Graduate Texts in Physics

Graduate Texts in Physics publishes core learning/teaching material for graduate- and advanced-level undergraduate courses on topics of current and emerging fields within physics, both pure and applied. These textbooks serve students at the MS- or PhD-level and their instructors as comprehensive sources of principles, definitions, derivations, experiments and applications (as relevant) for their mastery and teaching, respectively. International in scope and relevance, the textbooks correspond to course syllabi sufficiently to serve as required reading. Their didactic style, comprehensiveness and coverage of fundamental material also make them suitable as introductions or references for scientists entering, or requiring timely knowledge of, a research field.

More information about this series at http://www.springer.com/series/8431

Richard Osgood Jr. · Xiang Meng

Principles of Photonic Integrated Circuits

Materials, Device Physics, Guided Wave Design

Springer

Richard Osgood Jr.
Columbia University
New York, NY, USA

Boston University
Boston, USA

Xiang Meng
Columbia University
New York, NY, USA

ISSN 1868-4513   ISSN 1868-4521   (electronic)
Graduate Texts in Physics
ISBN 978-3-030-65195-4   ISBN 978-3-030-65193-0   (eBook)
https://doi.org/10.1007/978-3-030-65193-0

This Springer imprint is published by the registered company Springer Nature Switzerland AG
The registered company address is: Gewerbestrasse 11, 6330 Cham, Switzerland

Preface

The field of integrated photonics has advanced rapidly in the last two decades due to major advances in microfabrication, its design methods, its computer automation, and the expansion of our understanding of the physics of small-scale optical structures. Further, the growth in new commercial possibilities and actual devices has driven these advances through the realities of commercial needs. The excitement of this new technological world has led to the need to capture the field in a broadly scoped and full textbook for graduate students and those seeking to learn about this technology.

The materials from this book arose from a graduate course, which was taught for many years at Columbia University by R. Osgood. The intent of this course was to provide a full view of the overall field of integrated electro-optic devices. The authors worked to have a course not welded to the latest materials technology but rather to give a more pedagogical vehicle that showed how these devices may be simulated and best operated. Because of their small size and complex design, the course brings out unusual devices and new perspectives. Not unexpectedly the course involves an extensive range of ideas and physics, particularly in guided-wave optics.

The course builds not only on the most recent discoveries, and clearly has its foundations and insights from earlier classic monographs in this field. These prior classics include Tamir's highly respected monograph Guided Waves Opto-Electronics, the encompassing work on integrated opto-electronics, Optical Integrated Circuits by Nishihara, Haruna, Suhara, etc. Other texts with much broader pedagogical scope such as those by Bahaa and Teich, Yariv, or Haus are drawn upon for insights into new or unusual devices or their physics.

The text is organized and arranged so as to first present or review the basic electromagnetic fundamentals. It then shows how these electromagnetic "driver" equations are modified by interfaces and boundaries. In addition, the treatment of modal behavior and interaction via the coupled-mode equations and their modified version in the presence of adiabatic evolution during propagation in material media are then presented. These modified equations allow the initial formulation of elementary device structures, particularly those of interest for guided-wave phenomena to be developed and presented and their optical response examined. This introductory section is then followed by understanding of certain device types such as passive guided wave, i.e., waveguide crossings, Y-branches, bends, etc., grating-based devices, imaging devices, or switches and modulators. A major unique contribution of this book has been the inclusion of a chapter and illustrations of numerical methods for guided-wave problems. These methods are now the standards in areas where analytical methods had previously dominated. The precision and power of numerical calculations now even allow determining precise device behavior prior to fabrication.

Grateful Acknowledgments

Finally, this book has directly benefited from the input of Dr. Robert Scarmozzino and Prof. Miguel Levy in sections of earlier drafts of the book. The authors of the present text have used their substantial contributions and their encouragement in the years, in which later drafts were completed. In a like manner, Profs. Panoiu, Steel, and Dadap contributed a strong and exciting research environment, which gave new ideas and thinking in the content of the text. The text has also been the product of eager and hardworking graduate students, Zicong Huang, Rui Chen, Shijia Yan, and Songli Wang, as they moved from MS to Ph.D. in their studies at Columbia and helped in the preparation of the figures, proofreading, and scientific suggestions. Note that the book also has been generously helped by comments and suggestions by Prof. Paiella at Boston University, where R. Osgood spent the last several years of its preparation, and Prof. Bergman, Lipson, Teich, Herman, Kymissis, and Englund at Columbia and Prof. Willner of USC.

And finally it is a true joy to salute the grace and affection of Mrs. Alice Osgood and the Meng Family for their support of the authors during their day-to-day hard work. Without their work we would not have our tome.

New York, USA   Richard Osgood Jr.
                                                                          Xiang Meng

Contents

Chapter 1
Integrated Optics: An Overview

Abstract This chapter traces the history of integrated optical devices and gives an overview of its supporting technology. It also explains why integrated optics is so important in many areas of high-speed and high-density applications areas, which grow in significance with modern information systems. The recent growth in silicon photonics and its advances in fabrication technology is also emphasized.

1.1 Introduction

In the last two decades, the size and scale of photonics technology have been revolutionized by the same miniaturization process that has transformed electronic systems from collections of bulky solid state devices to nearly atomic-size arrays of transistors on single-crystal semiconductors. This development in photonics makes it possible to fabricate or integrate relatively complex optical subsystems on a single solid microchip. This book will focus on one form of integrated photonic systems, namely, those which are called photonic integrated circuits.

Table 1.1 makes the advantages of full microchip integration clear by comparing the optical components typically seen in three "formats" encountered in small-scale commercial photonic systems. Note that the advantages of integrated optics include greatly reduced size as scaled by the footprint of an optical waveguide, more robust optical alignment, i.e., that carried out by the fabrication process itself, and the lower cost made possible by the massive parallelism of planar processing. Of course, this lower cost is only realized when a large number of fabrication runs of devices or optical circuit are warranted.

The design of optical integrated circuits requires a fusion of design techniques that draw on contributions from three fields: materials, optical, and electronic engineering. Unlike much of the work on free-standing optical devices over the last few decades, integrated guided-wave designs require careful consideration of physics and performance of propagating optical waves in assessing device performance and properties. In particular, the designer must plan a design scheme, which focuses on the routing of optical waveguides and devices. Factors such as radiative loss, evanescent coupling, optical absorption, etc., must all be anticipated in the design of this

© Springer Nature Switzerland AG 2021
R. Osgood jr. and X. Meng, *Principles of Photonic Integrated Circuits*,
Graduate Texts in Physics,
https://doi.org/10.1007/978-3-030-65193-0_1

Table 1.1 Optical circuit technology

Technology	Macro-optics	Micro-optics	Integrated optics
Components	Lenses, mirrors	LEDs, LD fibers	PICs, single-mode LD
Alignment	needed	needed	fixed in fabrication
Connection size	cm	mm	μm
Electrode size	cm	mm	μm (fast, compact)

analog circuit. As a result, the design procedure for these circuits is drastically different from that of the usual digital integrated electronics circuits. Further, while design
methods are progressing at a very rapid pace in commercial venues, developing a
standard procedure and even a uniform choice of components for these designs is
still in a state of flux.

1.2   History of Integrated Optics

Integrated optics was first proposed in 1969 by Stuart Miller (then at Bell Labora-
tories), in the context of the rapidly growing field of integrated electronics and the
emerging interest in lightwave communications. This field grew sufficiently rapidly
such that by 1972 the first conference in integrated photonics was held: The Conference on Integrated and Guided-Wave Optics. Since then, integrated optics has
expanded into a field with not only a vigorous research program, but also with commercial markets in communications, cable television, fiber-optic sensors, and radar
control. Initially, these applications used optics with single or small arrays of devices,
but more recent interests in large integrated subsystems have arisen. This chapter is
intended to provide an understanding of the physics and methodology of design for
one class of optical systems that involving dominantly optical functions on a single
chip.

As in integrated electronic design, integrated optical systems have evolved in the
complexity of their design. At the same time, improvements in fabrication technology
have led to a steady increase in simplicity or uniformity in the choice of the basic
materials assembly for the optical chip. This progression caused "integrated" systems
to move from an assembly of diverse materials bonded together, generally by manual
alignment methods, to the present state, in which chips are fabricated by a succession
of large-scale planar processes on a single substrate chip. This evolution is illustrated
in Fig. 1.1, which shows the photonic chips by IBM as they were fabricated after many
years of development of an experimental single platform for optical systems.

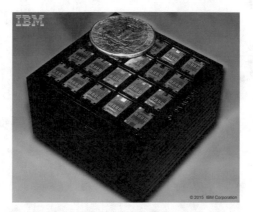

Fig. 1.1 IBM demonstrated its first fully integrated wavelength multiplexed silicon photonics chip. *Source* https://arstechnica.com/informationtechnology/2015/05/ibmdemos-first-fully-integratedmonolithic-siliconphotonics-chip/

1.3   Integrated Optical Circuits: Classification and Advantages

Integrated optical "Chips" have been used for various types of optical systems and have appeared in several different basic forms, each reflecting their intended application. These varying applications have been given several different names or classifications. The first form is basically a standard digital silicon or III-V IC with some provision for the detection or emission of light. Such a circuit might have, as its optical function, simply an onboard detector for receiving an optical data stream from a fiber-optic input. The second type, the opto-electronic integrated circuit (OEIC), is more clearly a photonic or optical circuit. This type usually involves significant electronic and optical components and circuits, and typically has important hybrid mounting methods. A good example is a fiber-optical transceiver, which contains a laser transmitter, its drive circuitry, a detector, and an amplifier. The integration level in OEICs has increased drastically over the last few decades; see Fig. 1.2.

Fig. 1.2 The dramatic change of integration level in OEICs in the last few years. The well-known Moore's Law is readily seen in these plots in this figure

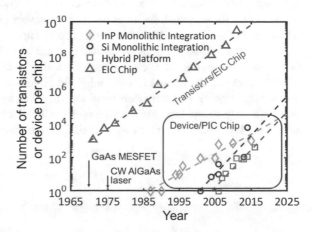

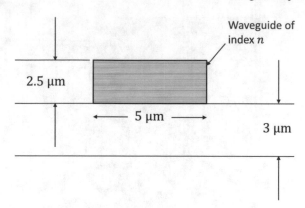

Fig. 1.3 The cross section of a waveguide showing an example of a typical size of the guided-wave elements in PICs

Finally, the concept of a photonic integrated circuit (PIC) has been developed over the last few decades. This circuit is in essence a miniaturized fully photonic system that has been confined to a planar geometry, thus making copious use of waveguides and two-dimensional optical elements. For PICs, the design is dominantly concerned with photonics functionality, since the number of electronic elements is small. Typically, most of these "circuits" use guided-wave elements, with waveguides acting as interconnections or "optical wiring," as well as being integral parts of both simple and complex devices, such as optical couplers and phase-delay routers, respectively. The transverse scale size of this "wiring" is small, e.g., 0.5 μm, but not as small as the wiring on electronic IC (see Fig. 1.3).

In this book, we will be concerned mainly with the design and understanding of PICs and to a lesser extent the optical portion of OEICs. Thus far, work in PICs has emphasized passive devices and circuits; this will be our emphasis as well, although we also present extensive discussions of modulators and switches. An example of such a PIC is shown in Fig. 1.4.

In terms of engineering integrated *optical* systems, PICs offer distinct advantages over hybrid systems, many of which are found only in part in other forms of optical systems. The first is an extremely broad spectral bandwidth, which can encompass a large number of optical channels. In practice, this means that many parallel channels are available for use, each coded at a separate wavelength. This wavelength parallelism can be used for a variety of network or sensor applications. A second advantage is found in the interfacing of many optical systems in an inexpensive robust manner. In this case, retaining the optical signal on a signal chip greatly reduces the cost of alignment and the optical loss for a hybrid system on a Si optical bench. It also provides greater thermal and mechanical stability than hybrid systems. These advantages are only examples; other advantages are available as well, including easier access to high modulation rates, more facile interconnect geometries, etc.

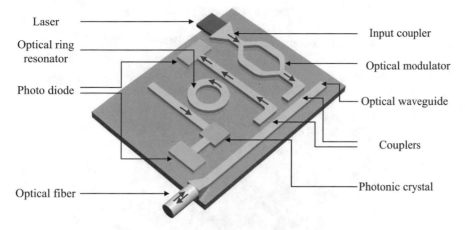

Fig. 1.4 A sketch showing the schematic of a photonic integrated circuit (PIC) (not to scale), showcasing different optical components

1.4 The Rise of Si and Si-Based Systems

Much of the challenge of realizing integrated optical systems is found in the difficulty of selecting a single universally effective material. For example, $LiNbO_3$ or other metal oxide optical materials make excellent electro-optical switching materials but suffer from a low refractive index and, hence, large characteristic bend radius. Other materials such as certain III-V crystals are superb as emitters and some passive devices but are relatively expensive. However, in trying to fix on the most desirable material, the Si-based materials system allows new possibilities for extensive integration of various materials (Soref 1993). After all, Si is the basis for the most successful and mature integrated systems, integrated electronics! In fact, it is reasonable to contemplate manufacture of photonic devices from the standard fabrication processes at a silicon-chip foundry. The recent major improvement in Si-based opto-electronic devices makes it worthwhile to focus this chapter on this specific system.

With these factors in mind, as well as the growing interest in using Si, like a miniature optical bench for glass waveguides (Henry et al. 1989) and other discrete components, it was suggested, in the 1980s by Soref (1993) and by others, such as Reed (2008), Jalali and Fathpour (2011) to employ guided-wave optical systems, which use patterned Si wafers as the light guiding material. Figure 1.5 shows a sketch of Soref's original concept for Si integrated photonic circuits. Much of the initial device work in this area focused heavily on the use of CMOS wafers (Jalali and Fathpour 2011), which were designed to have waveguides that were weakly guiding. Interestingly this early work did explore many integration issues that are still of interest and considered a key device concept such as free-carrier modulators, etc. Subsequent to this important pioneering work, approaches to build high confinement single-mode waveguides in Si were conceived and developed (Vlasov and McNab

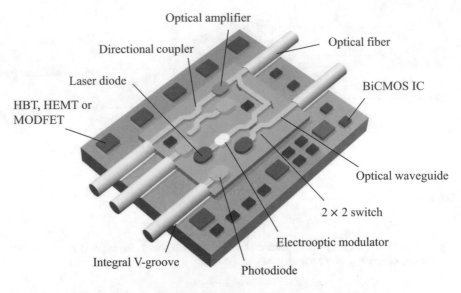

Fig. 1.5 A sketch of Soref's original concept for Si integrate photonic circuits

Fig. 1.6 An optical microscope image of PIC by PLCC showing the remarkably high-quality patterning on a silicon on insulator substrates (SOI) chip. *Source* https://www.extremetech.com/extreme/201163-ibm-to-demonstrate-first-on-package-silicon-photonics

2004). The most effective material for these devices was SOI (*silicon-on-insulator*) substrates, which were available in large-scale planar wafers and for which the methods for remarkably high-quality patterning had been developed. (see Fig. 1.6)

The most striking advance in Si photonics, however, occurred from the realization that guiding light in a pure silicon device layer made a very distinctive and potentially powerful approach to optical integration. This Si-base layer, which has a very high refractive index contrast, which enables compact bend waveguide systems, would make extremely small down-scaled photonic systems possible. In addition, while the array of allowed Si functionality has still certainly not yet been fully realized, it is possible to fabricate a wide array of active devices, including low-power modulators, high-sensitivity photodetectors, and high-efficiency switches (thermo and electro-optic). SOI is also an unsurpassed semiconductor material for patterning since it builds on the massive infrastructure that is in place for nanometer patterning of Si electronics. The degree of this infrastructure includes the full capability needed for the manufacturing of microsystems (Chrostowski and Hochberg 2015). Finally, the spatial resolution of Si patterning is such that spurious roughness is now no longer an issue.

The excitement of realizing photonics integration based on Si photonics has had a profound effect on the direction of integrated optics. The first of these directions is the development of radically new devices, which are enabled by the unique properties of Si (Xu and Lipson 2006) and its related Group IV material Ge (Liu et al. 2010). These advances include ultralow energy-per-bit modulators, typically based on micro-ring resonators, on-chip doped-Si or pure Ge detectors, new approaches to integration, and precisely defined, efficient grating couplers (Reed 2008); these devices will be discussed later in the book. The second area of recent advances is in achieving greater spectral reach for integrated devices and systems. Research is now emerging, which allows integration at Mid-IR wavelengths for sensing applications and in the near IR for quantum-computational applications. Finally, major growth is occurring in the area of large-scale integrated applications such as data systems or multi-wavelength transceivers (Shen et al. 2019). These systems are characterized by high speed and orthogonalities in wavelength, mode, and polarization.

The broad goal of one key application of Si photonics is to use the high bandwidth of optical technology for transmitting large amounts of data between computers, servers, and data centers. By integrating Si photonic devices into these systems, the costs of fabrication and electronic power consumption can then be driven down, and thus inexpensive, highly functional systems can be realized.

Silicon photonics can also improve the response time and allow ultrahigh-speed data processing. For example, one commercial vendor recently described a photonics chip, which employs four wavelengths in a single waveguide for data transfer. This system will enable "digitally sharing of six million images, or downloading an entire high-definition digital movie in just two seconds." (Green 2015) In this chip, for example, each wavelength provides an independent 25 Gb/s optical channel. Thus for an entire transceiver chip, the four wavelength channels would be multiplexed so as to obtain a total 100 Gb/s bandwidth in one Si wire waveguide. If the "wire" is designed to be single-mode, the bandwidth-distance product is no longer limited as it is in multimode links.

It is important to realize that the use of standard-thickness SOI makes it difficult to pattern and fabricate nanoelectronic and photonic systems on a single chip; see discussion in Sun et al. (2015). One solution to this problem is to place photonics and electronics on separate chips. This solution, while powerful, presents a fabrication issue in connecting between the two chips. Thus recently methods have been demonstrated to form a single-chip solution, which enabled the coplanar method for both electronic and photonic functionality, and methods have now been devised (Sun et al. 2015).

Finally, it is important to conclude this section with a cautionary note to point out that, as time has progressed, the emphasis on specific choices of photonic materials has shifted steadily. At present, Si appears to be an ideal solution for the most pressing of the current photonics system needs. However, a careful look at the commercial marketplace shows that glass, $LiNbO_3$, and III-Vs continue to be crucial for use in integrated optics. Thus to take the broadest of overviews, this book will consider photonic devices and systems based on each of these materials.

1.5 Contents of This Book

The goal of this book is to provide a summary of the essential devices and device concepts for photonic integrated circuits. Thus the first chapter lays out the properties and fabrication of major materials systems and the waveguide structures needed for these PICs. This book also targets achieving an understanding of the physics and analytic solutions to waveguiding structures. It is felt that presenting analytic solutions gives a deeper understanding of the operation principles of these structures. These analytic methods include a detailed discussion of coupled-mode theory since this perturbation-based method enables analytical analysis of many guided-wave structures. However, the advances in computational or numerical design methods and the computational power itself have been so substantial that they can add new design tools themselves and add capabilities such as highly accurate imaging. In all cases, we display many illustrated examples obtained from the numerical methods for an intuitive understanding of how the devices work. In addition, a review of the numerical methods used for photonic device and PIC design or simulation is presented in the text as well.

The book also includes examples of many basic devices. These devices include grating- and coupler-mediated devices, and adiabatic designs of the devices. In addition, devices for wavelength filtering and mux/demux are discussed and shown. The book is presented in a format, which may lend itself to a graduate class in integrated photonics or to a review for a working-level engineer or applied scientist.

1.6 Summary

The basic introduction and motivation for integrated optical circuits have been presented in this chapter. Clearly the ultimate driver of any commercial technology is the nature and extent of its marketplace, and this varies with time. But the basic argument for PICs is that integration reduces cost by allowing parallel processing of all devices on the wafer surface and by allowing alignment of the optical system with the lithography pattern tool.

Finally, note that parallel growth has occurred in the interest in potential devices for nonlinear-optical integrated photonic devices, particularly in the area of Si photonics

(Osgood et al. 2009). However, the focus of this book is on linear photonics, in part due to its more immediate commercial interest. The reader is referred to several recent reviews on the subject of integrated nonlinear optics (Atabaki et al. 2018).

References

Atabaki, A. H., Moazeni, S., Pavanello, F., Gevorgyan, H., Notaros, J., Alloatti, L., et al. (2018). Integrating photonics with silicon nanoelectronics for the next generation of systems on a chip. *Nature, 556*(7701), 349.

Chrostowski, L., & Hochberg, M. (2015). *Silicon photonics design: From devices to systems.* Cambridge: Cambridge University Press.

Green, W. (2015). Silicon photonics: The future of high-speed data. IBM Research Blog.

Henry, C. H., Blonder, G., & Kazarinov, R. (1989). Glass waveguides on silicon for hybrid optical packaging. *Journal of Lightwave Technology, 7*(10), 1530–1539.

Jalali, B., & Fathpour, S. (2011). *Silicon photonics for telecommunications and biomedical applications.* Boca Raton: CRC.

Liu, J., Sun, X., Camacho-Aguilera, R., Kimerling, L. C., & Michel, J. (2010). Ge-on-Si laser operating at room temperature. *Optics Letters, 35*(5), 679–681.

Osgood, R., Panoiu, N., Dadap, J., Liu, X., Chen, X., Hsieh, I.-W., et al. (2009). Engineering nonlinearities in nanoscale optical systems: Physics and applications in dispersion-engineered silicon nanophotonic wires. *Advances in Optics and Photonics, 1*(1), 162–235.

Reed, G. T. (2008). *Silicon photonics: The state of the art.* Hoboken: Wiley.

Shen, Y., Meng, X., Cheng, Q., Rumley, S., Abrams, N., Gazman, A., et al. (2019). Silicon photonics for extreme scale systems. *Journal of Lightwave Technology, 37*(2), 245–259.

Soref, R. A. (1993). Silicon-based optoelectronics. *Proceedings of the IEEE, 81*(12), 1687–1706.

Sun, C., Wade, M. T., Lee, Y., Orcutt, J. S., Alloatti, L., Georgas, M. S., et al. (2015). Single-chip microprocessor that communicates directly using light. *Nature, 528*(7583), 534.

Vlasov, Y. A., & McNab, S. J. (2004). Losses in single-mode silicon-on-insulator strip waveguides and bends. *Optics Express, 12*(8), 1622–1631.

Xu, Q., & Lipson, M. (2006). Carrier-induced optical bistability in silicon ring resonators. *Optics Letters, 31*(3), 341–343.

Chapter 2
Materials for PICs

Abstract The materials technology for integrated optics is complex and thus this chapter examines the optical properties, the electronic properties, the fabrication, and pattering of a set of the most common materials. These include polymers, crystalline silicon, epitaxial three-five thin films, and doped titanium dioxide. In the discussion, the relative importance of each material is described.

2.1 Introduction

Because the development of PICs is still evolving, the materials technology for photonic circuits also remains in a state of flux. Thus, a wide variety of new materials and materials combinations are still being used commercially and there is still major research into a wide variety of materials technologies. Many of these materials are ideally suited for a specific photonics application or for realizing one functionality but then fall short in other areas. Thus the development of a generally accepted and used materials system, such as is found in the case of SiO_2/Si for CMOS electronics has not occurred; research into Si photonics has become an increasingly focused choice for important areas of data communications. In fact, in some sense, the state of materials choice is even more uncertain than for the comparatively well-developed technology for microwave circuits, which rely on *four* different materials technologies: Au/GaAs, GaAs/AlGaAs, Si:Ge, and SiO_2/Si.

Nevertheless, within the last few years, most commercial PICs have generally used five basic materials systems: SiO_2, Si, III-V's (InP- or GaAs-based), polymers, and $LiNbO_3$. As a result this chapter will contain a discussion of each of these five materials areas.

An overall summary of these five materials systems is contained in Table 2.1. The table presents data on four characteristics: typical values of Δn and n, absorption coefficient α, and the typical planar dimension of the substrate of the five materials considered here. Note that in two cases (III-Vs and SiO_2/Si), two particular classes of the materials systems are discussed. Finally, the optical response, which is achievable with each material system and wavelength is also listed.

© Springer Nature Switzerland AG 2021
R. Osgood jr. and X. Meng, *Principles of Photonic Integrated Circuits*,
Graduate Texts in Physics,
https://doi.org/10.1007/978-3-030-65193-0_2

Table 2.1 Commonly used PIC materials systems and their approximate optical properties and available planar dimensions

Material	Refractive index n	Index contrast Δn	Absorption α/cm^{-1}	Typical available dimension l/cm	Active optical response
SiO$_2$/Si	1.46 (P-glass 1.3 μm)	7×10^{-3}	0.05	$\sim$10	Thermo-electric modulation
SiO$_2$/Si$_3$N$_4$/Si	1.97 (1.3 μm)	2–4×10^{-2}	0.3	$\sim$10	Optical amplification (with Er doping)
LiNbO$_3$	2.2(n_e) 2.24(n_o)	10^{-2}	5×10^{-3}	$\sim$2 $\times$ 6	Electro-optic modulation
Polymer	1.49	2×10^{-3}– 10^{-4}	0.02/0.1 (1.3/1.5 μm)	$\sim$30	Modulation, some emission
GaAs/AlGaAs	$\sim$3.6	5×10^{-2}	2×10^{-2} (at $n\sim10^{16}\,\text{cm}^{-3}$)	5	Modulation, Optical sources, Nonlinear response
InP/GaInAsP	$\sim$3.2	5×10^{-2}	2×10^{-2} (at $n\sim10^{16}\,\text{cm}^{-3}$)	5	

The reason for the particular choice of properties, presented in Table 2.1, is that these parameters control the physical scaling of the PIC. The index difference between the cladding and waveguide core, Δn, controls modal confinement in a waveguide. This confinement, in turn, establishes the widths of the waveguide mode as well as the minimum radius of bends in the waveguide that are achievable without significant radiative loss. The linear absorption coefficient, α, determines the insertion loss in the waveguide for a given length. Hence a large α will rule out fabrication of PICs with large linear dimensions. Finally the availability of large-area substrates of polymer and SiO$_2$ enable large-area circuits to be made of these materials in comparison to the much smaller size possible with Si or III-V materials.

Some of the optical properties in Table 2.1 are also important for the interfacing of PICs with an optical fiber link. Specifically, the absolute value of n controls the magnitude of Fresnel loss, i.e., that due to reflection, at the chip facets. The refractive index is also important, along with the change in core/cladding refractive index, Δn, in determining the mode matching between the fiber and the integrated- waveguide mode. The radiative loss, which is encountered in going on and off chip due to imperfect mode matching, can be substantial and is thus an extremely important quality in designing a practical PIC system. This radiative, mode-mismatch loss is obviously the same, regardless of chip size. These interfaces will be described more qualitatively in Chap. 7.

2.2 The SiO$_2$-Based Materials

An important passive PIC technology has been developed using the SiO$_2$-based materials system on a Si platform; this technology is sometimes called the Si optical bench technology (Henry et al. 1989). These "bench" materials are grown either by metal organic chemical vapor deposition (MOCVD), flame hydrolysis (also used for optical fibers), or by high pressure oxidation of a silicon substrate. Si$_3$N$_4$ layers can also be grown by low pressure chemical deposition (LPCVD). This technology is attractive because it is, in principle, compatible with many phases of Si IC manufacture. This compatibility is due to the fact that it makes use of silicon wafers and many of the chemical processing steps used in integrated circuit manufacturing. Despite the importance of this Si-based technology for passive devices, the absence of an electro-optical response and light-emission capability means that active Si-based optical circuits, except those based on thermo-optic control, are not possible. Of course in certain cases thermo-optical modulation, while relatively slow, does permit switching/modulation operations.

SiO$_2$-based PICs have a number of important advantages. First, this PIC is mounted on a Si platform and, thus, pig tailing of the fiber to the dielectric layer may be realized via Si V-groove technology, which is based on precise crystallographic etching (see Fig. 2.1). This capability makes optical packaging, which is generally a costly procedure, somewhat more straightforward. Second, because of the wide availability of Si wafers, the technology is in principle relatively inexpensive. In addition, the high thermal conductivity of the silicon substrate means that it is a practical platform for mounting active devices, such as laser diodes, which impose significant cooling requirements. Finally, the refractive index of this system is close to that of the fiber, making efficient coupling between fiber and integrated waveguide modes possible. The core of this waveguide is $\sim$4 μm $\times$ 7 μm (height and width) for a 1.55 μm wavelength system. This good mode matching has thus been used for a number of high-quality commercial components such as star couplers, filters, routers, and splitters.

In the remainder of this section, we will focus on SiO$_2$ on Si (silicon optical bench used initially by Bell Labs). In this system, there are three SiO$_2$-based materials systems of interest: undoped SiO$_2$, SiO$_2$:P (phosphorous doped silica), and Si$_3$N$_4$. These systems can be used to make the two types of waveguides shown in Fig. 2.2. In both types of guides: SiO$_2$ or low P-doped, the SiO$_2$ film functions as the cladding layer. In the case of SiO$_2$:P, the variable content of phosphorus dopant atoms can raise the index above that of the undoped SiO$_2$. This system (see Fig. 2.2) has indices which are most compatible with fiber I/O's. The Si$_3$N$_4$ film has a much higher index ($\sim$1.97), and typically can be used in circumstances where greater mode confinement is required. An excellent example of its use is in mode matching a waveguide interconnect to a semiconductor diode.

The properties of these films determine the comparative advantages of SiO$_2$ for waveguide materials. A typical refractive index step between the waveguide core and cladding is $\Delta n = 0.004$ (see Table 2.1). The absolute indices are those of glass

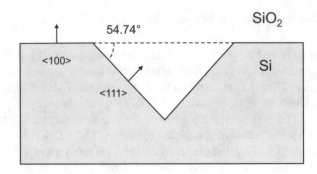

Fig. 2.1 Schematic representation of a v-groove

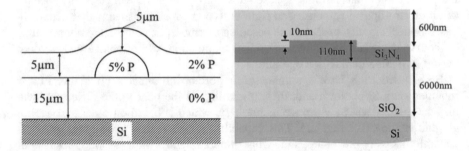

Fig. 2.2 Two typical waveguides using SiO₂-based materials. Left: SiO₂:P; right: Si₃N₄

except that prior to annealing, films can have a lower density than for typical glass. With such relatively small values of Δn, it is necessary to decouple the deposited thin film from the substrate using a ~15 µm layer. Once decoupled, waveguide losses are small, i.e.,, <0.05 dB/cm. The residual stress anisotropy in optical waveguides causes a small but important degree of birefringence in the deposited waveguide of ~5 × 10⁻⁴ for the TE versus the TM polarization, and where the TM polarization has a higher P-doped index. The nitride-core waveguides have higher index (~1.97) and loss (~0.3 cm⁻¹).

These glass layers are grown for waveguides in several ways. First, high pressure thermal oxidation in a steam ambient is used to form thick optical SiO₂ buffer layers on the Si wafer. This process is sometimes called HiPOx. If P-doped layers are desired, they can be grown from silicon and oxygen molecular precursors, with a phosphine dopant source, using LPCVD. A second approach, which yields a more conformal film, is based on LPCVD using tetroethylorthosilicone and ammonia. This film is called TEOS or, when doped with phosphorus from phosphine, P-TEOS. Flame hydrolysis has also been used to grow P-doped SiO₂ films. Annealing at higher temperatures can be used to relieve stress in the film. Finally, use of various nitrogen precursors in a LPCVD system will deposit a silicon nitride film. These films are limited by stress to less than 1500 Å in thickness.

Because of the commercial importance of SiO_2 structures, there have been extensive measurements of the optical properties of this materials system. A representative sample is provided in Fig. 2.2, which shows the measured index versus wavelength for different P concentration and for undoped SiO_2. Clearly, P-doping does increase the background index; however, note that the index contrast, Δn, between the various glasses remains nearly constant over the wavelength range. Also note that for undoped SiO_2 deposited using TEOS, annealing is needed to cause the index rise to reach the background index. Annealing changes the index due to glass densification.

2.3 Polymers

Polymers are a particularly interesting materials system for PICs because of the fact that large-area sheets of polymers can be prepared at very low cost. These thin sheets can then be readily patterned, by a variety of techniques, to form integrated optical circuits. At present, the use of polymers is well established for many different passive devices, such as filters, splitters, and star couplers. (Eldada and Shacklette 2000). In addition, polymers can, in principle, be a host for electro-optic functionality, as well as light emitters; this capability enables a nearly fully functional polymeric PIC to be made. In practice, however, the use of polymers for active components in PICs has been limited by thermal degradation, or the loss of the poling. This last issue means that it is very difficult to make a long-term stable electro-optic modulator since loss of poling eliminates the electro-optical coefficient.

In addition to research work, polymers are also finding wider use in many commercial large low-cost planar optical elements. For example, robust low-cost polymers have been used in the devices in aircraft fiber-sensor systems. In fact, many of these materials are being inserted commercially in various approaches to low-cost highly multimode optical interconnects.

A variety of polymeric materials have been utilized for these optical waveguide devices. The polymer classes include polyimides, olefins, and acrylates, including halogenated versions of these building-block molecules. These materials are typically applied in liquid form as monomers, polymers, or oligomers in solution. The materials can then be formed into thin sheets by cross-linking, either by exposure to light, thermal treatment, or desolvation in an oven, respectively. A list of some of the typical commercial polymeric waveguide materials is given in Table 2.2. The refractive indices of the materials listed in the table can range from 1.30 to 1.85.

Table 2.2 Examples of polymers used in passive integrated optical devices

Manufacturer	Polymer type or trade name	Patterning technique	Optical loss, dB/cm (at wavelength, nm)
NIT	Halogenated Acrylate	RIE	0.02 [830] 0.07 [1310] 1.7 [1550]
NIT	Deuterated Polysiloxane	RIE	0.17 [1310] 0.43 [1550]
Amoco	Fluorinated Polyimide [Ultradel^{TM}]	Photoexposure/wet etch	0.4 [1300] 1.0 [1550]
DuPont	Acrylate [Polyguide^{TM}]	Photolocking	0.18 [800] 0.2 [1300] 0.6 [1550]
Dow Chemical	Benzocyclobutene [Cyclotene^{TM}]	RIE	0.8 [1300] 1.5 [1550]
Dow Chemical	Perfluorocyclobutene [XU 35121]	Photoexposure/wet etch	0.25 [1300] 0.25 [1550]
JDS Uniphase Photonics	[BeamBox^{TM}]	RIE	0.6 [1550]
Allied Signal	Acrylate	Photoexposure/wet etch, RIE, laser ablation	0.02 [840] 0.2 [1300] 0.5 [1550]
Allied Signal	Halogenated Acrylate	Photoexposure/wet etch, RIE, laser ablation	<0.01 [840] 0.03 [1300] 0.07 [1550]

By mixing these materials on a single-material substrate such as a glass sheet, it is possible to "integrate" a wide range of different index polymers. However, chemical and physical incompatibilities of the different polymer units can set practical limits on the number of choices for mixing.

Examination of the table shows that a major issue in designing a high-quality optical polymer is the presence of undesired optical absorption. (Eldada and Shacklette 2000) This absorption is generally more significant at infrared than at visible wavelengths. Figure 2.3 shows that absorption loss in polyimide, a common prototypical polymeric material. The calculated points in the figure, which are connected to give an overall sense of the absorption envelope, show that the absorption maximum can be accounted for simply by counting the number of vibrational units times their individual absorption strength. The figure also shows that replacing the polymer hydrogen with increasingly heavier species, i.e., deuterium, fluorine, etc., shifts the absorption edge further into the infrared and, hence, lowers absorption in the visible and near-infrared regions. This strategy has led to low loss in commercial polymers for integrated optics.

A variety of other materials properties are important for optical applications, including detection of thermal aging and humidity sensitivity. First, partial thermal

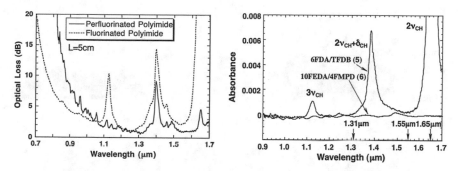

Fig. 2.3 Optical loss spectra of single-mode waveguides prepared by perfluorinated polyimide and partially fluorinated polyimide [S. Ando, 2004]

decomposition of polymers is a particularly severe problem for use at visible or ~800 nm wavelengths. It can be reduced to negligible amounts at these wavelengths by using heavily cross-linked polymers. In such a material, however, aging is not an issue for 1.3 and 1.5 μm wavelengths. Second, halogenation and cross-linking also have improved the humidity resistance of these polymers to the point where it is less of an issue.

Similarly, the optical properties of these polymers are now well categorized and are sufficiently favorable to be useful for a variety of optical applications. For example, refractive index dispersion is important for many multiple-wavelength-use devices (Eldada et al. 2000). The material dispersion in a well-designed polymer is roughly comparable to (albeit somewhat higher than) SiO_2. In the case of halogenated polymers, material birefringence is $<1 \times 10^{-6}$. This small value is a result of the non-oriented nature of the polymeric chain. Finally, polymers do have a larger temperature-dependent refractive index than SiO_2. For example, a value of $-3 \times 10^{-4}/°C$ has been reported for halogenated polymers. While this property can be detrimental in some applications, it has been used to make a series of very sensitive thermo-optic switches (Eldada et al. 2000). Note that if problematic, the detrimental aspects of the index variation with temperature can be ameliorated by use of a substrate with compensated thermal properties.

A variety of patterning techniques have been used to form devices or components in optical polymers. Three of the most promising commercial processes are photo-patterning, photodelineation, molding, and reactive-ion-etching (RIE). Photo-patterning typically uses a photolithographic technique for either polymerizing a monomer or for changing the index in a pre-existing polymer, a process sometimes called "photolocking." In the former technique, the monomer is "spun-on," as in a photoresist, and then after exposure developed with a solvent. These techniques yield structures with an index contrast between substrate/cladding and core of $\Delta n \sim 10^{-1}$. This index contrast in photodelineated waveguides is typically $\sim 1 \times 10^{-2}$. Schematic of polymer photo-patterning is shown in Fig. 2.4.

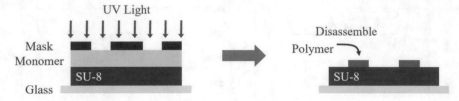

Fig. 2.4 Schematic of the photodelineation process for SU-8 photoresist

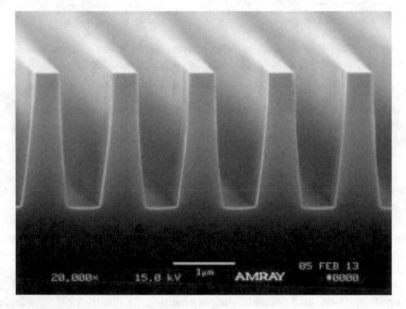

Fig. 2.5 SEM photo of an InP waveguide fabricated using RIE etching. *Source* https://www. plymouthgrating.com/about-pgl/pgls-technology/reactive-ion-etching/

The second approach to fabrication uses a molding or embossing tool which is transparent to UV radiation. In one approach, liquid monomers are pressed with the embossing tool, and then exposed through the tool with UV light. While this approach produces uniform surfaces, additional loss due to surface roughness still limits its applicability. The third patterning approach uses RIE etching of a pre-existing polymer substrate. Very high-quality devices have been made using this approach and it is thus useful for fabricating the most advanced devices (see Fig. 2.5).

Finally, as mentioned above, polymers can also be modified to permit their use as electro-optical and nonlinear-optical media. In their normal state, polymers are centrosymmetric amorphous materials and, thus, cannot exhibit an electro-optical response. However, "poling" of polymer materials with a high voltage can provide the molecular orientation for allowing induced dipolar properties. This response can be further enhanced by adding specific chromophores to the polymer. These chromophores can be tailored in functionality and varied in concentration. Polymer modulators based on poled polymers have recently achieved very high modulation speeds (Chen et al. 1997). The most important issue concerning the practical use of these polymers has been poling lifetimes. As mentioned earlier in this chapter, thus far, the results have been promising but not completely satisfactory.

2.4 Single-Crystal LiNbO$_3$

2.4.1 Overview

LiNbO$_3$ was one of the earliest choices for making relatively large integrated-photonics devices and circuits. It was selected because its low index allows low loss mode matching to a fiber I/O and because its combination of excellent and reproducible dielectric and electro-optic properties make it ideal for use as a modulator. In addition, LiNbO$_3$ is transparent over the wavelength range of most commercial fiber systems. Because it has been an important electro-optical integrated optical material for many years, the processing and optical properties of the material are extremely well characterized. As a result, LiNbO$_3$-based technology is in an advanced state, with many available commercial products. LiNbO$_3$ is a birefringent crystal, and thus two indices of refraction must be specified, namely,

$$n_o(\lambda) = 2.195 + 0.037/\lambda^2 \tag{2.1}$$

$$n_e(\lambda) = 2.122 + 0.031/\lambda^2 \tag{2.2}$$

where λ is in pm. Typically, birefringence is of minimal importance for LiNbO$_3$, even when using z-cut crystals. Note however, the crystal dependence of the index of refraction is very important for understanding or calculating the effect of electric field. One final important advantages of LiNbO$_3$ is that it can be grown in large crystals, with the size of the crystal being as large as 20 cm in diameter and 25 cm in length. This large size has allowed relatively complex multielement LiNbO$_3$ PICs to be made.

2.4.2 Waveguide Formation

LiNbO$_3$ waveguides may be formed by several methods; two of which will be discussed here: impurity doping and proton exchange.

Ti-Diffused Waveguides
Ti "in-diffusion" utilizes standard lithographic-based patterning of a deposited Ti thin film. These films are then in-diffused via a heat treatment similar to that used for semiconductor processing. The diffusion of Ti into the LiNbO$_3$ crystal causes a

Fig. 2.6 Ti-diffused LiNbO₃
concentration profile

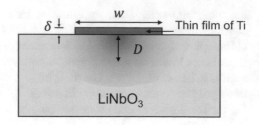

change in the index of the crystal. This method or procedure, which forms a graded-index structures with relatively low values of Δn, e.g., $1 - 4 \times 10^{-3}$. Because of the propensity of LiNbO$_3$ to lose volatile LiO$_2$ during heating, a number of important material chemistry innovations have been made to form reproducible structures. These are summarized in Korotky et al. (1987). Commercial dopant diffusion is done using uniform sample heating. As a result, the relative Ti-doping concentration (i.e., that percentage of the crystal composition) is given by an equation assuming local thermally equilibrated diffusion:

$$C(y) = \frac{2}{\sqrt{\pi}} \frac{\tau}{d} \exp\left(-\left(\frac{y}{d}\right)^2\right) \qquad (2.3)$$

where d is the diffusion distance into the crystal, y is the vertical coordinate in the crystal with $y = 0$ being at the crystal surface, and τ is the thickness of the deposited Ti film deposited on the surface of a wafer. The above equation assumes complete incorporation of the deposited film; i.e., no metal film remains on the surface after the heating step and that the deposited stripe width is much larger than the metal thickness, so that the "edge" effects from the stripe are negligible. The diffusion distance, d, in a time, t, is given by the usual diffusion equation, $d = 2\sqrt{Dt}$, where D is the diffusion coefficient. In general, D is a temperature-dependent diffusion coefficient, e.g., $D \sim 0.5 \times 10^{-12}$ cm^2/s at 1000 °C or $\sim 2 \times 10^{-12}$ cm^2/s at 1100 °C. Generally, D can be calculated with the simple equation

$$D = D_0 \exp(-T_0/T) \qquad (2.4)$$

where $T_0 = 2.5 \times 10^4$ K and $D_0 = 2.5 \times 10^{-4}$ cm^2/s. An example of the actual concentration gradient in a Ti-diffused waveguide is shown in Fig. 2.6. After complete in-diffusion (no free metal remaining on the surface) of the surface, the index profile is approximately Gaussian.

In fact, in many cases the actual waveguide diffusion profile is more exactly treated by a two-dimensional diffusion profile. For example, for typical single-mode waveguides which are defined by diffusion from a stripe of width, w, and thickness τ, where $w/d \sim 2 - 4$, and thus edge effects are important. In this case,

$$C(x, y) = C_0 X(x) Y(y) \qquad (2.5)$$

where

$$C_0 = \frac{2}{\sqrt{\pi}}\frac{\tau}{d}\mathrm{erf}\left(\frac{w}{2d}\right) \tag{2.6}$$

$$X(x) = \frac{1}{2\mathrm{erf}(w/2d)}\left(\mathrm{erf}\left[\frac{1}{d}\left(x+\frac{w}{2}\right)\right] - \mathrm{erf}\left[\frac{1}{d}\left(x-\frac{w}{2}\right)\right]\right) \tag{2.7}$$

and

$$Y(y) = \exp\left(-\frac{y^2}{d^2}\right) \tag{2.8}$$

where C_0 is the peak concentration and $X(x)$ and $Y(y)$ are the lateral- and depth-dependent concentration profile. In these equations, erf is the error function.

Doping changes the index of refraction as shown in Fig. 2.6. This change has been attributed to Ti-atom-induced strain in the crystal lattice. The data show that, conveniently, the change of index depends *approximately* linearly on the concentration, $\Delta n_0 \approx bC_0(y)$, where Δn_0 is the maximum in index change, i.e., from that at the surface, and b is a constant of proportionality. For $\lambda = 0.6\,\mu$m, for example, $b \approx 0.7$ for the extraordinary axis. In the case of the change in the ordinary index, a linear relation between Δn and concentration only holds at lower concentrations, i.e., <0.4%. For values above this concentration, the induced index change begins to saturate. This proportionality constant, b, also has a wavelength dependence, due to dispersion in the doped region:

$$b(\lambda) = 0.552 + 0.065/\lambda^2 \tag{2.9}$$

where λ is in μm. When the three-dimensional character of the waveguide is accounted for, then

$$\Delta n(x, y) = \Delta n_0 X(x) Y(y) \tag{2.10}$$

The relatively low index of the diffused doped region in Ti-diffused LiNbO₃ means that the guided mode has weak confinement in the waveguide, and hence it possesses a relatively large mode. Note also that the mode width is controlled to a large extent by the width of the diffused region. The large mode in Ti:LiNbO₃ is however not necessarily a disadvantage; in facet this large size means that there is typically good overlap between both these waveguides and that of the input fiber mode. This overlap is crucial for obtaining low loss pigtailed optical systems. Thus Burns and Hocker (1977) obtained a useful analytic expression for the overlap. Specifically he showed that the power coupling efficiency η between a fiber with a mode with a $1/e$ diameter a and a waveguide of width dx and depth dy can be written conveniently as

$$\eta = 0.93\left(\frac{4(d/a)^2}{[(d/a)^2 + r][(d/a)^2 + 1/r]}\right) \tag{2.11}$$

where $d = \sqrt{d_x d_y}$ and $r = d_x/d_y$.

In addition, often, a crystal is covered with a layer of SiO_2 on a $LiNbO_3$ surface, which optically decouples the surface from overlayer effects particularly metal electrodes. With good cladding or with no metal overlayer, a typical $LiNbO_3$ waveguide crystal loss is $\sim$0.05 dB/cm for high-quality crystals.

Proton-Exchange Waveguides

Proton exchange provides a technique for making a relatively high-index change in $LiNbO_3$. In this method, a patterned substrate containing a surface mask is suspended in hot benzoic acid, e.g., $\sim$240 °C. At this temperature, Li^+ ions on an unmasked region of the substrate are replaced with H+ ions from the acid. This compositional change leads to an index change of $\Delta n_e \sim 0.12$ and $\Delta n_o \sim -0.04$ for the extraordinary and ordinary index of refraction, respectively. As a result, these waveguides only guide TM waves. Typically, the index profile is more step-like and has a larger change in index than for Ti-diffused guides (see Fig. 2.6). It can be shown using expressions for loss in a bending waveguide that the high-index contrast also allows use of proton exchange possible for the high values required for tight bends in some PICs. Typical losses for proton-exchange guides are $\sim$0.5 dB/cm.

The properties of these proton-exchange guides can be altered by adjusting the processing procedure. For example, post-exchange annealing leads to a reduced index gradient. In addition, use of a mixed benzoic and lithium-benzoate solution increases the stability of the index profile in the presence of an electric field more than for neat benzoic acid. This lack of stability for the neat acid still limits the use of proton-exchange waveguides for modulators. The use of annealing (annealed proton exchange) reduces loss and improves the otherwise-degraded electro-optic coefficient in proton-exchange $LiNbO_3$.

2.5   GaAs/AlGaAs or InP/InGaAsP

2.5.1   Overview

III-V materials are well developed for electro-optic and integrated optic applications. This materials system has several important advantages, including the fact that a III-V PIC can incorporate a high degree of functionality and easily tailorable index profiles. Note that increased functionality includes the fact that the standard laser and LED optical sources are made in III-V materials. However, integration of fully functional PICs does require complex growth steps, typically involving regrowth or selective-area epitaxy. In addition, the size of most III-V wafers is relatively small.

2.5.2 Index Changes

Two techniques can be used for changing the refractive index difference in epitaxial layers of these III-V semiconductors: doping and hetero-layers (Leonberger and Donnelly 1988). However, while doping may occasionally be useful, it suffers from an attendant loss due to free-carrier absorption. In fact, optical loss due to background doping is an important consideration in designing III-V PICs. Note that free-carrier effects are also present in voltage-induced changes in carriers as well; these will be discussed in regard to certain optical modulator types.

Doping-induced changes in semiconductors are due to free-carrier-induced change in the dielectric constant. This negative change can be obtained using a simple free-carrier plasma calculation of the dielectric constant. For doped semiconductors, the index is given by $n = n_0 - \Delta n$ where n_0 is the index of the undoped material and

$$\Delta n = \frac{n_c e^2}{8\pi^2 \epsilon_0 n_0 v^2 m^*} \tag{2.12}$$

where v is the optical frequency, m^* is the effective mass of carriers in the semiconductor, e is the charge of an electron, n_c is the number density of carriers, and ϵ_0 is the dielectric constant of free space. For example, for a semiconductor with $m^* = 1$ and $n_0 = 3.5$, $\Delta n \sim 3 \times 10^{-20} n_c \lambda_0^2$, where λ_0 is the free-space wavelength. Thus for GaAs the index goes down with doping. Unfortunately, this index change comes with a price! The physical effect causing Δn also causes a concomitant change in α, the optical absorption, in fact this change is invariably an increase. In particular, for doped GaAs,

$$\alpha = \frac{g e^3}{4\pi^2 m^{*2} \mu n_0 c \epsilon_0} \tag{2.13}$$

where $g \sim 1$–3 and is related to carrier scattering times, and μ is the carrier mobility. This loss causes the guide loss in a simple single-mode homojunction guide to be, at a minimum, 2.5 dB/cm, if the substrate is doped to $n_c \sim 10^{18}$ cm^{-3} and $\lambda_0 = 1.3\,\mu$m. Note that in this case, the losses result only from the evanescent tail of the optical mode in the substrate. A fully confined mode in such a substrate would have a loss of 4.3 dB/cm, also at $\lambda_0 = 1.3\,\mu$m.

Other doping-induced absorption effects are present near the band edge. These are due to deep impurity levels, excitons, band filling, and interband absorption, such as the near edge absorption features in InP, which is doped to different levels of impurity concentration and led to a significant free-carrier absorption increase at longer wavelengths as well as a decreasing region just below the band edge due to interband absorption.

A very important III-V technology that is evolving, which competes with Si photonics (see below) is that based on InP photonics. This technology has been the subject of a recent very thorough review article by a European group, Smit et al. (2014).

2.6 Single-Crystal Si and Amorphous Si

2.6.1 Overview

As discussed in the introduction to this book, Si has, over the last 1–2 decades, become a major materials platform for integrated optics and PICs in particular. The high-quality fabrication that is possible using silicon CMOS electronics planar processing technology is simply unparalleled for realizing the very high spatial image resolution and large-scale size needed for commercial PICs. Two solid phases of Si material have been suggested for Si photonics. The first is that of crystalline silicon and is typically based on silicon-on-insulator (SOI) wafers. The second uses amorphous Si, which has attractive important advantages for nonlinear Si optics but has consistently shown degradation in its optical properties over time in working devices. (Foster 2015)

Several important and pressing applications are drivers for the development of silicon photonics. Thus Si based on SOI is developing into the key materials technology for truly large-scale and high-performance integrated optics for communications and computer interconnects (Shen et al. 2019). This development has grown rapidly, due to recent demonstration of fabricating essential and basic integrated-photonics Si building blocks, including modulators, lasers, detectors, and passive optics. In addition, other building-block active devices have been developed including current driven switches, very low-energy modulators, simple ultrafast thermo-optical switches. The scaling down of these devices to the ultrasmall ($<0.1 \, \mu m^2$-area) dimensions for single-mode waveguides, called Si nanowires (see Chaps. 3 and 4) has given rise to a new waveguide technology. This technology has been made possible by using the Si fabrication tools and infrastructure used in modern CMOS Si fabrication foundries for integrated electronics.

The unique properties of silicon-wire photonic devices are determined by the high refractive contrast available in the silicon-on-insulator (SOI)-materials platform. This high contrast permits extremely tight confinement in two dimensions, hence the name Si wires. (Driscoll et al. 2015) This high contrast allows deep scaling of single-mode devices so as to have ultrasmall Si waveguide cross sections ($<0.1 \, \mu m^2$) and consequently allows for design of compact device areas. Note that the utility of high-index contrast has been noted in earlier generations of devices using III-V materials for integrated optical technologies; see, for example, Levy et al. (1999), Huan et al. (2000). The device size reduction yields a set of more subtle but very important advantages, such as the possibility for engineering of waveguide dispersion, high optical-field densities, reduced group velocities and thus group-velocity-enhanced effects, and unique optical properties of the supported guided modes such as a strong longitudinal electric field component.

Fig. 2.7 Refractive index
versus wavelength for silicon
at room temperature

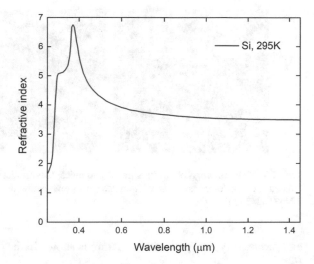

2.6.2 Materials Properties of SOI

State-of-the-art silicon devices have generally been made of silicon based on the silicon-on-insulator (SOI) materials platform. This platform enables large-scale wafers to be fabricated using ion-induced film separation followed by wafer bonding. Silicon has a very large refractive index (n = 3.5), which, in conjunction with a low-index waveguide cladding (n = 1 for air or n = 1.45 for silica), permits tight confinement of light to sub-wavelength dimensions. Because of the ultrasmall dimensions of the oxide, geometrical dispersion can have a major effect on its effective (see Chap. 3) refractive index. Both of these materials also exhibit significant chromatic dispersion. The dispersion in the refractive index for Si (Green 2008) is shown in Fig. 2.7.

As introduced above Si photonic wires are waveguides typically patterned on a SOI wafer or chip with a 1–3 μm-thick buried oxide layer (BOX), 100–300 nm Si device layers. Typically many of the photonics devices based on SOI fabrication employ a 220 nm layer because of its modal properties. The thickness of the underlying oxide layer is important to reduce optical coupling to the substrate. In fact, in some cases sufficiently low optical loss can only be obtained via under etching of the buried oxide layer. Finally note that the electrical properties of the relatively thick SOI can severely limit its use if significant electrical performance is required. In fact, its thickness does not allow co-integration of CMOS and SI photonics in most cases. Thus there is currently a major research thrust in using bulk Si for projects needing full integration.

The crystal orientation of the chip is important for certain fabrication steps for the chip. The direction of the propagating optical wave is typically oriented so that the waveguide is aligned along the [1 1 0] crystallographic axis of the Si device layer (Driscoll et al. 2015). This choice of [1 1 0] orientation is selected to place the waveguide facet along a cleavage plane of Si for the ready formation of high-quality

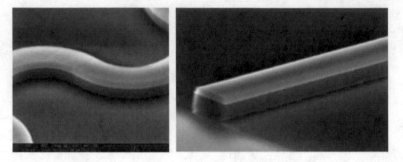

Fig. 2.8 SEM micrograph of a silicon photonics waveguide, with an overlay of the calculated TE mode-profile *Source* https://www.osapublishing.org/oe/fulltext.cfm?uri=oe-23-25-32452&id=333314

end facets. A typical patterned structure is shown in the SEM micrograph shown in Fig. 2.8.

2.6.3 Loss and Patterning

Because of the high quality of Si lithography patterning, the optical loss at C-band wavelengths in Si-wires is largely determined by optical scattering loss induced by etch-induced sidewall roughness of the etched waveguide. This loss, including a useful expression for the upper bound of the loss, is discussed in Yamada (2011) and is generally a result of etch-induced defects. Thus scattering loss can be reduced by using advanced dry-etching methods such as hydrogen bromide reactive-ion-etching. In its current state, state-of-the-art waveguide losses for channel waveguides using such etching and e-beam patterning are typically measured to be $\alpha \sim 1.0 - 3.0$ dB/cm for the 1550 nm range. If the operating wavelength is in the mid-IR region scattering losses can be smaller but substrate absorption will increase (Liu et al. 2010). The high confinement of wavelength-scale features due to Si's high refractive index difference reduces optical loss at bends. This effect enables tight folding of Si PICs and leads to much higher packing density than previously possible.

The origin of loss due to wall roughness on Si wires has been examined using analytic coupled-mode theory as well as numerically using finite-difference time-domain (FDTD) computations. In addition it is important that, as mentioned above, the buried oxide layer sufficiently thick to prevent evanescent coupling to the bulk Si substrate (Poulton et al. 2006).

2.6.4 Doping-Induced Index Changes and Loss

Si does not possess an intrinsic χ^2 and as a result it is not possible to use a simple intrinsic layer of Si for electro-optical modulation or switching. As a result, alternative approaches to modulation have been used. The most common electro-optical approach involves doping of the Si to form a local concentration of free carriers, which exhibit the well-known carrier physics seen in basic electronic device books and discussed earlier in this chapter for the case of compound semiconductors, such as GaAs or InP.

2.6.5 Temperature Variation of the Si Index of Refraction

In many photonics applications, temperature stability, or conversely, the ability to change phase with temperature is an important quantity in device design. Crystalline Si has a significant temperature induced-refractive index shift that is due to changes in its electronic structure and shifts in carrier physics. This shift, dn_0/dT, can be normalized to the value of n_0, which for Si at $\lambda = 1.55\,\mu m$ is $5.2 \times 10^{-5} K^{-1}$. This quantity will be seen later in this text to be important for thermo-optical switches and phase stability in resonant and non-resonant devices.

2.6.6 Amorphous Silicon

Many applications in Si photonics require interconnects, which are deposited on nonplanar and fragile substrates; a particular example is the formation of 3D optical interconnects. This deposition capability is not possible with crystalline Si waveguides, but it is important for low cost, flexible, or even very dense-device fabrication.

In principle it should be possible to use standard polycrystalline Si application. However pure α-Si cannot be used because of its poor optical properties, particularly its high optical absorption due to its high concentration of free dangling bonds. On the other hand, if these free bonds are reacted with hydrogen, optical absorption drops dramatically. This hydrogenated amorphous silicon (α-Si:H) is deposited using low temperature ($\sim$100–400 °C, compared to poly) plasma-enhanced chemical vapor deposition (PECVD). In addition to its flexibility in deposition its use can also reduce propagation loss (Burns and Hocker 1977) for at least one form of noncrystalline Si (Foster 2015). As an example of a loss of 3.46 dB/cm has been achieved for amorphous Si in a 480 nm × 220 nm Si wire at 1550 nm (TE mode) (Soref 1993). This result indicates that the propagation loss in α-Si:H waveguides are approaching that of the best Si wire waveguides, i.e., $1 - 3$ dB/cm (Kimerling et al. 2004; Reed 2004). Do note however that the crystalline silicon of silicon-on-insulator substrates is the only possibility for high-index contrast waveguiding.

In addition, α-Si:H waveguides have a high thermal stability, which is useful in stacked waveguides. It has been reported that these α-Si:H waveguides can be stable even at room temperature (Soref 1993) and their propagation loss does not increase after rapid annealing at 550 °C (Burns and Hocker 1977). In contrast the thermal budget of in CMOS is limited to that of the back end process of $\sim$450 °C.

2.7 Summary

This section of our text has focused on a short overview of the materials properties for five choices of integrated optical platforms. Each of these has been used in commercial devices and thus we have chosen to review each. In our review we have also attempted to give brief mention of the devices that are important for each material type. We have also given the typical properties for each material class. Note that the choice of materials is determined by the PIC application.

Problems

1. Coupling to a fiber-optic input/output for a photonic chip is important for a low loss optical system.
   Find the FWHM of the mode from a single-mode commercial fiber; what is the approximate shape of the mode? You will have to do some searching (web, references, books).
2. Determine the Al Composition, i.e.,, x, needed to obtain a $\Delta n = 0.04$ for GaAs/Al$_x$Ga$_{1-x}$As structure at $\lambda = 1.3\,\mu$m and $\lambda = 0.9\,\mu$m, respectively.
3. A Ti diffused waveguide is made in LiNbO$_3$ by annealing the following structure at 1050 °C. If the Ti stripe thickness is 900 Å, how long will it take to change the index by 1×10^{-3}? The stripe width is 8 μm. Sketch the index profile at $x = 0$, along Y, $\lambda = 1.3\,\mu$m (Fig. 2.9).
4. Derive the wave equation for $\vec{E}_x$, propagating along $+z$, in the homogeneous medium having ε, μ_o.
5. It's important to get a sense of how the waveguides are actually manufactured using different materials and techniques. Look into this chapter and search online to answer the following questions about fabrication techniques:

   (a) Sketch and explain how a GaAs/AlGaAs slab waveguide is made.
   (b) Sketch and explain how a LiNbO$_3$ channel waveguide is made.

Fig. 2.9 Cross-sectional view of a Ti-diffused Waveguide

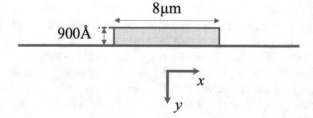

References

Burns, W. K., & Hocker, G. (1977). End fire coupling between optical fibers and diffused channel waveguides. *Applied Optics, 16*(8), 2048–2050.

Chen, D., Fetterman, H. R., Chen, A., Steier, W. H., Dalton, L. R., Wang, W., et al. (1997). Demonstration of 110 GHz electro-optic polymer modulators. *Applied Physics Letters, 70*(25), 3335–3337.

Driscoll, J. B., Osgood, R. M., Grote, R. R., Dadap, J. I., & Panoiu, N. C. (2015). Squeezing light in wires: Fundamental optical properties of Si nanowire waveguides. *Journal of Lightwave Technology, 33*(14), 3116–3131.

Eldada, L., & Shacklette, L. W. (2000). Advances in polymer integrated optics. *IEEE Journal of Selected Topics in Quantum Electronics, 6*(1), 54–68.

Eldada, L. A., Beeson, K. W., Pant, D., Blomquist, R., Shacklette, L. W., & McFarland, M. J. (2000a). Polymeric components for all-optical networks. In *Optoelectronic integrated circuits IV* (Vol. 3950, pp. 78–90). International Society for Optics and Photonics.

Foster, A. C. (2015). Advances in amorphous silicon waveguides for nonlinear optical signal processing. In *2015 IEEE Summer Topicals Meeting Series (SUM)* (pp. 90–91). IEEE.

Green, M. A. (2008). Self-consistent optical parameters of intrinsic siliconat 300 K including temperature coefficients [J]. *Solar Energy Materials and SolarCells, 92*(11), 1305–1310.

Henry, C. H., Blonder, G., & Kazarinov, R. (1989). Glass waveguides on silicon for hybrid optical packaging. *Journal of Lightwave Technology, 7*(10), 1530–1539.

Huan, Z., Scarmozzino, R., Nagy, G., Steel, J., & Osgood, R. (2000). Realization of a compact and single-mode optical passive polarization converter. *IEEE Photonics Technology Letters, 12*(3), 317–319.

Kimerling, L., Dal Negro, L., Saini, S., Yi, Y., Ahn, D., Akiyama, S., Cannon, D., Liu, J., Sandland, J., Sparacin, D., et al. (2004). Monolithic silicon microphotonics. In *Silicon photonics* (pp. 89–120). Berlin: Springer.

Korotky, S., Alferness, R., & Hutcheson, L. (1987). Integrated optical circuits and components.

Leonberger, F., & Donnelly, J. (1988). Semiconductor integrated optic devices. In *Guided-wave optoelectronics* (pp. 317–395). Berlin: Springer.

Levy, D. S., Park, K. H., Scarmozzino, R., Osgood, R. M., Dries, C., Studenkov, P., et al. (1999). Fabrication of ultracompact 3-dB 2 x 2 MMI power splitters. *IEEE Photonics Technology Letters, 11*(8), 1009–1011.

Liu, X., Osgood, R. M, Jr., Vlasov, Y. A., & Green, W. M. (2010). Mid-infrared optical parametric amplifier using silicon nanophotonic waveguides. *Nature Photonics, 4*(8), 557.

Poulton, C. G., Koos, C., Fujii, M., Pfrang, A., Schimmel, T., Leuthold, J., et al. (2006). Radiation modes and roughness loss in high index-contrast waveguides. *IEEE Journal of Selected Topics in Quantum Electronics, 12*(6), 1306–1321.

Reed, G. T. (2004). Device physics: The optical age of silicon. *Nature, 427*(6975), 595.

Shen, Y., Meng, X., Cheng, Q., Rumley, S., Abrams, N., Gazman, A., et al. (2019). Silicon photonics for extreme scale systems. *Journal of Lightwave Technology, 37*(2), 245–259.

Smit, M., Leijtens, X., Ambrosius, H., Bente, E., Van der Tol, J., Smalbrugge, B., et al. (2014). An introduction to InP-based generic integration technology. *Semiconductor Science and Technology, 29*(8), 083001.

Soref, R. A. (1993). Silicon-based optoelectronics. *Proceedings of the IEEE, 81*(12), 1687–1706.

Yamada, K. (2011). Silicon photonic wire waveguides: Fundamentals and applications. In *Silicon photonics II* (pp. 1–29). Berlin: Springer.

Chapter 3
Dielectric Slab Waveguide

Abstract The origin of guided-wave behavior is presented for the case of slabs and for thin films of dielectric materials. The discussion examines the origin of the guided wave using a classical optics approach and that of Maxwell's equations, as well as the matching of fields at the interfaces of dielectric layers. The chapter also includes various approximations and graphical methods for waveguides with abrupt and diffused geometry.

3.1 Introduction

Waveguides are the wires of photonic integrated circuits. They transport light just as metallic wires transport electrons. Unlike wires, however, waveguides or an assembly of waveguides can be used as functional passive elements in a circuit. There are many different forms of waveguides, including those with a variety of geometries and materials. Generally, however, a waveguide typically has a light-confining structure obtained by surrounding a high-index material with a low index cladding. Two examples are shown in Fig. 3.1. Such waveguide functions by virtue of confining the light in the core through the phenomenon of total internal reflection (TIR).

3.2 "Thin-Film" Waveguides

3.2.1 Slab Waveguide (2D)

Most integrated waveguides are formed from thin-film structures. The structure fabrication starts with a substrate, to which is added a higher index film and lower index cover (which could be air); see Fig. 3.2. In the simplest of these structures, which is termed a "slab" waveguide, total internal reflection confines light vertically, but not horizontally. Since a slab waveguide confines light only vertically, it is in itself not typically very useful for routing light within a PIC. A slab waveguide, however,

© Springer Nature Switzerland AG 2021
R. Osgood jr. and X. Meng, *Principles of Photonic Integrated Circuits*,
Graduate Texts in Physics,
https://doi.org/10.1007/978-3-030-65193-0_3

Circular Symmetry

Rectangular Symmetry

n_{core}

$n_{cladding}$

n_{core}

$n_{cladding}$

(Easy to analyze due to symmetry)   (Cannot be analyzed exactly)

Fig. 3.1 The cross section of two important waveguide shapes: circular and rectangular waveguides. Despite its difficulty for analysis, the rectangular waveguide is the most commonly used type for analysis

Fig. 3.2 The cross section of a typical thin-film-based structure. In general the index of the cover and substrate region are not identical

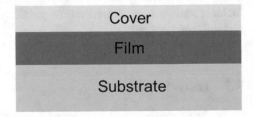

Cover

Film

Substrate

has the advantage of being able to be exactly analyzed: it thus can serve as a model structure for understanding the basic physics and technology of waveguiding.

In practical PIC and OEIC chips, however, light must also be confined laterally in two dimensions in order to achieve the two-dimensional routing, that is needed for the working surface of the light circuit. Lateral confinement is achieved by introducing changes in the lateral geometry or in the material index of refraction. As a result, a physics analysis of PICs requires understanding of light propagation in these 3D waveguides; see Fig. 3.3. Three-dimensional guides are surprisingly difficult to analyze, although they may be fabricated relatively easily. Introduction and analysis of different types of 3D waveguides will be considered in the next chapter.

A full description of waveguiding requires consideration of several areas of physics. These include general waveguide phenomena, such as lightwave polarization, guided modes, and radiation modes. In addition, the relation of waveguide geometry to the properties of the guided wave, such as modal "shape" and its relation to guided-wave loss, and coupling, need to be considered.

In this chapter, the analysis of basic slab waveguides will be demonstrated. Note, however, that much of the behavior of 3D guides can be anticipated via a thorough understanding of 2D waveguides.

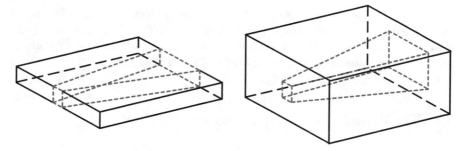

Fig. 3.3 Waveguides that lack (left) and that incorporate (right) lateral wave confinement. The 2D or "slab" waveguide spreads by diffraction as the waveguide beam propagates along in the x-direction. For the 3D waveguide, index guiding typically leads to a confined structure

3.3 Slab Waveguide: Ray Optics Picture

One of the most readily understandable approaches to slab waveguides is that obtained from ray optics, combined with the known interaction of a plane wave with a dielectric interface. This approach is useful for understanding both the phase conditions for transverse modes to exist, as well as the variation in phase velocity with mode number. To apply ray optics, we start with a plane wave incident at some angle and reflecting within the film. The confinement by the waveguide is determined by the behavior of the plane wave as it strikes the interface.

There are three possible trajectories for a ray, which travels within the slab structure. In the first, there is total internal reflection (TIR) at the cover and substrate interface, and a trapped or guided ray is then possible. In the second instance, there is TIR at the cover interface only, while the ray scatters at the substrate interface into the "substrate." Finally, in the third case, there is no TIR at either interface and radiative loss occurs into both the substrate and the cladding.

Since total internal reflection at either a top or bottom interface is important, we define two critical angles:

$$\theta_c \equiv \sin^{-1} \frac{n_c}{n_f}, \quad \theta_s \equiv \sin^{-1} \frac{n_s}{n_f} \tag{3.1}$$

where $n_c \leq n_s < n_f$, $\theta_c \leq \theta_s$, and where $n_{c,s,f}$ are the index of refraction in the cladding, substrate, and core region. Now, for a ray to radiate through the top or bottom surface of the waveguide, θ_t must be a real angle in the range $0 \leq \theta_t \leq 90°$; this implies $\sin \theta_t \leq 1$. Thus, for the substrate the existence of refracted beam requires, by Snell's law, that $\theta_i \leq \sin^{-1}(n_s/n_f) \equiv \theta_{\text{crit}}$. When $\theta_i > \theta_{\text{crit}}$, we have the condition of total internal reflection, in which condition, light is trapped in the waveguide.

It is possible to continue to use the ray optics approach in conjunction with the known interfacial phase shifts for a plane wave so as to analyze the dielectric slab waveguide, and to obtain expressions for modal dispersion, etc. (Kogelnik 1988). This analysis shows that a mode exists when the total phase shift for a ray, in traversing the waveguide from the substrate to cladding and back to the substrate, is an integral number of 2π, say $2m\pi$, where m is an integer. However, a more useful and complete approach is obtained by directly using Maxwell's equations.

3.4   The Wave Equation for a Slab Waveguide

3.4.1   The Wave Equation and Its Boundary Conditions

The modal fields of a slab waveguide, as shown in of Fig. 3.4, are obtained directly by using the wave equation to solve for the relative fields in the dielectric slab and then applying the boundary condition in the transverse directions at the slab boundaries. In this derivation, a simple harmonic, $e^{i\omega t}$, time dependence, is assumed, in part, since monochromatic light signals are typically used in PICs. Next, the waveguide is assumed to have a medium, which is uniform within, but of different indices, inside and outside of the waveguide, and of nonmagnetic character. As seen in the sketch in Fig. 3.4, the slab consists of three layers, each with its owe indices, n_c, n_s, and n_f. Application of Maxwells equations, thus,

$$\nabla \times \vec{E} = -i\omega\mu_0\vec{H} \tag{3.2}$$

$$\nabla \times \vec{H} = +i\omega\epsilon_0 n^2 \vec{E} \tag{3.3}$$

where

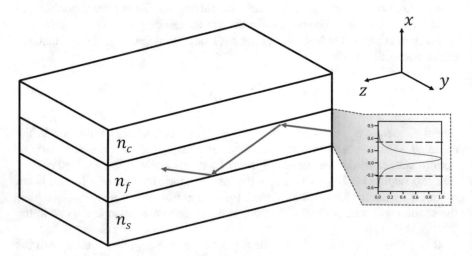

Fig. 3.4 Light confined and propagating in a slab waveguide

$$n^2 \equiv \frac{\epsilon}{\epsilon_0} \tag{3.4}$$

Then gives the wave equation

$$\nabla^2 \vec{E} + k^2 n^2 \vec{E} = 0 \tag{3.5}$$

where k, the wavenumber in vacuum, is

$$k = \omega \sqrt{\epsilon_0 \mu_0} = \frac{2\pi}{\lambda} \tag{3.6}$$

This well-known form of the wave equation is called the Helmholtz equation; it is used repeatedly in waveguide problems.

The geometry of the slab waveguide is unconstrained in the longitudinal or z-direction, and hence waves in the z-direction are propagating waves. As a result, the z-dependence is of the form $e^{-i\beta z}$ for forward-traveling waves, where β is the propagation constant in the z-direction.

The magnitude of the propagation wavenumber in a free dielectric medium, which forms the core of the waveguide, is kn. In a waveguide, transverse, i.e., top and bottom, confinement causes the allowed propagating wave modes to have wavevector components in the z- and x-direction, such that

$$k_x^2 + \beta^2 = n^2 k^2 \tag{3.7}$$

One goal of solving the waveguide problem is thus to determine the mode-dependent values of k_x and β.

The Helmholtz equation as shown above assumes a uniform medium; thus it only holds within each layer but not at the interfaces. At the interfaces, boundary conditions, which are derived from Maxwell's equations, must be applied. These boundary conditions are the continuity of the transverse components of $\vec{E}$ and $\vec{H}$. In addition, for a slab waveguide, it is assumed that the slab waveguide solutions are uniform in y, the second lateral direction as shown in Fig. 3.4, and thus, $\frac{\partial \vec{E}}{\partial y} = 0$.

3.4.2 Effective Index

Since the z-dependence of the propagating wave in the guide is determined by β, it is important to examine the implications of having $\beta < nk$. For example, since the effective wave velocity along z is defined as follows:

$$v = \frac{\omega}{\beta} = \frac{ck}{\beta} = \frac{c}{N_{\text{eff}}} \tag{3.8}$$

the wave velocity in the z-direction can be viewed as resulting from an "effective" index,

$$N_{\text{eff}} \equiv \frac{\beta}{k} \qquad (3.9)$$

which describes the modification of the wave velocity by the waveguide. Note that this definition encompasses both material- and geometry-dependent effects, through n and d, respectively, where d is the waveguide thickness.

The allowed values of β are determined by the indices of refraction of the three layers. Specifically, recall that guided modes exist in the range $\theta_s < \theta < 90°$; thus

$$n_s < \frac{\beta}{k} \, (\text{or } N_{\text{eff}}) < n_f \qquad (3.10)$$

and similarly, we have

$$\frac{\beta}{k} < n_s \qquad (3.11)$$

for any radiation modes.

3.4.3   Polarization: Why TE and TM

Slab waveguide modes are normally classified as either transverse electric (TE) or transverse magnetic (TM). This choice of terminology is also used for the two possible polarizations of the fields. TE-polarized light has its electric field only in a plane parallel to that of the waveguide slab, while for the TM polarization, the magnetic field lies in this plane. Note that substitution in Maxwell's equations shows that these two polarizations are independent of each other. In addition, because of the waveguide confinement in the x-direction, it can also be shown that the magnetic field for the TE polarization and the electric field in the TM polarization each possesses a non-zero component along the z or propagation axis. Thus the waves deviate from the fully transverse behavior of simple plane waves; this "deviation" can ultimately be seen to originate from the transverse confinement of the lightwave.

Specifically, if we solve the wave equation for the slab waveguide geometry, along with the appropriate boundary conditions, we find that, by convention, there is one solution in which the longitudinal electric field is zero, which implies that E is only transverse, the TE mode. Similarly, the other solution has a zero longitudinal magnetic field, the TM mode. These two solutions result from the application of the two different boundary conditions. If E_z and H_z are given, the other components can be determined, and, in fact, the longitudinal components of $\vec{E}$ and $\vec{H}$ determine the optical field entirely. The wave equations for the two separate polarization cases are summarized in Table 3.1.

In the following section, we derive the modal properties for each of these two cases. This derivation is accomplished by solving the appropriate wave equation

Table 3.1 Summary of the wave equations for TE and TM modes

	Wave equation	Boundary condition
TE mode ($E_z = 0$)	$\dfrac{\partial^2 E_y}{\partial x^2} + (k^2 n^2 - \beta^2) E_y = 0$	E_y continuous
		H_z continuous $\Longrightarrow \dfrac{\partial E_y}{\partial x}$ cont.
TM mode ($H_z = 0$)	$\dfrac{\partial^2 H_y}{\partial x^2} + (k^2 n^2 - \beta^2) H_y = 0$	H_y continuous
		E_z continuous $\Longrightarrow \dfrac{1}{n^2} \dfrac{\partial E_y}{\partial x}$ cont.

with boundary conditions given in Table 3.1. This approach yields both the field profiles and the dispersive properties, i.e., β versus ω for the two cases.

3.4.4 TE Guided Modes

3.4.4.1 Dispersion Relation

The simplest case to examine is that for the TE guided mode, i.e., for E_y. Then, a solution is sought for E_y, which is confined within the slab in the x-direction. Confinement in the x-direction will yield a mode which is sinusoidal inside the slab with wavenumber in lateral direction of $\kappa \equiv k_x$, and has exponential decay outside as governed by δ and γ. Thus we write

$$E_y = \begin{cases} C e^{-\delta x} & x > 0 \\ A \cos \kappa x + B \sin \kappa x & -d \leq x \leq 0 \\ D e^{+\gamma x} & x < -d \end{cases} \tag{3.12}$$

where δ, κ, γ, A, B, C, and D are parameters to be determined from the wave equation, with application of the appropriate boundary conditions.

Substitution of this form of E_y into the Helmholtz equation in each region yields the following relationships:

$$\begin{aligned} -\delta^2 &= k^2 n_c^2 - \beta^2 \\ -\gamma^2 &= k^2 n_s^2 - \beta^2 \\ \kappa^2 &= k^2 n_f^2 - \beta^2 \end{aligned} \tag{3.13}$$

note that the last equation of (3.13) is identical to (3.7). Further, the boundary condition on the continuity of E_y across the interface gives $C = A$ and $D e^{-\gamma d} = A \cos \kappa d - B \sin \kappa d$.

Thus,

$$E_y = \begin{cases} Ae^{-\delta x} & x \geq 0 \\ A\cos\kappa x + B\sin\kappa x & -d < x < 0 \\ (A\cos\kappa d - B\sin\kappa d)e^{\gamma(x+d)} & x \leq -d \end{cases} \qquad (3.14)$$

Now, applying the continuity of H_z or $\dfrac{\partial E_y}{\partial x}$ across the interface then yields the following:

$$-\delta A = \kappa B \qquad (3.15)$$

and

$$\gamma(A\cos\kappa d - B\sin\kappa d) = \kappa(A\sin\kappa d + B\cos\kappa d) \qquad (3.16)$$

Equation (3.15) gives $B = -(\delta/\kappa)A$; inserting this relation into (3.16) then yields the following relation between δ, γ and κ:

$$\tan\kappa d = \frac{\kappa(\gamma + \delta)}{\kappa^2 - \gamma\delta} \qquad (3.17)$$

Expressing the relations of δ, γ, and κ in terms of β and k, i.e., (3.13), and inserting them into (3.17), we obtain an eigenvalue equation for the allowed β's. In fact, a careful analysis of (3.17) shows that this equation leads to an equivalent way of expressing the fact that the phase accumulation of a ray in the waveguide during one complete transverse path must be as an integer number times 2π.

Equation 3.17 can be solved numerically using commercial mathematical solvers, or alternatively the equation can be solved graphically.

The particular value of κd given in (3.18) as follows:

$$\kappa d = kd\sqrt{n_f^2 - n_s^2} \qquad (3.18)$$

is such an important physical quantity in designing waveguides that it is often called the "normalized frequency" of the guide, V see, for example, Nishihara et al. (1989). Thus,

$$V = kd\sqrt{n_f^2 - n_s^2} \qquad (3.19)$$

Note that it is a dimensionless quantity. As a practical matter, V is also directly proportional to the numerical aperture, NA, of the waveguide, since $NA = V/kd$.

The allowed modes, the cross points of lift and right-hand side of (3.17), are shown in Fig. 3.5. Note that as the waveguide V increases, the number of modes also increases; that is, large d (compared to λ) and higher index contrast, i.e., $\Delta n = n_f - n_s$ lead to a larger number of modes. As V decreases (thus implying a narrower core, a smaller n, a longer wavelength, or a low optical frequency), the number of modes decreases and, in fact, a width will be reached where no modes can exist in the case of an asymmetric slab waveguide. This condition is termed waveguide "cutoff".

Fig. 3.5 A plot of the functions in the left and right-hand side of (3.17). The cross points are the discrete solutions in κd for the allowed guided mode

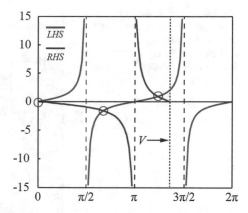

A more explicit equation displaying the mode number m can be obtained as an alternately written version of (3.17). This equation expresses the fact that the boundary-condition equations can have a non-trivial solution if

$$\kappa d = (m+1)\pi - \tan^{-1}\left(\frac{\kappa}{\gamma}\right) - \tan^{-1}\left(\frac{\kappa}{\delta}\right) \qquad (3.20)$$

where $m = 0, 1, 2\ldots$ This equation is termed the dispersion relation for the waveguides, since it specifies the allowed values of β and ω (or k) for the waveguide geometry and its optical properties. The equation also expresses the fact that the total lateral phase shift for each mode is an integral value of 2π, since the phase shifts at the cladding-core and substrate-core interface are

$$\phi_s = \tan^{-1}\left(\frac{\kappa}{\gamma}\right) \qquad (3.21)$$

and

$$\phi_c = \tan^{-1}\left(\frac{\kappa}{\delta}\right) \qquad (3.22)$$

respectively, and the phase shift in passing through the waveguide is κd. This dispersion relation (3.20), which can also be obtained using the ray optics approach, is plotted in Fig. 3.6 for the first three modes of a symmetric waveguide, i.e., $n_c = n_s$.

The dispersion relation (3.20) implies that a slab waveguide will possess a mode-dependent group velocity, $v_g = \partial\omega/\partial\beta$. This velocity can be obtained by taking the total derivative directly from the dispersion relation after expressing κ, δ and γ in terms of k and β. The velocity can be seen graphically as the slope in the dispersion at a specific ω for each mode as shown in Fig. 3.6.

In practical waveguide (and fiber optic) design, recourse is typically made to a useful graph which contains quantities that are normalized and thus generally applicable to any slab waveguide. First, recall that we have previously introduced

Fig. 3.6 The normalized dispersion curve between wave vector and frequency. This plot enables rapid design of single-mode behavior

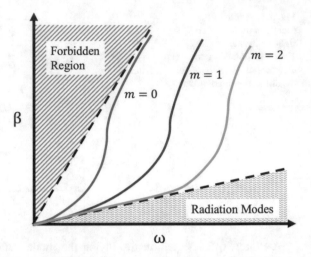

Fig. 3.7 The dispersion relation between normalized frequency and normalized guide index for a slab waveguide operating on three different spatial different modes

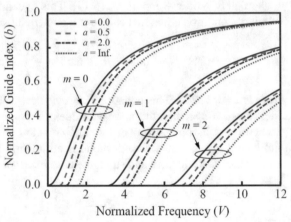

one such quantity, i.e., V, as defined above. In addition, two other quantities, "a" and "b," can be introduced. The quantity "a" is termed the waveguide asymmetry, since it quantifies how much the cover and substrate index, n_c and n_s, differ from each other, where

$$a \equiv \frac{n_s^2 - n_c^2}{n_f^2 - n_s^2} \tag{3.23}$$

If $n_c = n_s$, then $a = 0$. The quantity b is typically termed the normalized waveguide index and is defined by

$$b \equiv \frac{\left(\frac{\beta}{k}\right)^2 - n_s^2}{n_f^2 - n_s^2} = \frac{N_{\text{eff}}^2 - n_s^2}{n_f^2 - n_s^2} \tag{3.24}$$

Table 3.2 Summary of parameters used to describe the behavior of slab waveguides for TE and TM light

	TE	TM
Normalized thickness	$V \equiv kd\sqrt{n_f^2 - n_s^2}$	$V \equiv kd\sqrt{n_f^2 - n_s^2}$
Asymmetry factor	$a = \dfrac{n_s^2 - n_c^2}{n_f^2 - n_s^2}$	$a_m \equiv \dfrac{n_f^4}{n_c^4}a$
Normalized guide index	$b \equiv \dfrac{(N_{\text{eff}})^2 - n_s^2}{n_f^2 - n_s^2}$	$b_m = \left(\dfrac{n_f}{n_s q_s}\right)^2 b,$
		where $q_s =$ $\left(\dfrac{N_{\text{eff}}}{n_f}\right)^2 + \left(\dfrac{N_{\text{eff}}}{n_s}\right)^2 - 1$

Thus when β/k or $N_{\text{eff}} \to n_s$, $b \to 0$; and when β/k or then $N_{\text{eff}} \to n_f$, then $b \to 1$. These three normalized quantities are summarized in Table 3.2.

The definitions in Table 3.2 allow the dispersion relation (3.20) to be written in a normalized form:

$$V\sqrt{1-b} = m\pi + \tan^{-1}\sqrt{\frac{b}{1-b}} + \tan^{-1}\sqrt{\frac{b+a}{1-b}} \qquad (3.25)$$

This expression is often alternatively written as

$$V\sqrt{1-b} = (m+1)\pi - \tan^{-1}\sqrt{\frac{1-b}{b}} - \tan^{-1}\sqrt{\frac{1-b}{b+a}} \qquad (3.26)$$

One can solve this equation numerically and obtain a general graph of its solutions; this graph is shown in Fig. 3.7. This curve is called the normalized dispersion curve, since it plots the variation of the normalized guide index, b, versus normalized frequency, V. The diagram also provides a convenient means for determining modal cutoff.

The graph in Fig. 3.6 shows that cutoff occurs sequentially for higher order modes as the waveguide is made thinner or as waveguide confinement is reduced. This cutoff condition is reached when V is less than its cutoff value V_c^m for a specific mode m. Using (3.25), the normalized frequency at cutoff can be found to occur at

$$V_c^m = m\pi + \tan^{-1}\sqrt{a} \qquad (3.27)$$

Notice that, for a symmetric waveguide, $a = 0$ and the lowest order mode, $m = 0$, mode does not have a "cutoff" value. In addition, (3.27) shows that for a waveguide with normalized frequency V, the number of allowed modes, m, is given by $m \approx V/n$.

3.4.4.2 Approximate Solutions

Very useful approximate, but analytic forms for the above equations can be obtained from the normalized dispersion curve near certain limiting regions. For example, near cutoff, $b \to 0$, is found that

$$V\sqrt{1-b} = m\pi + \sqrt{\frac{b}{1-b}} + \tan^{-1}\sqrt{a} \tag{3.28}$$

or

$$V\sqrt{1-b} = V_c\sqrt{1-b} + b \tag{3.29}$$

Finally note that other expressions are found in other limiting regions.

3.4.4.3 Normalization of Waveguide Modes

In order to understand how the shape of a waveguide mode varies with its material properties, the modes local field structure needs to be normalized. Thus recasting (3.14), as

$$E_y = \begin{cases} Ae^{-\delta x} & x > 0 \\ A\left(\cos\kappa x - \dfrac{\delta}{\kappa}\sin\kappa x\right) & -d \le x \le 0 \\ A\left(\cos\kappa d + \dfrac{\delta}{\kappa}\sin\kappa d\right)e^{\gamma(x+d)} & x < -d \end{cases} \tag{3.30}$$

where $B = -\dfrac{\delta}{\kappa}A$ has been inserted in (3.14).

The arbitrary constant, A, can now be specified by fixing the power carried by the mode as a constant value, P. In order to do this, integrate the z-component of the Poynting vector,

$$S_z = \frac{1}{2}\text{Re}\{\vec{E} \times \vec{H}^*\}\cdot\hat{z} \tag{3.31}$$

over the transverse cross section of the guide. Thus, after using Maxwell's equations to obtain $\vec{H}$ from the form of E_y determined above, an expression for the power can be found:

$$\frac{\beta}{|\beta|}P = -\frac{1}{2}\int_{-\infty}^{\infty} E_y H_x^* dx = \left(\frac{\beta}{2\omega\mu_0}\right)\int_{-\infty}^{\infty} |E_y|^2 dx \tag{3.32}$$

where the presence of factor $\beta/|\beta|$ ensures that the power (P) is always positive. Substitution for $\beta/|\beta|$ then yields the amplitude of the lowest order mode,

Fig. 3.8 A notional plot of the effective width, T_{eff}, of a waveguide versus its physical width, d

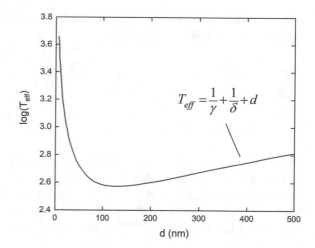

$$A^2 = \frac{4\kappa^2 \omega \mu_0 P}{|\beta|(\kappa^2 + \delta^2)\left(d + \dfrac{1}{\gamma} + \dfrac{1}{\delta}\right)} \tag{3.33}$$

This relation shows that the peak intensity of the waveguide mode depends on the confinement of the mode in the waveguide by the quantities $1/\gamma$, $1/\delta$ and d. This leads to a definition for an effective width of the waveguide, T_{eff}:

$$T_{\text{eff}} \equiv \frac{1}{\gamma} + \frac{1}{\delta} + d \tag{3.34}$$

It is instructive to determine how the effective width of a waveguide varies as its physical width is reduced. This variation is shown in a plot of normalized variables in Fig. 3.8. Notice that reducing the waveguide width,d, reduces the width of the mode only up to a certain point. For guides smaller than that, the effective width of the mode begins to expand. While this behavior implies a lack of confinement, it has important uses such as to improve mode matching in fiber-waveguide interfaces.

3.4.4.4   Modal Confinement

The adoption of normalized units is also useful for specifying practical engineering properties for waveguides. For example, it is often important, in diode laser design, to specify the "confinement factor" of a waveguide. This factor is the ratio of the power in the waveguide core to the total power in the waveguide. For a slab waveguide, this quantity (Hutcheson 1987) is

$$\Gamma = \frac{d}{t_{\text{eff}}}\left[1 + \frac{\sqrt{b}}{V} + \frac{\sqrt{b+a}}{V(1+a)}\right] \tag{3.35}$$

where the second and third terms are due to the asymmetric shape of the mode. For a symmetric waveguide, Γ reduces to

$$\Gamma = \frac{V\sqrt{b} + 2b}{V\sqrt{b} + 2} \tag{3.36}$$

3.4.4.5 Symmetric Slab

Symmetric slab waveguide are useful for easily capturing the essence of the properties of waveguide. Their dispersion relations are particularly simple, thus allowing a simple physical picture for modes with odd and even mode numbers. In this case, when $n_c = n_s$, $\delta = \gamma$ and the dispersion relation becomes $\tan \kappa d = (2\kappa\gamma)(\kappa^2 - \gamma^2)$. The use of trigonometric identities leads to two classes for the modes:

$$\tan \frac{\kappa d}{2} = \frac{\gamma}{\kappa} \quad \text{for even } m \tag{3.37}$$

and

$$\tan \frac{\kappa d}{2} = \frac{\kappa}{\gamma} \quad \text{for odd } m \tag{3.38}$$

In particular, rewriting the expression for the field in the slab by shifting the coordinate origin to the center of the slab (see Fig. 3.9), we see that even, symmetric ($A \cos \kappa x$), and odd, antisymmetric ($B \sin \kappa x$), modes exist. Recall also that for a symmetric slab ($n_c = n_s$), $a = 0$, and there is no cutoff for $m = 0$, or the lowest order mode in a symmetric guide, since we cannot have $V < 0$, which is not physical!

The field distributions in a symmetric waveguide structure are a simplification of the earlier expressions for the general structure:

$$E_y = \begin{cases} A e^{-\delta x} & x > 0 \\ A \left(\cos \kappa x - \dfrac{\delta}{\kappa} \sin \kappa x \right) & -d \leq x \leq 0 \\ A \left(\cos \kappa d + \dfrac{\delta}{\kappa} \sin \kappa d \right) e^{\gamma(x+d)} & x < -d \end{cases} \tag{3.39}$$

Finally, note that if the waveguide is not symmetric, the electric fields at the substrate-core (E_s) and cladding-core (E_c) interfaces are no longer the same. It can then be shown via the boundary condition that these ratios are

$$E_c^2 = E_f^2 \frac{n_f^2 - N_{\text{eff}}^2}{n_f^2 - n_c^2} \tag{3.40}$$

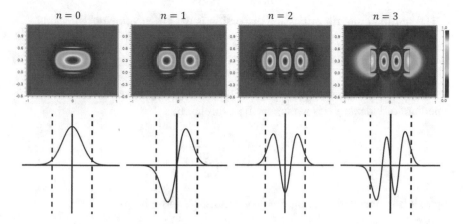

Fig. 3.9 The electrical field distribution calculated for four different TE modes, using a typical waveguide similar to that shown in Fig. 3.2

$$E_s^2 = E_f^2 \frac{n_f^2 - N_{\text{eff}}^2}{n_f^2 - n_s^2} \qquad (3.41)$$

where E_f is the maximum field amplitude in the core. Notice that the smaller the value of n_c, n_s, the smaller the electric field at the cladding, substrate interface.

3.4.4.6   TM Modes

In the derivation above, we have ignored discussion of TM modes. In fact, the derivations and results follow very closely to those for the TE modes. However, the normalized quantities "a" and "b" are different, although practically speaking "b" is often identical to that defined for the TE mode if n_s and n_f are close, that is, if the waveguide has only a small index contrast. The expression for "a" for TM modes is typically significantly different from for TE modes and thus is designated by the symbol a_m. The expression for a_m is given in Table 3.2. In addition the normalized dispersion curves for TM modes will then have a much more complicated form than that of the TE mode; however, in practice, particularly for low index contrast materials, the curves generally have a very similar shape.

3.4.4.7   Dispersive Properties of Waveguides

Waveguides are dispersive elements because of the sensitivity of propagation constants to their material properties, waveguide geometry, and mode number. These dispersive properties can be of direct importance in several different PICs, including optical delay lines and wavelength routers.

The starting point for discussing dispersion is the group velocity, defined as

$$\nu_g = \frac{d\omega}{dk} \tag{3.42}$$

This is the velocity at which a pulse of light travels. Using the expression for k in a medium of index n gives a more useful expression,

$$\nu_g = \frac{c}{n + \omega \left(\dfrac{dn}{d\omega} \right)} \tag{3.43}$$

where it is convenient to define the group index, N_g, as

$$N_g = n + \omega \frac{dn}{d\omega} = n - \lambda \frac{dn}{d\lambda} \tag{3.44}$$

When several operating wavelengths are present in a device, it is important to account for the dispersion of the group velocity with wavelength. This dispersion leads to a difference in arrival time, Δt, for pulses centered between two wavelengths separated by $\Delta \lambda$,

$$\Delta t = \frac{L}{c} \frac{dN_g}{d\lambda} \Delta \lambda \tag{3.45}$$

where L is the difference in path lengths and $\dfrac{dN_g}{d\lambda}$ is the group index dispersion.

The group index dispersion can be rewritten in terms of the second (wavelength) derivative of the refractive index,

$$\frac{dN_g}{d\lambda} = -\lambda \frac{d^2 n}{d\lambda^2} \tag{3.46}$$

thus in transparent materials the dispersion can be defined as

$$D = \frac{-\lambda}{c} \frac{d^2 n}{d\lambda^2} \tag{3.47}$$

Examples of these quantities for the case of SiO_2 are given in Fig. 3.10.

As suggested by an earlier discussion in this chapter, dispersion also can occur because of the vertical confinement in waveguides. This dispersion arises due to the sensitivity of β to the wavelength of light in a waveguide, a result apparent from the waveguide eigenvalue equation. Specifically, the group velocity in the waveguide is thus given by

$$\nu_g = \frac{d\omega}{d\beta} \tag{3.48}$$

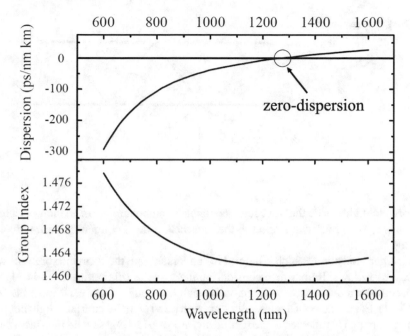

Fig. 3.10 The group index and the dispersion relation for the case of bulk SiO_2

where β has replaced the quantity k used in the previous equation. The group index, N_g, is then

$$N_g = \frac{d\beta}{dk} \tag{3.49}$$

As the wavelength of the light changes, this variation causes a change in the confinement of the mode. For example, considering the right-hand region of Fig. 3.5, as k is varied, the mode samples more of the cladding at low k and more of core at high k. As a result, the effective index is seen by the light changes with λ.

3.4.5 TE Radiation Modes

The number of guided modes is finite; thus there must be other solutions of Maxwell's equations in order to provide the complete set of modes, which must be used to describe an arbitrary initial-field configuration. In fact, such an arbitrary field distribution has to include both *guided* and *radiative* modes.

Radiative modes were encountered earlier in this chapter when discussing the slab waveguide using the ray optics picture. For example, recall that for $\theta_c < \theta < \theta_s$, or $n_c < \beta/k < n_s$, there are substrate radiation modes, or for $0 < \theta < \theta_c$, or $0 < \beta/k < n_c$, there are substrate and cladding radiation modes present. For example, if

Fig. 3.11 Ray picture of
radiation modes

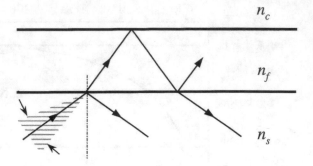

the original field inside the waveguide contains k components with an incident angle,
which is too shallow with respect to the interface, it must contain, in part, radiation
modes.

Another viewpoint, which is helpful for understanding the form the solutions will
take, is as follows. Rather than considering the wave as originating from inside the
slab, consider a plane wave incident on the slab from one side, say the substrate (see
Fig. 3.11). For an incident angle shallow enough relative to the interface, light couples
into the film and is reflected at the second air interface. At this second interface, there
is a decaying or evanescent field in the air. But also note that there is then a reflected
wave going into the substrate. The combination of incident and reflected waves in
the substrate leads to a (partial) standing wave, and this will be reflected so as to the
form of a solution for this case.

3.4.5.1   Substrate Radiation Modes

To solve for the radiation modes of the waveguide, the equations and boundary
conditions are the same as used earlier. However, the range of β is such that the
parameters γ and δ may no longer be real and the field then oscillates along the
y-direction; that is, radiation may propagate or radiate away from the waveguide.

This approach leads to developing a set of continuous radiation modes, labeled
by parameter ρ, with ρ in the range:

$$0 \leq \rho \leq k\sqrt{n_s^2 - n_c^2} \tag{3.50}$$

$$k\sqrt{n_s^2 - n_c^2} \leq \rho \leq n_s \tag{3.51}$$

However, ρ is also allowed in the range $n_s \leq \rho \leq \infty$, ($\beta$ is imaginary), and the
"substrate-cladding" form of the equations holds true. Note that these modes are
evanescent in z and are particularly needed for describing the field around imperfections.

The guided and radiation modes that are then found form a complete set. As a result, any arbitrary field distribution in the waveguide can be written as a superposition of these modes.

3.5 Graded-Index Waveguides

Thus far in our presentation of slab waveguides, we have only considered abrupt-index guides. However, many practical guides are made using fabrication techniques which lead to smoothly merging index profiles. These methods include diffusion, ion implantation, and optical exposure. Each of these techniques leads to confinement by a graded-index gradient.

Because of their relative complexity, several approaches have evolved for dealing with slab waveguides which are confined by a graded-index profile. These include analogies to well-known problems of ray optics, quantum mechanics, and the WKB method (see Kogelnik 1988). We will consider only the first here.

3.5.1 Ray Optics Approach for Diffused Waveguides

A ray optics approach can be applied to finding the normalized dispersion curve for a graded-index waveguide, using the same basic methods as for an abrupt-index slab waveguide (Nishihara et al. 1989). In this method, a ray trajectory is first "followed," and its accumulated phase shifts along the path and summed at the interfaces. The principal value of this method, which is very nicely discussed by Nishihara et al. (1989) is that it provides physical insight into the trapping of light in the waveguide (see Fig. 3.12). If the waveguide is formed such that the index, $n(x)$, decreases with x, the distance into the cladding-core interface, a ray propagating both in the x- and z-direction undergoes a continual change in trajectory as measured by the angle θ_c such that $\cos \theta_i = N_{\text{eff}}/n(x)$, where i denotes the ith increment of the ray's path. The angle, θ_i, is the angle between the ray and the z-direction and N_{eff} is the effect index of the waveguide. At the end of the downward trajectory, $x = x_t$ and $\theta_i = 0$ when $n(x_t) = N_{\text{eff}}$. At this point, the wave turns upward toward the guide, where x_t is the turning point of the ray. Thus x_t can be defined as the *effective thickness* of the diffused waveguide.

In this waveguide class, the waveguide modes are obtained by quantizing the cumulative phase change in units of 2π as the ray undergoes one complete path length of the graded-index film, i.e.,

$$\sum \phi_i + 2\phi_0 + 2\phi_{x_t} = 2\pi m \tag{3.52}$$

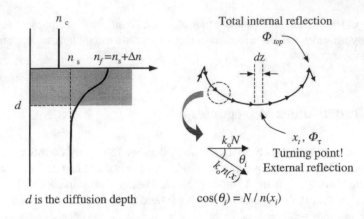

d is the diffusion depth $\cos(\theta_i) = N / n(x_i)$

Fig. 3.12 Ray picture of propagation in a graded-index waveguide

where $\phi_0 \equiv \phi_{x=0}$ and $\phi_T \equiv \phi_{x=x_t}$. These values are obtained from the Fresnel equations for reflection at a dielectric interface. In this case, using standard formulae for such a phase-shift and the assumption that $\Delta n \ll n_s$ gives the phase at the top and the bottom of the trajectory,

$$\phi_{x=0} = \frac{\pi}{2} \tag{3.53}$$

$$\phi_{x=x_t} = \frac{\pi}{4} \tag{3.54}$$

In addition, phase accumulates during each segment of the path,

$$\phi_i = k_{x_i} \Delta x_i \tag{3.55}$$

or, substituting for k_{x_i}

$$\phi_i = k n(x_i) \sin \theta_i \Delta x_i$$
$$= k \sqrt{n^2(x_i) - N_{\text{eff}}^2} \Delta x_i \tag{3.56}$$

where we have used the vector relation between k_z and k_{x_i} as shown in the insert of Fig. 3.12. The total phase change for a ray going through one entire trajectory is then given by

$$\int \phi_i = 2k \int_0^{x_t} \sqrt{n^2(x_i) - N_{\text{eff}}^2} \, dx \tag{3.57}$$

It is now possible to derive a normalized curve for a graded-index guide just as seen for an abrupt waveguide. The equation is obtained by using (3.52), along with the values for each of the phase terms. To obtain a useful expression, it is necessary to give the depth-dependent index in terms of a characteristic distribution function $f\left(\dfrac{x}{d}\right) = f(\zeta)$, where we have defined $\zeta \equiv \dfrac{x}{d}$ for simplicity. Then

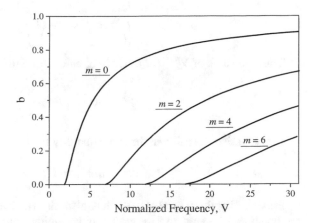

Fig. 3.13 The normalized dispersion curve for graded-index waveguide

$$n(x) = n_s + \Delta n \cdot f\left(\frac{x}{d}\right) = n_s + \Delta n \cdot f(\zeta) \qquad (3.58)$$

where d is the depth for which the distribution decreases to $1/e$ of its peak value; this is sometimes called the diffusion depth.

For low index contrast, this can be written as

$$n^2(x) \approx n_s^2 + (n_f^2 - n_s^2) f(\zeta) \qquad (3.59)$$

where n_f is defined to be the maximum index in the waveguide. Thus $n_f \equiv n_s + \Delta n$.

Then, defining the normalized frequency for a graded-index (typically a diffused) waveguide as

$$V^g = kd\sqrt{n_f^2 - n_s^2} \qquad (3.60)$$

We can write the phase equation (3.52), as

$$2V^g \int_0^{x_t/d} \sqrt{f(\zeta) - b}\, d\zeta - \frac{3}{2}\pi = 2m\pi \qquad (3.61)$$

In addition we can define b, i.e., the normalized index, its form is the same as in (3.24). This equation can be solved analytically for several cases of specific interest for a diffused waveguide-dopant distribution. For instance, in the case of Ti-diffused LiNbO$_3$, the index has a half-Gaussian distribution. In this case $\Delta n \sim e^{-\zeta^2}$, where again $\zeta \equiv x/d$. For this case, Fig. 3.13 shows the normalized waveguide quantities obtained by numerical solution of the normalized equation for the existence of modes (3.61). This diagram is valid for both TE and TM waves, when $\Delta n \ll n_s$, because of the small birefringence of LiNbO$_3$ crystals.

At cutoff, $b = 0$ and if we assume a large ratio $x_t/d \gg 1$, then

$$\int_0^{+\infty} e^{\frac{-\zeta^2}{2}} \, d\zeta = \sqrt{\frac{\pi}{2}} \tag{3.62}$$

Thus the value of $V_c^{g,m}$ for mode m at cutoff for a diffused guide is

$$V_c^{g,m} = \sqrt{2\pi} \left(m + \frac{3}{4} \right) \tag{3.63}$$

In comparison, recall that the same quantity for the wave in a slab guide is

$$V_c^m = m\pi + \tan^{-1} \sqrt{a} \tag{3.64}$$

Finally, Nishihara et al. (1989) has also shown that an analogous ray-tracing argument can be applied to a graded-index guide with a Gaussian distribution, a geometry comparable to that seen for the transverse direction in a Ti-diffused channel guide. This analysis yields the normalized frequency for the full Gaussian, V^{2g},

$$V^{2g} \int_0^\zeta \sqrt{f(\zeta) - b} \, d\zeta = \left(m + \frac{1}{2} \right) \pi \tag{3.65}$$

for the normalized dispersion curve across a region of halfwidth d. Cutoff for the mth mode of this Gaussian guide, $V_c^{g,m}$, occurs when $b = 0$, or

$$V_c^{2g,m} = \sqrt{\frac{\pi}{2}} \left(m + \frac{1}{2} \right) \tag{3.66}$$

3.5.2 Numerical Solutions

The advent of compact powerful digital computers has made numerical solutions to the mode equation a practical and widely used approach. We will discuss the numerical approaches later in the text in Chap. ?? and in addition, we give many examples of computer solutions to integrated optical problems throughout this text.

3.6 Conclusion

We have accomplished several goals in this chapter. First, we have introduced the physics and terminology of several forms of optical waveguides, including abrupt and diffused guides. To keep the explanations clear, we have concentrated exclusively on two dimensional or slab waveguides, since they have simple, closed-form solutions. Secondly, we have, in the process, obtained several useful formulae and normalized graphs The normalized graphs (and approximate formulae) are particularly useful for simple approximate analytical solutions. However, perhaps the most important consequence of this chapter is that it provides the theoretical underpinning and insight for the far more complex but more realistic problem of three-dimensional waveguides, which will be presented in the next chapter.

Problems

1. Consider a symmetric waveguide with $y = 0.1$ and $\lambda = 0.84\,\mu m$ (Fig. 3.14)

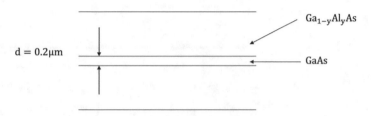

Fig. 3.14 Cross-sectional view of the symmetric waveguide

Find n_f, n_c, V, b, γ, for the TE$_0$ mode

2. You fabricated a very wide Ti:diffused slab waveguide by starting with $t = 80\,nm$ thick Ti and annealing for 6 h at 1000 °C

   (a) Calculate the index profile at 0.6 μm.
   (b) How many modes can this slab guide support at $\lambda = 1.3\,\mu m$.

3. You make the following slab waveguide (Fig. 3.15):

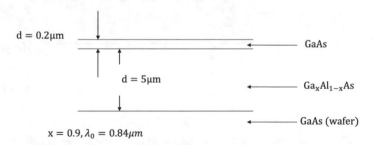

$x = 0.9, \lambda_0 = 0.84\mu m$

Fig. 3.15 Cross-sectional view of a slab waveguide

   (a) What are n_s, n_c and n_f?
   (b) Sketch the shape of the lowest order modes
   (c) How many modes are supported?
   (d) How much of the lowest order mode is "in" the GaAs wafer?

4. Derive (3.23)

$$V\sqrt{1-b} = m\pi + \tan^{-1}\sqrt{b/(1-b)} + \tan^{-1}\sqrt{(b+a)/(1-b)}$$

starting with (3.15).
5. Using simple physics intuitions, estimate the number of modes of the slab waveguide below (Fig. 3.16).

Fig. 3.16 A slab waveguide

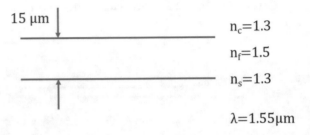

15 µm

$n_c = 1.3$

$n_f = 1.5$

$n_s = 1.3$

$\lambda = 1.55\mu m$

6. Using the modal propagation quantization equation, write an approximate expression for β_m of the highest order mode, m. Assume the waveguide has $n_f \gg n_c$, $\lambda = \lambda_0$ and width w.
   Hint: first determine the transverse propagation constant k_x as a function of the mode number m and width w.
7. A symmetric slab waveguide is given with the parameters as shown in the following figure (Fig. 3.17).

Fig. 3.17 A symmetric slab waveguide

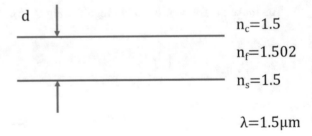

d

$n_c = 1.5$

$n_f = 1.502$

$n_s = 1.5$

$\lambda = 1.5\mu m$

   (a) Design the waveguide with a thickness that is 50% of the cutoff for $m = 1$ mode. Find the width d.
   (b) What are effective-index n_{eff} and propagation constant β for the $m = 0$ mode at the value of d you found in (a)?
   (c) What is the evanescent decay length in the cover region?

8. Design a single-mode waveguide, i.e., 10% below $m = 1$ cutoff, using a buried-channel waveguide architecture as below. Basically, you need to find out the value of core thickness d and width w (Fig. 3.18).

Fig. 3.18 A buried-channel waveguide

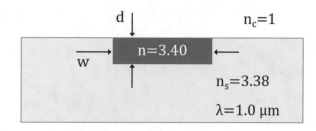

References

Haus, H. A. (1984). *Waves and fields in optoelectronics*. Upper Saddle River: Prentice-Hall.

Hocker, G., & Burns, W. (1975). Modes in diffused optical waveguides of arbitrary index profile. *IEEE Journal of Quantum Electronics, 11*(6), 270–276.

Hutcheson, L. D. (1987). Integrated optical circuits and components: Design and applications. In *Optical engineering*, No individual items are abstracted in this volume (Vol. 13, 417 p). New York: Marcel Dekker, Inc.

Kogelnik, H. (1988). Theory of optical waveguides. In *Guided-wave optoelectronics* (pp. 7–88). Berlin: Springer.

Nishihara, H., Haruna, M., & Suhara, T. (1989). *Optical integrated circuits* (Vol. 1). New York: McGraw Hill Professional.

Pollock, C. R., & Lipson, M. (2003). *Integrated photonics*, (Vol. 20). Berlin: Springer.

Saleh, B. E., Teich, M. C., & Saleh, B. E. (1991). *Fundamentals of photonics* (Vol. 22). New York: Wiley.

Snyder, A. W., & Love, J. (2012). *Optical waveguide theory*. Berlin: Springer Science & Business Media.

Chapter 4
Three-Dimensional Waveguide

Abstract In general most PICs use three-dimensionally confined waveguides for optical transport on an optical chip. Thus this chapter focuses on understating and controlling the properties of these waveguides and on using these waveguides in device-like applications. Although the basic principles of 3D waveguides are straight forward to understand, precise analytic calculations of their properties are not easy to obtain. However, a set of useful approximations and approaches have been developed, including designing for single-mode operation in both transverse directions, the influence of waveguide cross section and materials choice on waveguide dispersion, and low loss designs. These design approximations can lead to many important device ideas including mode-transformation devices and precise delay lines.

4.1 Introduction

While slab waveguides such as described in Chap. 3 are important for their analytic insight and are even used in some practical diffractive components, most PICs use three-dimensional or channel waveguides for optical transport on the chip. In addition, channel guides also form the heart of many complex PIC devices, e.g., precise delay lines or mode-transformation devices. Although the basic principles of channel guides are readily grasped, exact analysis is much more complex than for two-dimensional waveguides. While this chapter is meant to be very general in discussions, of applications, references to specific materials types will be made in several of the sections below.

There are several topics that are recurrent in the analysis of 3D waveguides including realization of single-mode operation in both transverse directions, the influence of waveguide cross section and materials choice on waveguide dispersion, and designs that result in reduction in waveguide loss. Each of these topics will be covered in this chapter.

© Springer Nature Switzerland AG 2021   57
R. Osgood jr. and X. Meng, *Principles of Photonic Integrated Circuits*,
Graduate Texts in Physics,
https://doi.org/10.1007/978-3-030-65193-0_4

4.2 Types of Channel Guides

There are several different types of channel waveguide geometries, all of which
clearly have three-dimensional characteristics. Some of the most common step-index
varieties are shown in Fig. 4.1; the comparative advantages of each step-index type
are given in Table 4.1. As the waveguide types move from strip to rib to strip-loaded,
the conditions for optical lateral confinement of the mode become progressively
more subtle. Also notice that in general $n_c \leq n_s < n_f$ for all waveguides, while in a
strip-loaded guide, $n_c < n_l < n_f$.

In addition to these 3D guides with abrupt rectangular geometries, three-
dimensional waveguides have also been made using graded-index structures. These
guides include Ti:diffused and proton-exchanged LiNbO$_3$ and some SiO$_2$—and
polymer-based structures. The smooth features in graded-index guides reduce the
wall-scattering loss that occurs in many lithographically formed features.

In addition, in the last decade, extremely small (<1 μm wide) crystalline Si and
III–V waveguides have been fabricated using modern pattern-transfer methods; these
are termed "wires" because of their small lateral dimensions. While these guides can
have relatively high lineal loss, their tight confinement and high-index contrast can
allow them to be used in the fabrication of small-area PICs. In addition, improvements

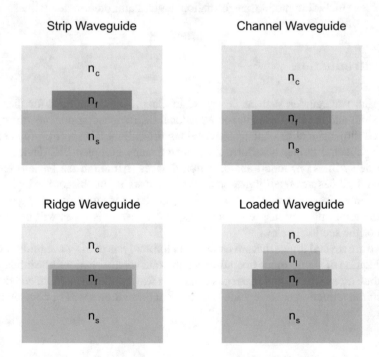

Fig. 4.1 Cross section of four of the more common step-index waveguides. Subscripts on the
indices are "s" substrate, "c" cladding, "f" film, and "l" loading strip

Table 4.1 Comparative advantages of certain common channel waveguides

Buried	Ridge	Strip loaded
• Smooth surfaces possible • Formed via ions, light or diffusion	• Strong confinement • Wall roughness • Low bend loss	• Weak confinement: $10^{-5}, 10^{-4}$ • Substantial bend loss

in fabrication methods and new waveguide structures can be expected to make lineal loss issues less important.

4.3 Modal Analysis of Three-Dimensional Waveguides

The analysis of three-dimensional or channel waveguides is very complex because their modes are not pure TE or TM: they are hybrid, i.e., TEM. As a result, the nomenclature for modes in a channel guide is more complex than in slab guides. A mode having its electric field dominantly oriented along x, i.e., TM-like, is labeled by E_{pq}^{x}, while a mode lying dominantly along y, i.e., TE-like, is labeled by E_{pq}^{y}. In this notation, p and q denote the number of modes in the x- and y-direction, respectively. An example of the hybrid nature for a common rib waveguide is shown in Fig. 4.2, where it is seen that electric field components lie along both x and y, although E_{x} is clearly dominant. Despite the hybrid nature, however, the simple approximate terms TM and TE are usually retained, because there is typically one dominant polarization, which allows a single polarization approximation.

The two possible polarization components in channel waveguide modes result from the fact that the channel waveguide geometry couples the two transverse directions in the wave equation. As is shown below, approximate solutions using the scalar-wave equation can be employed successfully if the mode is far from cutoff or

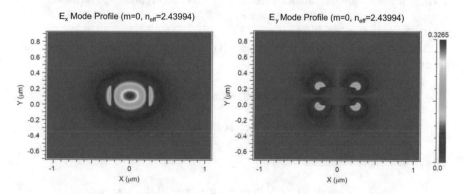

Fig. 4.2 Calculated plots of left: E_x Mode Profile; right: E_y Mode Profile

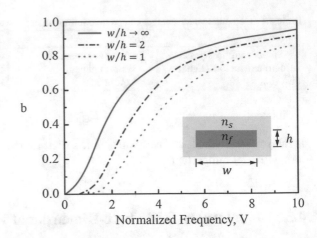

Fig. 4.3 The normalized dispersion curve between guide index (**b**) and normalized frequency for different w/h. This plot shows that when the ratio between width and thickness, w/h, becomes larger, the normalized dispersion curve approaches that for a slab waveguide

if the waveguide width, w, is greater than the thickness, t, or $w/h > 1$. The criterion of $w/h > 1$ is a reasonable approximation since clearly, the greater this ratio, the more the channel waveguide resembles a slab waveguide. This behavior is illustrated by the results of a numerical calculation on a buried-channel waveguide of height t and width w, shown in Fig. 4.3. Notice that when $w/h > 2$, the normalized dispersion curve approaches that for a slab waveguide, i.e., that with $w/h \to \infty$.

The root of the difficulty in analyzing three-dimensional waveguides is the fact that $\nabla \epsilon$ has both y and x gradients. These gradients introduce additional complexities into the vector-wave equation by coupling both transverse components of the $\vec{E}$ and $\vec{H}$ fields. In the presence of such a spatially varying dielectric, the wave equations then become

$$\nabla^2 \vec{E} + \nabla(\vec{E} \cdot \nabla \ln \epsilon) + \omega^2 \epsilon \mu_0 \vec{E} = 0 \tag{4.1}$$

$$\nabla^2 \vec{H} + \nabla(\nabla \ln \epsilon) \times (\nabla \times \vec{H}) + \omega^2 \epsilon \mu_0 \vec{H} = 0 \tag{4.2}$$

These wave equations cannot be reduced to scaler forms and must be used in their vectorial form. In essence, the (transverse) gradients in the waveguide prevent purely TE- or TM-polarized guided waves.

Dividing $\vec{E}$ and $\vec{H}$ into their transverse (t) and longitudinal (z) components and substitution in the wave equations yields

$$\nabla^2 \vec{E} + \nabla(\vec{E} \cdot \nabla \ln \epsilon) + (\omega^2 \epsilon \mu_0 - \beta^2)\vec{E} = 0 \tag{4.3}$$

$$\nabla^2 \vec{H} + \nabla(\nabla \ln \epsilon) \times (\nabla \times \vec{H}) + (\omega^2 \epsilon \mu_0 - \beta^2)\vec{H} = 0 \tag{4.4}$$

Substitution yields

$$\nabla^2 \vec{E}_t + \nabla(\vec{E}_t \cdot \overline{\nabla} \ln \epsilon) + (\omega^2 \epsilon \mu_0 - \beta^2)\vec{E}_t = 0 \tag{4.5}$$

$$\nabla^2 \vec{H}_t + (\nabla \ln \epsilon) \times (\nabla \times \vec{H}_t) + (\omega^2 \epsilon \mu_0 - \beta^2)\vec{H}_t = 0 \qquad (4.6)$$

where the complete del, ∇, can be used, since all operations involving ∇_z are zero. Careful examination of these results shows that the gradient in ϵ couples only the two transverse components of $\vec{E}$ and $\vec{H}$; no longitudinal component appear. Further for many waveguide types, the terms containing $\nabla \ln \epsilon$ are negligible. Once the transverse components are obtained, the longitudinal components are readily found to yield Maxwell's equations

$$j\beta E_z = \nabla \cdot \vec{E}_t + \vec{E}_t \cdot \nabla \ln \epsilon \qquad (4.7)$$

and

$$j\beta H_z = \nabla \cdot \vec{H}_t \qquad (4.8)$$

Note that once $\vec{E}_t$ (or $\vec{H}_t$) is determined, $E_z(H_z)$ is found, thus, the complete $\vec{E}$ and $\vec{H}$ field is available.

4.4 Approximate Methods—Generally for Slab-Like (Rectangular) Waveguides

Several approaches are used to obtain approximate solutions for the dispersion and fields of channel waveguides, including vector perturbation, separation of variables, method of field shadows, and the effective-index method. We will concentrate on the latter two methods in this section because of their immediate and widespread applicability. More recently, numerical techniques have been developed, which are extremely accurate and are generally widely applicable; they will be discussed in Chap. 14 and we will see examples of their use throughout the book. In such computations, the effective-index method is also frequently used to convert a 3D numerical calculations into a much simpler and faster 2D case. Numerical methods also have the advantage of providing exact effectively solutions.

4.4.1 Method of Field Shadows

In principle, for certain geometries, it should be possible to apply separation of variables to solve the 3D wave equation. This approach is useful in the limited cases in which the permittivities, i.e., the square of the refractive index, can be shown to be separable into two different orthogonal variables, e.g., $n^2(x, y) = n_0^2 + n_x^2(x) + n_y^2(y)$. For example, it can be used to obtain an exact solution for the fields and modes in the presence of a parabolic index distribution for which $n^2(x, y) = n_f^2(1 - (x^2/x_0^2 - (y^2/y_0^2)))$. However, this method is not widely applicable to most practical

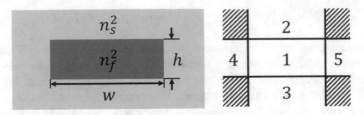

Fig. 4.4 A plot depicting the basic geometry of the Method of Field Shadows from Marcatili (1969). By ignoring the shadow region, the separation of variables can be realized

waveguide problems, for which this condition on the permittivity does not hold. Thus, a more useful approach would be to develop a method which, while approximate, allows separation of variables to be used more extensively. The "Method of Field Shadows" is such an approach.

The basic geometry used in the method of field shadows (or as it is sometimes called, "Marcatili's Method") is depicted in Fig. 4.4. The basic idea is to find the geometric regions around waveguides that can be ignored, so as to allow the desired separation of variables. In the rectangular geometry of Fig. 4.4, this separation can be realized by ignoring the "shadow region." This is a good assumption if light is confined well in a waveguide, which means the waveguide is far from cutoff. This can be achieved either by a high enough index contrast between the core and cladding, or by a large enough transverse geometry. Note that this approach provides only an fully scalar solution and, thus, is *prima facie* not exact.

This technique is illustrated by considering a rectangular buried channel (see Fig. 4.4). In this case, the waveguide can be decomposed into two slabs in order to seek two independent solutions, each slab corresponding to one of the two orthogonal variables, x and y,

$$E(x, y) = X(x)Y(y) \tag{4.9}$$

where $X(x)$ is the field in the x-slab guide and $Y(y)$ is the field in the y-slab guide.

This separation can be done in two steps: first, ignore the fields in the shadows, and then divide geometrically the guide into two orthogonally oriented waveguides, namely, waveguides lying along the x- and y-directions. The electromagnetic fields in each of these two separate slab waveguides may be found exactly as indicated earlier in Chap. 3. Specifically, the fields for each of the waveguides are cosinusoidal in the central region, labeled 1 in Fig. 4.4,

$$X(x) = A \cos(\kappa_x x + \phi_x) \tag{4.10}$$

$$Y(y) = B \cos(\kappa_y y + \phi_y) \tag{4.11}$$

and exponentially decaying in regions 2, 3, 4, and 5 with an appropriate decay constant γ_i for region i. For the solutions of region 1 to be valid, it is required that

$$\beta^2 = k^2 n_1^2 - \kappa_x^2 - \kappa_y^2 \tag{4.12}$$

again, notice that the orthogonal slab waveguide approach is only an approximate solution, since the fields in the corner regions are not then properly accounted for.

In the second step, divide the refractive index of the channel guide to allow separating out of the two slab waveguides. At this point, in order to provide a more specific example, consider the case of an embedded symmetric channel waveguide. Thus, regions 2, 3, 4, and 5 have the same index, n_s, and the central region has index $n_1 = n_f$. In this case, two independent slab waveguides can then be constructed if each of the waveguides has an embedded slab with $n_f^2/2$ and cladding/substrates of $n_s^2 - n_f^2/2$. When summed, these two decomposed guides yield that of the original buried rectangular guide, except for the corner or shadow regions. However, the fact that these regions contain relatively small fields in many cases allows this to be a useful method.

Using these two artificial index slabs, eigenvalues of the wave equation can be solved, just as for the standard slab waveguides. Because the solution is now a product of $X(x)$ and $Y(y)$, $\beta^2 = k^2 n_f^2 - \kappa_x^2 - \kappa_y^2$, and thus $\beta^2 = \beta_x^2 + \beta_y^2$ and $N^2 = N_x^2 + N_y^2$, where β_x, β_y, N_x, N_y are the propagation constant and the effective index for the x- and y-slabs, respectively. This suggests that it is possible to use a variant of the normalized dispersion variables, including its graphical presentation, that used in 2D waveguides. The normalized guide frequency of the two component slab waveguides is then

$$V_x = hk\sqrt{n_f^2 - n_s^2} \equiv V \tag{4.13}$$

and

$$V_y = k\omega\sqrt{n_f^2 - n_s^2} = \frac{\omega}{h}V \tag{4.14}$$

Notice that by convention, the thinnest waveguide dimension, denoted by h here, is typically used to specify the overall guide V number. This choice is reasonable, since the thinnest dimension sets the lowest guide modal cutoff.

The normalized guide index can then be written in both the x- and y-direction as

$$b_x = \frac{N_x^2 - n_s^2 + n_f^2/2}{n_f^2 - n_s^2} \tag{4.15}$$

and

$$b_y = \frac{N_y^2 - n_s^2 + n_f^2/2}{n_f^2 - n_s^2} \tag{4.16}$$

where the convention has been adopted that in this chapter, that the symbol N implies effective index. Since the above relations for V and b are for two slab waveguides, these waveguides can be designed using the normalized dispersion curve provided in Chap. 3.

Fig. 4.5 The normalized
dispersion curve between
guide index (**b**) and
normalized frequency for
different values of m. Note
that the waveguide shows
cutoff before $V = 0$

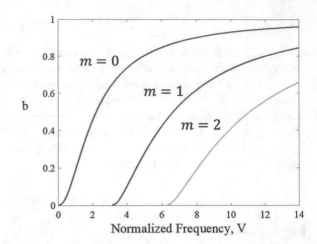

Then, since

$$b = \frac{N^2 - n_s^2}{n_f^2 - n_s^2} \tag{4.17}$$

and

$$N_x^2 + N_y^2 = N^2 \tag{4.18}$$

we find

$$b = b_x + b_y - 1 \tag{4.19}$$

Thus the normalized dispersion curve for the two slab waveguides can be used
to find the quantity, b, for the channel waveguide. Once the vertical slab $V(=V_x)$ is
found, it is also possible to construct a normalized waveguide curve for the channel guide; however, note that unlike the curve for a simple vertical slab, this curve
is specific to the channel guide geometry. A plot of b versus V for this symmetric channel waveguide, given in Fig. 4.5, shows that this waveguide exhibits cutoff
before $V = 0$. Since the waveguide is symmetric, we anticipate, on the basis of our
slab waveguide results, that such a symmetric waveguide with proper design can
always allow the fundamental mode to operate. This result indicates that for low V,
Marcatili's method is no longer accurate. Further comments on its accuracy will be
provided in Sect. 4.4.3.

As mentioned earlier, this method allows the field distribution to be obtained as
well! These fields are obtained by solving for the slab modes for each of the two
waveguides separately and then taking the product to find the final mode shape. For
example, for the above problem, the field distributions of $H_y(x, y)$, namely, the E_{00}^x,
and E_{01}^x mode (TM-like), appear as shown in Fig. 4.6.

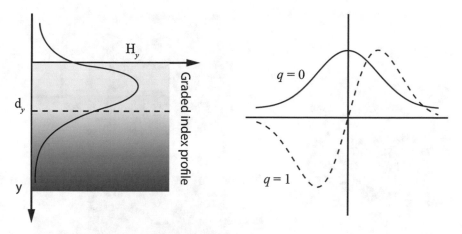

Fig. 4.6 The calculation of field distribution for $H_y(x, y)$, E_{00}^x, and E_{01}^x. These distributions are obtained by solving the slab modes for each of the two waveguides separately

4.4.2   Effective-Index Method

The effective-index method is easily implementable even for comparatively unusual guides. It is, thus, in many ways the "workhorse" technique for designing channel waveguides. Again, the approach is to decompose a three-dimensional guide into two intersecting two-dimensional guides. The essential idea is to replace the indices of the sometimes complex layered waveguide structure with a single, effective index. The specific implementation of this idea varies significantly from one waveguide geometry to another, but the basic approach is the same. To be specific, let us start with an example, using an abrupt waveguide geometry.

The steps in the approach are shown in Fig. 4.7 for a buried strip guide. As we will find, the method works best for $\dfrac{(n_f - n_s)}{n_f} \ll 1$ (i.e., weak confinement) and away from cutoff. The assumption of weak confinement also allow us setting $b_{TM} = b_{TE}$ for TM modes.

In step I, the same procedure as for any slab waveguide is used; for example, for the TM-like mode (i.e., dominant E_x), V_I is found from $V_I = hk\sqrt{n_f^2 - n_s^2}$. The normalized dispersion curve, along with the appropriate asymmetry factor, $a_m = (n_f/n_c)^4(n_s^2 - n_c^2)/(n_f^2 - n_s^2)$ may then be used to obtain b_I. The value of b_I could also be determined by using the explicit expression for the normalized TM dispersion curve for a slab waveguide. The effective index can be determined directly from b_I

$$N_I = \sqrt{n_s + b_I(n_f^2 - n_s^2)} \tag{4.20}$$

where the definition of normalized index, given earlier, has been used.

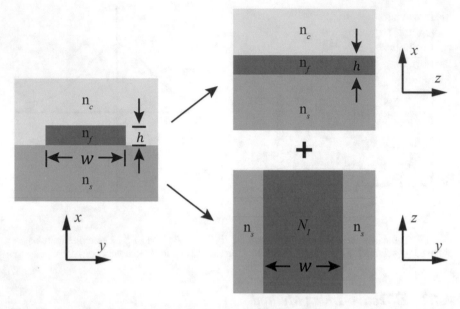

Fig. 4.7 An example of the basic steps of implementing the effective-index method on a buried strip guide. The material quantities are those defined earlier in the book

In step II, a "new" slab waveguide core of index N_I and with cladding of n_s is constructed. The TM-like mode for guide I is now TE-like for this second waveguide. Again, the normalized dispersion curve is used to derive V_{II}, then b_{II}. For this symmetric slab guide with TE modes, $a = 0$ and the effective index in the slab is given by

$$V_{II} = k\omega\sqrt{N_I^2 - n_s^2} \tag{4.21}$$

At this point in the analysis, b_{II}, the normalized thickness of the slab, can be found either by substituting in the normalized eigenvalue equation for a symmetric waveguide,

$$V_{II} = \frac{1}{\sqrt{1 - b_{II}}}\left((m + 1)\pi - 2\tan^{-1}\sqrt{\frac{1 - b_{II}}{b_{II}}}\right) \tag{4.22}$$

or by using the graphical form b_{II}. N_{II} (or $\beta = kN_{II}$), is determined by

$$N_{II} = \sqrt{n_s^2 + b_{II}(N_I^2 - n_s^2)} \tag{4.23}$$

Since the effective-index guide is symmetric, when cutoff occurs at $b_{II} = 0$, $V_{II} = q\pi$.

A very complete tabulation of the effective-index method for a wide variety of channel waveguide structures is presented in Kogelnik (1988) and a selection of these waveguides are given here in Table 4.2. For example, the application to ridge waveguide structures reveals the calculation of the effective index for the vertical slab in both the ridge and adjacent waveguide regions. The equivalent b and V of the horizontal slab waveguide is then calculated using the effective indices from this calculation.

The effective-index method is an excellent technique for obtaining the dispersion relation for a channel waveguide. In addition, the method can also be used to find the field distribution using the two orthogonal waveguides. However, since a simple "slab" approximation is being made to otherwise complex geometries in some cases, e.g., the ridge waveguide, the detailed shape of the distribution will sometimes be *at variance with the actual modal shape*. Still, the "effective-index" fields will give at least some indication of the average spatial extent of the fields.

4.4.3 Accuracy of Approximate Techniques for Channel Waveguides

It is useful to compare the accuracy of these methods with the generally exact calculations done with modern numerical techniques. A very interesting study in that regard has been described by Hocker and Burns (1977) for a buried rectangular channel waveguide and is shown in the two panels of Fig. 4.8. These graphs in these panels plot a comparison of calculations based on "exact" numerical techniques with results obtained using the field shadow and the effective-index methods for two different ω/t ratios. The buried-channel waveguide was chosen to allow direct comparison with the results given earlier in Fig. 4.3 for both slab and channel waveguides.

Fig. 4.8 The comparison of the accurate calculations and the two approximate techniques. The solid blue line represents the exact calculation while the red and green dashed lines follow the field-shadows method and the effective-index method, respectively

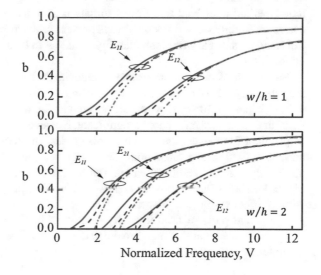

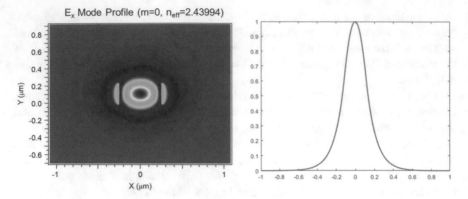

Fig. 4.9 Left: Mode profile calculated using the numerical Beam Propagation Method (BPM). Right: Mode profile calculated with effective-index method

In Fig. 4.8, the effective-index method uses $V = V_I$; it also employs the results given in Table 4.2 for the expression for the overall b of a buried-channel waveguide given in Table 4.2, namely, $b = b_I b_{II}$. In addition, the figure also shows the results of the field-shadows method that uses the normalized index of the two orthogonal slabs to obtain b, i.e., $b = b_x + b_y - 1$.

The results show that the effective-index method is the more accurate of the two approximate techniques; however, the field-shadows method becomes increasingly more accurate as ω/t increases. Note also that both methods are most accurate far from cutoff and for low mode numbers. The effective-index method tends to overestimate the normalized index, while the field-shadows method underestimates the same geometry. This underestimation occurs because the field distribution spreads over a significant portion of the corner or "shadow" regions. Although the effective-index method does lead to an overestimation of b near cutoff, it at least gives the correct reason that one mode is always present in a symmetric channel waveguide. This fact, plus its easy implementability, causes it to be a common tool in integrated optics design.

Finally, all of these approximate techniques do have one important inadequacy: they are incapable of determining any waveguide modulation of the transverse mode field distribution. In some waveguides, such as the buried-channel waveguide, this field "fine" structure might not be significant. However, in a surprisingly large number of practical 3D waveguides, such as rib guides, the modulation in the vertical or horizontal field can be substantial. The understanding of this structure is crucial for predicting even the "zeroth-order" performance of devices such as couplers or filters. Figure 4.9 shows the simulation results of this effect in the field of a rib waveguide with its effective-index equivalent.

The actual plots of field intensity cannot, in fact, be obtained by approximate methods. Instead, careful numerical calculation must be used. These methods will be presented in Chap. 14.

Table 4.2 Effective index parameters for channel guides

Structure	Guide height	Effective index	$N_I^2 - N_{II}^2$	Channel guide index b
General	$V_I = kh\sqrt{n_f^2 - n_s^2}$ $V_{II} = kh\sqrt{n_I^2 - n_f^2}$	$N_I^2 = n_s^2 + b_f(n_f^2 - n_s^2)$ $N_{II}^2 = n_s^2 + b_f(n_I^2 - n_s^2)$	$b_f(n_f^2 - n_s^2) - b_I(n_I^2 - n_s^2)$	$b_f b_{II} + b_I(1 - b_{II})a_{ch}$
Buried	$V_I = kh\sqrt{n_f^2 - n_s^2}$	$N_I^2 = n_s^2 + b_f(n_f^2 - n_s^2)$ $N_{II} = n_s$	$b_f(n_f^2 - n_s^2)$	$b_f b_{II}$
Raised	$V_I = kh\sqrt{n_f^2 - n_s^2}$	$N_I^2 = n_s^2 + b_f(n_f^2 - n_s^2)$ $N_{II} = N_c$	$(n_s^2 - n_c^2) + b_f(n_f^2 - n_s^2)$	$b_f b_{II} - (1 - b_{II})a$
Rib	$V_I = kh\sqrt{n_f^2 - n_s^2}$ $V_{II} = kl\sqrt{n_f^2 - n_s^2}$	$N_I^2 = n_s^2 + b_f(n_f^2 - n_s^2)$ $N_{II}^2 = n_s^2 + b_l(n_f^2 - n_s^2)$	$(b_f - b_l)(n_f^2 - n_s^2)$	$b_f b_{II} + b_l(1 - b_{II})$
Embedded	$V_a = kh\sqrt{n_f^2 - n_s^2}$	$N_I^2 = n_s^2 + b_f(n_f^2 - n_s^2)$ $N_{II} = n_s$	$b_f(n_f^2 - n_s^2)$	$b_f b_{II}$
Ridge	$V_a = kh\sqrt{n_f^2 - n_s^2}$ $V_{bII} = kl\sqrt{n_f^2 - n_s^2}$	$N_I^2 = n_{s1}^2 + b_f(n_f^2 - n_{s1}^2)$ $N_{II}^2 = n_{s2}^2 + b_l(n_{s2}^2 - n_{s2}^2)$	$(1 - b_l)(n_{s1}^2 - n_{s2}^2) + b_f(n_f^2 - n_{s1}^2)$	$b_{II}(1 + b_f a_{ridge}) + b_l(1 - b_{II})$

4.5 Wavegiude Dispersion for Channel Guides

Thus far, we have been using the scalar-wave equation, plus the matching of boundary conditions, to obtain the field amplitudes in the waveguide. Note that within this scalar approximation, there should be no difference in modal propagation constants because this equation, i.e.,

$$\nabla^2 \psi(x, y) + (n^2 k^2 - \beta_0^2)\psi(x, y) = 0 \tag{4.24}$$

where ψ is the scalar field, does not distinguish the polarization state of the field amplitude. Instead, dispersion arises from the inclusion of the $\nabla(\vec{E} \cdot \nabla ln\epsilon)$ term in the wave equation:

$$\nabla^2 \vec{E} + \nabla(\vec{E} \cdot \nabla \ln \epsilon) + (n^2 k^2 - \beta^2)\vec{E} = 0. \tag{4.25}$$

While including this transversely varying dielectric constant in the wave equation makes its solution much more difficult, it is possible to use a perturbation method to gain insight into its behavior. For example, perturbation theory can be used to obtain the shift in propagation constants, β, due to waveguide dispersion. The perturbation treatment finds the dispersive shift in β by comparing the initial scalar-wave equation, i.e., that without the vector perturbation, with the perturbed full vector-wave equation. In the presence of this perturbation, $\vec{E} = \vec{E}_0 + \vec{E}_1$, where $\vec{E}_0$ is the solution to the scalar-wave equation.

$$\beta_{TE}^2 = \beta_0^2 + \Delta\beta_{TE}^2 \tag{4.26}$$

$$\beta_{TM}^2 = \beta_0^2 + \Delta\beta_{TM}^2 \tag{4.27}$$

where β_0 is obtained from the scalar-wave equation, i.e., with no $\nabla \ln \epsilon$ term. Using the difference between the scalar and perturbed equation and (4.26) and (4.27), and neglecting the second-order perturbation effects, the following shifts are obtained for the TE- and TM-like modes:

$$\Delta\beta_{TE}^2 = \frac{1}{2} \frac{\int \psi^2 \frac{\partial^2}{\partial y^2}(\ln \epsilon)\, dx dy}{\int \psi^2\, dx dy} \tag{4.28}$$

$$\Delta\beta_{TM}^2 = \frac{1}{2} \frac{\int \psi^2 \frac{\partial^2}{\partial x^2}(\ln \epsilon)\, dx dy}{\int \psi^2\, dx dy} \tag{4.29}$$

where $\psi(x, y)$ is the modal field from the scalar equation. As an example, the application of this approach to a parabolic dielectric profile, characterized by half widths of d_x and d_y yields

$$\beta_{TE}^2 - \beta_{TM}^2 \left(\frac{\lambda}{\pi n_f}\right)^2 \left(\frac{1}{d_x^4} - \frac{1}{d_y^4}\right). \tag{4.30}$$

Finally recent developments in Si and Si fabrication (see Chap. 2) allow Si nanowire waveguides to be made. These waveguides have extremely high-index contrast at edges and this high contrast causes additional factors in waveguide design to be considered; this phenomenon has particularly important effects for dispersion in the waveguide (Chen et al. 2006). Thus, the strong quartic dependence of the square of the propagation constant seen in (4.30)suggests that structural dependence is a major issue for tightly confined waveguides. Thus, structural dispersion is particularly important in the case of Si-wire waveguides, due to the tight confinement of the mode in these guides. An example of this effect is seen in Fig. 4.2, which compares the material and structural contributions to the dispersion of a Si-wire waveguide for a wire with a cross section of $0.2 \times 0.5\,\mu m$. As is readily seen, the dispersion is dominated by the structural component. The dominance of structural dispersion is typical for Si wires. In addition, its importance causes small variations in the width or height of the wires during fabrication to affect the chromatic response of many Si waveguide devices, particularly devices with delay lines. These dispersion effects also are of major concern when designing waveguides for nonlinear optical experiments, for which phase matching is crucial.

4.6 The Design of Single-Mode Channel Waveguides

The performance of an integrated optical circuit often depends critically on maintaining single-mode operation in the waveguiding system. For example, higher order modes cause enhanced loss in waveguide bends and alter the performance of a waveguide coupler. The procedure for designing single-mode devices usually proceeds by decomposing the problem into two dimensions through the use of the effective-index method. This approach has been nicely discussed and presented by Nishihara et al. (1989).

In order to illustrate the design process, consider a simple step-index rectangular waveguide such as discussed in earlier sections. This waveguide geometry is, for example, very important in the case of very high contrast Si waveguides such as the Si wires, mentioned above, and for the related high-index clad waveguides in some III–V materials.

To design such a waveguide structure, *both* TM- and TE-like modes must be considered. As mentioned in previous chapters, this process is made more difficult since, in principle, the normalized quantities b and a are much more complex for the case of TM-like modes. If, however, the waveguide has low index contrast, then it suffices simply to substitute the quantity a_m for a in the equation for the dispersion relation, (3.23), since $b_{TE} = b_{TM}$.

Fig. 4.10 A plot of the
solution to the normalized
dispersion curve for
single-mode operation of
w/h versus V_I normalized
frequency. This graph shows
that a single-mode
waveguide can be made
inside the shadow area

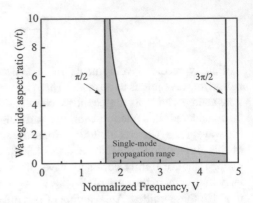

To proceed with an example, consider the buried step-index guide shown earlier in Fig. 4.7 for the case of a TM-like mode, i.e., E_{00}^x mode. The first step consists of adjusting the height of the vertical slab, waveguide I, so that it is above cutoff for $m = 0$ and below that for $m = 1$, that is, V_I is such that

$$\pi + \tan^{-1} \sqrt{a_m} > V_I > \tan^{-1} \sqrt{a_m} \tag{4.31}$$

where $V_I = hk\sqrt{n_f^2 - n_s^2}$. In cases where the lateral adjacent regions also have waveguiding properties, these regions may or may not be in themselves single mode; it is only important that the central "core" region be single mode for this design procedure.

In the second step, the "effective-index" symmetric slab is considered, and thus $a = 0$. In this case,

$$0 < V_{II} < \pi \tag{4.32}$$

Substitution then gives

$$0 < \frac{w}{h} < \frac{\pi}{\sqrt{b_I V_I}} \tag{4.33}$$

Notice that this equation shows that guide I, and hence, b_I and V_I, determine the width-to-thickness ratio for single-mode behavior. In applying this equation, the dispersion give for single lowest order mode related the value of b_I for each value of V_I. The solution to single-mode operation for w/h versus V_I has been given in graphical form by Nishihara et al. (1989) see Fig. 4.10.

4.7 Graded-Index Channel Waveguides

Thus far we have considered only step-index waveguides, however, channel guides can also be made in a graded-index medium, with a frequently used example being Ti-diffused waveguides. In this case, the effective-index method can again be used. In our

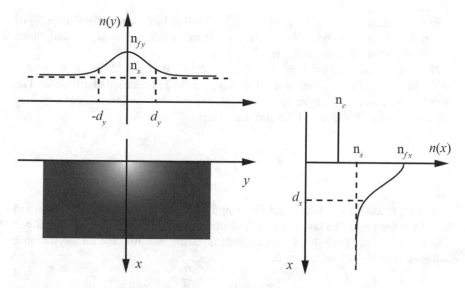

Fig. 4.11 A plot of index vs. spatial location for a 3D graded-index waveguide, the quantity $d_{x,y}$ indicates the diffusion depth in x- and y-direction. As in the case for 3D abrupt-index waveguides, the effective-index method can still be used

discussion here, we will again follow the excellent discussion given by Nishihara et al. (1989). The method decomposes the problem into vertical and lateral waveguides, just as in the case of 3D abrupt-index waveguides; see Fig. 4.11. First, a laterally uniform but vertically graded slab waveguide with a Gaussian index distribution is assumed in order to find the normalized frequency V_I^g of this vertical guide and, then, the conditions for single-mode confinement.

To determine the modal properties of the guide, the eigenmode equation is used for a graded-index guide given in Chap. 3. Specifically, for single-mode behavior, the value of V_I^g must be greater than the cutoff value for $m = 0$ and less than the same for $m = 1$, or, using (3.62),

$$\frac{3}{4}\sqrt{2\pi} < V_I^g \leq \frac{7}{4}\sqrt{2\pi} \tag{4.34}$$

Once V_I^g is fixed, the guide diffusion depth, d_x, is found from the usual definition of V_I^g,

$$d_x = \frac{V_I^g}{k\sqrt{n_{fx}^2 - n_s^2}} \tag{4.35}$$

In addition, this value of V_I^g can be used in (3.22) to find the effective-index N_I of the vertical guide, if this quantity is needed for further waveguide design calculations, e.g., computation of dispersion, etc.

Next, the dimension for the *lateral* guide, shown in Fig. 4.11, has to be found. To do this, the guide is approximated with a symmetric Gaussian distribution. The lateral waveguide diffusion depth can be written in terms of the normalized frequency V_{II}^{2g} and the effective index of the vertical guide:

$$d_y = \frac{V_{II}^{2g}}{k\sqrt{N_I^2 - n_s^2}} \tag{4.36}$$

This equation shows clearly that the design of the vertical waveguides couples to that of waveguide II, the lateral guide. The condition for a single propagating mode is then given by the graded-index eigenmode equation, but this time for a symmetric Gaussian distribution, with $m = 0$.

$$\sqrt{\frac{\pi}{8}} < V_{II}^{2g} \le 3\sqrt{\frac{\pi}{8}} \tag{4.37}$$

This equation can be rewritten by dividing through by V_I^g and inserting b_I to obtain an equation which gives the range of the ratio of the lateral, d_y, to vertical, d_x, distance needed for a specific vertical waveguide design specified by either b_I or V_I^g:

$$\sqrt{\frac{\pi}{8b_I}}\frac{1}{V_I^g} < \frac{d_y}{d_x} \le 3\sqrt{\frac{\pi}{8b_I}}\frac{1}{V_I^g} \tag{4.38}$$

Alternately, this inequality can also be used to determine V_I^g if the ratio d_y/d_x is given. As a specific example, consider the important case of $d_y/d_x = 1$, i.e., diffusion from a single "line source." Substituting this value in (4.38) and manipulating the equation gives

$$\sqrt{\frac{\pi}{8}} < \sqrt{b_I V_I^g} < 3\sqrt{\frac{\pi}{8}} \tag{4.39}$$

This equation, or its equivalent for other ratios of d_y/d_x, thus provides a second set of contrasts on V_I^g above that given in (4.37). To find the values of V_I^g given in this new inequality, we use the normalized dispersion curves shown in Fig. 4.12 to determine which value of the product $\sqrt{b_I V_I^g}$ is valid. This gives the solution

$$V_I^g \approx k_0 dx \sqrt{2\Delta n} \tag{4.40}$$

where $\Delta n \equiv n_f - n_s$, then we can rewrite this inequality as the following criteria for a single-mode 3D waveguide:

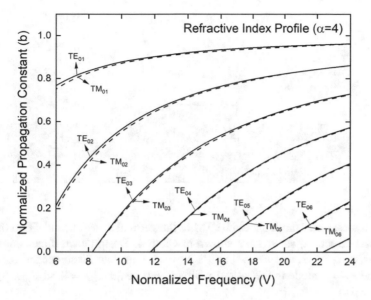

Fig. 4.12 Normalized dispersion curve for graded-index waveguide

$$0.26 < \left(\frac{d_x}{\lambda}\right)\sqrt{n_s \Delta n} \leq 0.39 \tag{4.41}$$

Notice that Δn and d_x play an essential role in setting the single-mode condition.

4.8 Summary

The complexity of 3D waveguide analysis, even for relatively simple geometries, provides a clear motivation for eventual use of numerical analysis. These techniques will be described in Chap. 14. However, the methods used in this chapter, especially the effective-index method, are surprisingly useful even in the case of numerical analysis since their employment can accelerate the design of complex integrated circuits.

At this point, the text will begin to analyze the optical properties of multiple waveguides both in simple coupled systems and in more complicated devices.

Problems

1. Use the effective-index method to determine the dimensions needed for single-mode operation in the following hetero-structure waveguide:

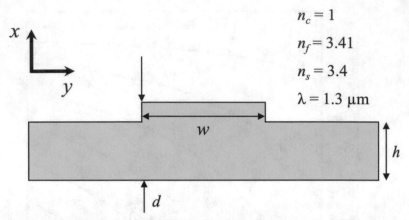

$$n_c = 1$$
$$n_f = 3.41$$
$$n_s = 3.4$$
$$\lambda = 1.3 \ \mu m$$

The lowest order mode should be TM in the vertical direction and TE in the horizontal direction, as we have shown in Chap. 4. Design each dimension in the figure (i.e., w, d, and h) so that it's 20% smaller than those needed for cutoff of the $m = 1$ mode in the y- and x-directions (i.e., $d_{design} = 0.8 d_{cutoff}$). Assume $h = 2 \ \mu m$.

2. The same as for *Problem 1*, except using a strip Si$_3$N$_4$/SiO$_2$ structure as below.

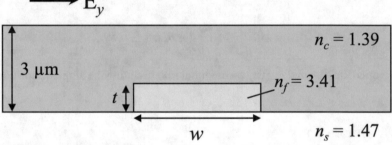

3. Consider a Ti-diffused waveguide with two Gaussian distributions in the x- and y-directions (a half and a full), as shown below.

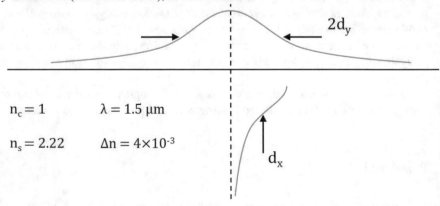

$n_c = 1$ $\lambda = 1.5 \ \mu m$

$n_s = 2.22$ $\Delta n = 4 \times 10^{-3}$

Design for single-mode operation (find d_x and d_y) using the same 20% rule and polarization as in *Problem 1*. As a reminder, the index of Ti-diffused waveguide satisfies following relation:

$$\begin{cases} n_x = n_s + \Delta n \cdot e^{-\frac{x^2}{d_x^2}} \\[2mm] n_y = n_s + \Delta n \cdot e^{-\frac{x^2}{d_y^2}} \end{cases}$$

4. Show that the following expression is valid under conditions of low index contrast:

$$N_{\text{eff}} \approx n_s + b(n_f - n_s).$$

References

Chen, X., Panoiu, N. C., & Osgood, R. M. (2006). Theory of raman-mediated pulsed amplification in silicon-wire waveguides. *IEEE Journal of Quantum Electronics*, *42*(2), 160–170.

Hocker, G., & Burns, W. K. (1977). Mode dispersion in diffused channel waveguides by the effective index method. *Applied Optics*, *16*(1), 113–118.

Kogelnik, H. (1988). Theory of optical waveguides. *Guided-wave optoelectronics* (pp. 7–88). Berlin: Springer.

Marcatili, E. A. (1969). Dielectric rectangular waveguide and directional coupler for integrated optics. *Bell System Technical Journal*, *48*(7), 2071–2102.

Nishihara, H., Haruna, M., & Suhara, T. (1989). *Optical integrated circuits* (Vol. 1). New York: McGraw Hill Professional.

Chapter 5
General Introduction to Coupled-Mode Theory

Abstract This chapter presents a formalism for showing that small changes in the structure of an optical waveguide structure can couple optical fields, as they propagate through the structure. This approach is an effect akin to the perturbation theory seen, say, in quantum physics; formally this approach is termed as coupled-mode theory. Coupled-mode theory can be used with a wide variety of geometries or hetero-materials and has thus been used to analyze periodic structures or a small isolated surface dent or defect. When waveguide coupling is present, a new set of normal modes of the coupled system are formed, which are termed as system modal fields. This coupling can be expressed in terms of the uncoupled fields and the parameters of waveguide geometry. In addition, this chapter will also consider a variation of the coupled-mode theory, which is written for a longitudinally varying system.

5.1 Introduction: A Simple Example of Mode Coupling

Coupling of modes is illustrated clearly by the example of the optical fields in two adjacent waveguides. Coupling occurs because the modal field in one waveguide overlaps with the dielectric medium of the other. Specifically, as shown in Fig. 5.1, the field, E_a, of waveguide a is perturbed by the dielectric step, which forms waveguide b. Coupling between modes allows power to flow from one waveguide to another. The power transfer is maximized when the two modes are nearly synchronous with each other, i.e., when propagation velocities along the waveguides are comparable.

When waveguide coupling is present, a new set of normal modes of the coupled system are formed; Fig. 5.1 shows these "system" modal fields, $E_{o,e}$. These can be expressed in terms of the uncoupled fields E_a and E_b and the parameters of waveguide geometry. For symmetric waveguides, the power in each waveguide varies sinusoidally with distance (see Fig. 5.2). This behavior can also be interpreted as a periodic interference between the fields of a set of even and odd system modes, E_e and E_o, each propagating with slightly different propagation constants, namely, β_e and β_o.

Finally note that our discussion thus far appears superficially to focus on a simple waveguide structure. However, in fact, the coupled-mode equation applies to a much

© Springer Nature Switzerland AG 2021
R. Osgood jr. and X. Meng, *Principles of Photonic Integrated Circuits*,
Graduate Texts in Physics,
https://doi.org/10.1007/978-3-030-65193-0_5

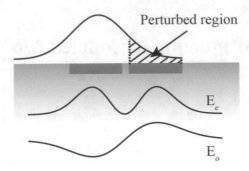

Fig. 5.1 When the modal field in one waveguide overlaps with the dielectric medium of the other, coupling occurs. This figure shows the case that one waveguide couples with another identical waveguide and forms two new modal fields E_o and E_e. Here, "o" and "e" stands for "odd" and "even," respectively

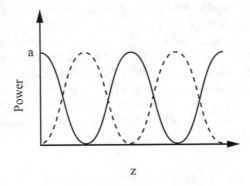

Fig. 5.2 Plot showing symmetric waveguides, the power in each waveguide varies sinusoidally with distance

wider variety of guided-wave structures such as diffraction gratings or waveguide arrays. We will discuss some of these applications later in this chapter and elsewhere in this book.

5.2 General Coupled-Mode Formulation

The coupled-mode equations relate the first derivative of one mode amplitude to the amplitude of a second mode via a coupling coefficient, typically denoted by Greek letter κ. These equations have been derived via several routes. One particular common approach is to use the Helmholtz equations in conjunction with the slowly varying wave approximation (Saleh et al. 1991; Hutcheson 1987).

However, a more rigorous and general approach to the coupled-mode equations, using modal expansions of the fields and vector manipulation of Maxwell's equations, has been described in the three references (Marcuse 1974; Kogelnik 1988; Nishihara et al. 1989). Elements of this straightforward derivation will be touched on and outlined here; more extensive references are found in the references. The starting point in the derivation is to examine a guided-wave structure and choose the canonical structures, the modes of which are only "perturbed" in the presence of the complete structure with perturbations, even major changes. For example, in Fig. 5.1, the canonical structures are the waveguides and the complete structure is the coupler. The transverse field of the canonical structure, labeled 1, may then be expanded in modes of the guide,

$$\vec{E}_{1t} = \sum (a_p^+ + a_p^-)\vec{E}_{tp} \tag{5.1}$$

$$\vec{H}_{1t} = \sum (a_p^+ + a_p^-)\vec{H}_{tp} \tag{5.2}$$

where a_p^+ and a_p^- are functions of z (along the guide) and denote the amplitudes of the forward- and backward-running field, respectively. The functions $\vec{E}_{tp}$ give the transverse field, i.e., x distribution. It is typical in many integrated-optic problems to normalize the waveguide modes such that the power is set equal to unity, which in our case will be 1W/cm^2. This is a particularly important point when calculating practical quantities such as the coupling coefficient.

Mode coupling occurs through the presence of a change or perturbation in the dielectric constant. The change may be scalar,

$$\vec{P} = \Delta\epsilon\vec{E} \tag{5.3}$$

or tensoral

$$\vec{P}_i = \Delta\epsilon_{ij}\vec{E}_j \tag{5.4}$$

The former allows coupling among like (e.g., TE) modes. The latter allows mixing between TE and TM modes as well. Mode conversion via a tensor dielectric change is important in magneto- and electro-optical devices, and for nonlinear optical effects.

The coupled-mode equations may then be written in a form such as

$$\frac{da_q^+}{dz} + j\beta_q a_q^+ = \kappa_{qp} a_p^+ \tag{5.5}$$

where for simplicity, the case of coupling only between two forward-moving modes, q and p, has been considered. A more complete expression including both forward and backward waves can be obtained (Kogelnik 1988). In many practical cases, only coupling between two modes is sufficient to provide a rigorous treatment of the device operation. However, clearly the inclusion of only a limited number of modes could, in some cases, limit the accuracy of the solution.

Below, we will show that on the basis of simple considerations, relations between certain coupling coefficients can be derived.

5.3   Practical View of Coupling of Modes

The coupled-mode equations given above, i.e., in the form such as (5.5), allow the use of a very simple and more practical approach for solving coupling of modes in waveguide structures. Before examining this approach, it is important to lay out an explicit expression for the mode amplitude with a complete expression for its field. The complete expression for the harmonic field at frequency ω, say in one mode, is

$$\vec{E}(x, y, z, t) = a'_t(z)e^{-j\beta_1 z} \vec{f}_1(x, y)e^{j\omega t} \tag{5.6}$$

where $a'_t(z)$ is the (slowly varying) wave amplitude of the wave labeled by 1, β_1 is its propagation constant, and $\vec{f}_1(x, y)$ is the transverse distribution of the mode. Notice, in this case, that a purely transverse field has been assumed. This assumption will be followed throughout this chapter and, except where it is noted or obvious, throughout the book. In some cases, it is more instructive to lump the amplitude and the r-dependent-phase term into one term, which is designated by $a_1(z)$; note that the field is designated without a prime superscript in this case: notice that this notation was used in the previous section. In this formulation, the slowly varying spatial dependence of the modal power can be obtained through $|a_1|^2$, etc.

If modes are labeled by 1 and 2, and they are weakly coupled via coupling coefficients κ_{12} and κ_{21}, two coupled-mode equations will describe their intertwined spatial variation,

$$\frac{da_1}{dz} = -j\beta_1 a_1 + \kappa_{12}a_2 \tag{5.7}$$

$$\frac{da_2}{dz} = -j\beta_2 a_2 + \kappa_{21}a_1 \tag{5.8}$$

The normalization conditions are such that for weak coupling, and $|a_1|^2$ and $|a_2|^2$ represent the powers carried by these particular waves or modes.

Haus (1984) has introduced the notation which gives the total power flow, P, from these two modes as

$$P = p_1|a_1|^2 + p_2|a_2|^2 \tag{5.9}$$

where $p_{1,2} = \pm 1$ depending on whether the power flow is forward or backward. This notation thus allows the expectation that the coupled modes can be copropagating or counterpropagating. Each of these cases will be considered below. In the absence of power loss or gain, the total power flow of the modes should be such that

$$\frac{dP}{dz} = 0 \tag{5.10}$$

and thus

$$p_1\frac{d|a_1|^2}{dz} + p_2\frac{d|a_2|^2}{dz} = 0 \tag{5.11}$$

The fact that

$$\frac{d|a_1|^2}{dz} = a_1 \frac{da_1^*}{dz} + a_1^* \frac{da_1}{dz} \tag{5.12}$$

plus the coupled-mode equations for $\dfrac{da_1^*}{dz}$ and $\dfrac{da_1}{dz}$ can then be used to show that

$$p_1 \kappa_{12} + p_2 \kappa_{21}^* = 0 \tag{5.13}$$

Thus power conservation yields a fixed relation between κ_{12} and κ_{21}^*, the two coupling coefficients appearing in (5.7) and (5.8), and depends on the direction of propagation through the factor p.

5.3.1 Propagation Constants for the Coupled Modes

The coupled-mode equations, (5.7) and (5.8), can now be solved to find the new propagation constants, $\beta'_{1,2}$, of the coupled system. For example, in the coupled waveguide example given in Sect. 5.1, these "system" modes would contain even and odd modes shown in Fig. 5.1, having propagation constants $\beta' \equiv \beta_e$ or β_o. Setting the determinant of the coupled-mode equations equal to zero, we find

$$(\beta - \beta_1)(\beta - \beta_2) + \kappa_{12}\kappa_{21} = 0 \tag{5.14}$$

This equation yields the solutions, or eigenvalues, for the propagation constants of the system modes,

$$\beta'_{1,2} = \frac{\beta_1 + \beta_2}{2} \pm \sqrt{\frac{(\beta_1 - \beta_2)^2}{2} - \kappa_{12}\kappa_{21}} \tag{5.15}$$

Consider first the case of copropagating waves. In this case, $p_1 p_2 = +1$ and then (5.13), plus some algebraic manipulations, show that $\kappa_{12}\kappa_{21} = -|\kappa_{12}|^2$. Thus, β'_1 and β'_2 are always real, and two propagating solutions are obtained. Coupling between the waveguides is only substantial if $\beta_1 \approx \beta_2$, or, specifically, coupling is important when $|\beta_1 - \beta_2| \approx |\kappa_{12}|$. This is the condition for the two waves to be close to phase matching or synchronous. It is of interest to examine how the system or coupled modes vary as the difference in the propagation constants of the isolated waveguide modes, β_1 and β_2, evolve. For example, if the isolated waveguide β_1 and β_2 vary slowly and linearly with ω, near the point where $\beta_1 = \beta_2$, then the system modes $\beta'_{1,2}$ will behave as shown in Fig. 5.3. See also the discussion concerning (5.26) and (5.27), below. Note that there is an avoided "crossing" in the dispersion plot of the two system modes as a result of the modal coupling term. Also the propagation constants approach the uncoupled or isolated waveguide values for $|\beta_1 - \beta_2| > |\kappa_{12}|$. This

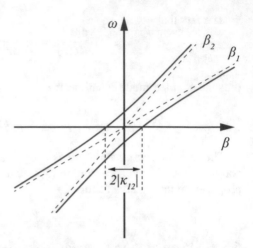

Fig. 5.3 The dispersion relation when β_1 and β_2 vary slowly and linearly with ω, near the point where $\beta_1 = \beta_2$. The blue lines represent β_1' and β_2'

behavior is due to the fact that in the limit where the waveguides are sufficiently asynchronous that their fields do not interact. The use of coupling, over several propagation lengths, between two waveguides is important in a variety of integrated optical devices such as switches and routers.

For the case of counterpropagation modes, the same eigenvalue equation (5.14) can be applied. In addition, $p_1 p_2 = -1$ and then $\kappa_{12}\kappa_{21} = -|\kappa_{12}|^2$. By inspection of (5.15), this constraint implies that the two waves will have a complex propagation constant for certain values of $|\beta_1 - \beta_2|$, i.e., when

$$\frac{\beta_1 - \beta_2}{2} < |\kappa_{12}| \tag{5.16}$$

In this range, the two modes will exhibit attenuation as they propagate into the coupling region; that is, propagating waves do not exist. The dispersion curve for counterpropagating modes, having different uncoupled group velocities is shown in Fig. 5.4. In the center region of this figure, propagation does not occur and thus an "optical" bandgap in the grating structure exists. This behavior is also typical of periodic structures such as those used in Bragg reflectors. Photonic crystals are the same type of periodic structure, but have a larger coupling constant, κ.

5.3.2 Power in the Coupled Systems

The power in coupled systems can be found by solving the appropriate coupled-mode equations, while taking into account the appropriate initial conditions. For example, for the case of the copropagating waves, we use the coupled-mode equations given in (5.7) and (5.8) as well as the modal propagation constants given in (5.15). With

Fig. 5.4 The dispersion curve for counterpropagating modes, which have different uncoupled group velocities

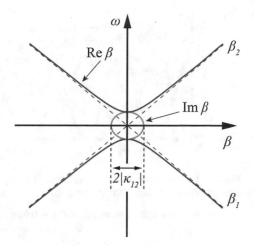

this starting point, we can show that for the case of excitation of wave 1 only, i.e., $a_1(0) = 1$, $a_2(0) = 0$, then

$$\frac{|a_1(z)|^2}{|a_1(0)|^2} = 1 - \left(\frac{\kappa}{\beta_c}\right)^2 \sin^2 \beta_c z \qquad (5.17)$$

and

$$\frac{|a_1(z)|^2}{|a_1(0)|^2} = \left(\frac{\kappa}{\beta_c}\right)^2 \sin^2 \beta_c z \qquad (5.18)$$

where

$$\beta_c = \sqrt{\kappa^2 + \left(\frac{\beta_1 - \beta_2}{2}\right)^2} \qquad (5.19)$$

and thus

$$\left(\frac{\kappa}{\beta_c}\right)^2 = \frac{1}{1 + (\Delta/\kappa)^2} \qquad (5.20)$$

If $\Delta \equiv (\beta_2 - \beta_1)/2$. In these equations, we have also assumed that $\kappa_{12} = \kappa_{21}$ and that κ is real.

Plots of the power coupling curve in the two guided waves is shown in Fig. 5.5. The use of $|a_{1,2}(z)|^2$ removes the rapidly varying, β-dependent phase from the length variation. Notice that the power coupling is repetitive over a characteristic length given by $L_c = \dfrac{\pi}{2\beta_c}$. For example, for synchronous waveguides, $\beta_1 = \beta_2$ and $L_c = \dfrac{\pi}{2\kappa}$. Also, the maximum possible power transfer, from $|a_1|^2$ to $|a_2|^2$, is given by the ratio $(\kappa/\beta_c)^2$; this term is unity at the phase-matching point, $\beta_1 = \beta_2$, and 0 for $|\beta_1 - \beta_2| \gg \kappa$.

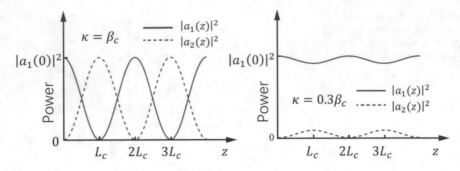

Fig. 5.5 Plots of the power coupling curve in two guided waves. Notice that the coupling power is repetitive over a characteristic length given by $L_c = \pi(2\beta_c)$. The left is when $\beta_1 = \beta_2$ and thus $|a_1|^2 = |a_2|^2$. The right is when $|\beta_1 - \beta_2| = 6.36\kappa$ and thus $\dfrac{|a_1|^2}{|a_2|^2} = (\kappa/\beta_c)^2 = 0.09$

In the case of counterpropagating waves, a different set of equations is obtained. Further, because the waves are oppositely directed, it is not possible to have phase matching in the presence of a simple non-periodic coupling term, as was the case for copropagating modes. If, however, the coupling does have a periodic spatial dependence, phase matching can be obtained for counterpropagating waves if this spatial periodicity has a Fourier component equal to $|\beta_1 - \beta_2|$. Specifically, if the waves are coupled through a spatial periodic structure, introduced by, say, a surface diffraction grating on a single waveguide, then the coupled-mode equation is written as

$$\frac{da_1}{dz} = -j\beta a_1 + \kappa_{12}a_2 e^{-jKz} \tag{5.21}$$

$$\frac{da_2}{dz} = j\beta a_2 + \kappa_{21}a_1 e^{+jKz} \tag{5.22}$$

and where $K \equiv 2\pi/\Lambda$, Λ is the spatial period of the structure, and thus K is the wavenumber of the periodic coupling element.

In these equations, it is assumed for the sake of simplicity that $\beta_1 = \beta_2 = \beta$; that is, we examine coupling between the forward and backward waves in a single-mode synchronous structure fabricated on the waveguide. This simple scheme is common in many practical counterpropagating devices, for example, in a structure where a grating is etched into the surface of a waveguide. The analogous coupled-mode equations for the case in which β_1 and β_2 are different but can also be solved; it is, however, a straightforward but tedious exercise.

These equations can be reduced to a form analogous to those for the copropagation coupling by the substitution

$$\begin{cases} a_1(z) = A_1(z)e^{-jKz/2} \\ a_2(z) = A_2(z)e^{-jKz/2} \end{cases} \tag{5.23}$$

where $A_1(z)$ is the field amplitude that varies even more slowly than $a_1(z)$ by removing any field variations at the grating period. This substitution gives

$$\frac{dA_1}{dz} = -j\left(\beta - \frac{K}{2}\right)A_1 + \kappa_{12}A_2 \tag{5.24}$$

$$\frac{dA_2}{dz} = j\left(\beta - \frac{K}{2}\right)A_2 + \kappa_{21}A_1 \tag{5.25}$$

Notice that these equations are identical to those encountered earlier for codirectional coupling (see (5.7) and (5.8)), with the substitution of $\beta_{1,2} \to \pm(\beta - K/2)$. Also note that for counterpropagating waves, hence by (5.13), $\kappa_{12} = \kappa_{21}^*$. Thus by analogy, phase matching occurs when $2(\beta - K/2) = 0$, or when $\beta = K/2$. Note also that the coupling coefficient κ_{12} does not include the grating spatial variation. The fact that we have removed the grating periodicity by adopting (5.23) also means that in calculating the coupling coefficient, κ_{12}, we must also remove the grating spatial variation in the modal fields.

Depending on the device, either β or K can be tuned to achieve a specific functionality, i.e., switching, filtering, etc. A simpler notation can be used which will be important for examining the behavior of a periodic structure in the presence of a wavelength-tuned optical source. Specifically, let β be expanded in a Taylor's series about the resonant or synchronous frequency for the grating,

$$\beta \simeq \beta(\omega_0) + \frac{d\beta}{d\omega}(\omega - \omega_0) \tag{5.26}$$

$$= \beta(\omega_0) + \frac{\omega - \omega_0}{v_g} \tag{5.27}$$

where we have used the fact that $d\omega/d\beta = v_g$ is the group velocity. We then introduce the detuning of β from the resonant spatial period of the periodic structure,

$$\delta \equiv \beta - \frac{K}{2} \tag{5.28}$$

$$\equiv \frac{\omega - \omega_0}{v_g} \tag{5.29}$$

This substitution allows the coupled-mode equations to be rewritten as

$$\frac{dA_1}{dz} = -j\delta A_1 + \kappa A_2 \tag{5.30}$$

$$\frac{dA_2}{dz} = j\delta A_2 + \kappa A_1 \tag{5.31}$$

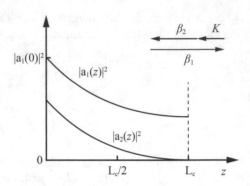

In these equations, A_1 is the amplitude of the forward-propagating mode and A_2 the amplitude of the backward-propagating mode. Again, solving the determinant of these equations gives the propagation constants of the coupled modes:

$$\beta'_{1,2} = \pm\sqrt{\delta^2 - |\kappa|^2} \tag{5.32}$$

The coupled-mode equations can then also be used to obtain the expressions for the modal power as a function of distance along a periodic coupler of length L.

To obtain these expressions, we assume $A_1(z = 0) = 1$ and $A_2(z = L) = 0$, and find for the power of the forward and backward waves,

$$\frac{|A_1(z)|^2}{|A_1(0)|^2} = \frac{1 + (\kappa/\beta_d)^2 \sinh^2[\beta_d(z - L)]}{1 + (\kappa/\beta_d)^2 \sinh^2 \beta_d L} \tag{5.33}$$

and

$$\frac{|A_2(z)|^2}{|A_1(0)|^2} = \frac{(\kappa/\beta_d)^2 \sinh^2[\beta_d(z - L)]}{1 + (\kappa/\beta_d)^2 \sinh^2 \beta_d L} \tag{5.34}$$

where $\beta_d = \sqrt{\kappa^2 - \delta^2}$.

Figure 5.6 shows a plot of these functions for several values of the normalized length $\beta_d L$ at $\delta = 0$. It shows that as the length increases, the efficiency of the transfer of power in the backward direction increases. Also, notice that along the propagation path, i.e., from left to right in the figure, both A_1 and A_2 decrease along the periodic structure, since A_1 is gradually attenuated by reflection and A_2, the reflected wave, is the sum of all reflections along z. The reflected power or power transfer, $|A_2(0)|^2/|A_1(0)|^2$, from A_1 to A_2 is 99% when $L = \pi/\kappa$; in general, the reflected power for no propagation mismatch $\delta = 0$ is

$$\frac{|A_2(0)|^2}{|A_1(0)|^2} = \tanh(|\kappa|L) \tag{5.35}$$

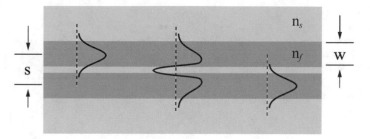

Fig. 5.7 Copropagating mode coupling between the two symmetric slab waveguides

Thus a characteristic coupling length can be defined in the contradirectional case as well: $L_c = \pi/2\kappa$.

Finally, note that while grating-assisted coupling is usually considered in a contradirectional context, a grating can also be used to compensate for differing propagation constants in codirectional coupling. In that case, codirectional coupling can be accomplished with a grating of spatial frequency K, if $\beta_1 - \beta_2 \approx K$. Thus codirectional coupling requires a larger spatial period than contradirectional coupling. Interestingly, as mentioned earlier in the chapter, the κ for codirectional coupling is typically stronger than for counterdirectional coupling because of the sign of the small but finite κ_z term, which acts against κ_t for contradirectional coupling. As will be discussed in a subsequent chapter, the power transfer characteristics for codirectional coupling in a grating are similar to those for a directional coupler.

5.4 Two Specific Examples of Coupling Coefficients

We can illustrate the calculation of the coupling coefficient for two of the more important cases of the application of coupled mode theory: coupled waveguides and the coupling of forward and backward waves in a grating.

Consider first copropagating mode coupling between the two symmetric slab waveguides shown in Fig. 5.7. Using the fields for slab waveguides obtained from Chap. 3, it can be shown that for these two slab waveguides,

$$\kappa_{12} = \frac{j2\gamma k_x^2}{\beta(k_x^2 + \gamma^2)(d + (2/\gamma))} \exp(-\gamma(h - d)) \tag{5.36}$$

In these expressions, we have used the expression k_x rather than κ as used in Chap. 3 to denote the transverse wavevector in the waveguide and to avoid confusion with the coupling coefficient. Also, γ is the evanescent transverse wavevector in the slab cladding. Notice the exponential dependence of the coupling on the separation of waveguides, $h - d$; this strong, nonlinear dependence makes precise-value couplers hard to make.

Fig. 5.8 A grating
waveguide with a simple
spatial variation
$d(z) = d_0 + \Delta d \cos Kz$,
where $K = 2\pi/\Lambda$

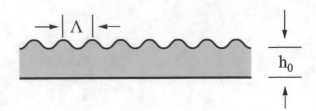

For a grating, the index perturbation is spatially periodic. Thus, let us assume here a simple spatial variation $d(z) = d_0 + \Delta d \cos Kz$ as shown in Fig. 5.8. As mentioned earlier, this spatial periodicity allows the phase-matching condition (i.e., the waves are synchronous) $2\beta \approx K$ to be achieved for forward and backward moving waves in the grating coupler.

To obtain the coupling coefficient, we use the known field distributions for a slab waveguide, given in Chap. 3, and assume TE radiation. The presence of this surface corrugation alters locally the dielectric constant around $x = d$ such that if $d(z) > d_0$, the effect is to add a small volume of dielectric with $\Delta\epsilon = \epsilon_0(n_f^2 - n_c^2)$, while if $d(z) < d_0$, the dielectric change is negative and of opposite magnitude $\Delta\epsilon = \epsilon_0(n_c^2 - n_f^2)$.

These fluctuations are substituted in (5.22) and (5.23) along with the appropriate modal fields. Since the field is TE, the longitudinal coupling coefficient of (5.23) is zero, i.e., $\kappa^z = 0$. However, the transverse coupling coefficient is given by

$$\kappa^t = \iint \Delta\epsilon E_y^2 \, dx \, dy \tag{5.37}$$

If the grating height corrugation is small compared to γ_c^{-1} in the cover or k_x^{-1} in the waveguide dielectric, then $E_y(d(z)) \approx E_y(d_0)$, a constant independent of integration. The coupling coefficient, including the sinusoidal z dependence, κ', is then

$$\kappa' = \kappa^t = j\omega\epsilon_0 E_y^2(d_0)(n_f^2 - n_c^2)\Delta d \cos Kz \tag{5.38}$$

However, as mentioned earlier in this chapter, the usual formulation of the coupled-mode equations removes the grating-induced rapid spatial variation from the modal fields (see Sect. 5.3.2), to obtain the usual coupled-mode κ for gratings. In addition, the value of E_y in (5.38) can be related to the maximum field of the waveguide mode via the relations derived earlier in Chap. 3, e.g. (3.36),

$$E_f^2(n_f^2 - N_{eff}^2) = E_c^2(n_f^2 - n_c^2) \tag{5.39}$$

where E_f is the maximum field in the guided mode and E_c is the field at the cover interface and equals $E_y(d_0)$. Recall from the initial discussion of the modal expression in Sect. 5.2 that the electric fields used are normalized by setting the waveguide power equal to 1. This power is obtained from the Poynting vector:

$$P = N_{eff} \sqrt{\frac{\epsilon_0}{\mu_0}} E_f^2 T_{eff} \qquad (5.40)$$

where the average effective thickness of the slab waveguide is $T_{eff} = d_0 + \frac{1}{\delta} + \frac{1}{\gamma}$.
Then, setting

$$\kappa' \equiv \kappa e^{-jKz} \qquad (5.41)$$

an expression for the coupling coefficient is realized,

$$\kappa = j \frac{\pi}{\lambda} \frac{\Delta d}{T_{eff}} \frac{n_f^2 - N_{eff}^2}{N_{eff}} \qquad (5.42)$$

5.5 An Alternate Approach to Obtaining the Coupling Coefficient

The general coupled-mode theory given earlier in Sect. 5.2 made it possible to derive a rigorous and very general expression for the coupling coefficient. In order to gain more insight in the physical nature of the coupling, we consider a more direct but simplified approach to the equation for the coupling coefficient. This approach yields the coupling coefficient, κ_{ij} via a derivation based on general power flow arguments; the discussion of the approach is made more accessible by considering the specific geometry of two symmetric coupled waveguides (Haus 1984). A sketch of this waveguide geometry is shown in Fig. 5.9. Physically, the coupling from waveguide 1 into waveguide 2 is due to the polarization current of the dielectric in waveguide 2 which is caused by the presence of a field in waveguide 1. If $\vec{f}_1(x, y)$ is the normalized transverse field distribution of waveguide 1 and a_1 its amplitude (see (5.6)), then the cross waveguide polarization current $\dfrac{d\vec{P}_{21}}{dt}$ is, by definition,

$$\frac{d\vec{P}_{21}}{dt} = \frac{d(\vec{P}_{21}(x, y) \cdot e^{-j\omega t})}{dt} = -j\omega \vec{P}_{21} = j\omega(\epsilon_f - \epsilon)a_1 \vec{f}_1(x, y)e^{-j\omega t} \qquad (5.43)$$

where ϵ is the cladding dielectric outside of the waveguides, ϵ_f is the dielectric in guide 2 and where we have subtracted off the current, which would exist in the absence of waveguide 2.

Using the Poynting vector, the power transferred can then be determined to be

$$\frac{1}{4} \int_A \vec{E}_2^* \cdot (-j\omega) \vec{P}_{21} \, dx \, dy = \frac{j\omega}{4} a_1 a_2^* \int_A (\epsilon_f - \epsilon) \vec{f}_1 \cdot \vec{f}_2^* \, dx \, dy + c.c. \qquad (5.44)$$

where A is the area of the waveguide. But we know from coupled-mode theory (see above) that

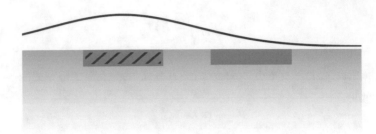

Fig. 5.9 A sketch of two symmetric coupled waveguides: the mode propagates in left waveguide overlap with the right waveguide

$$\frac{d|a_2|^2}{dz} = \kappa_{21}a_1a_2^* + \kappa_{21}^*a_1^*a_2 \tag{5.45}$$

Equating the left-hand sides of (5.44) and (5.45) thus shows that

$$\kappa_{21} = \frac{j\omega}{4}\int_A (\epsilon_f - \epsilon)\vec{f}_1 \cdot \vec{f}_2^* \, dx \, dy \tag{5.46}$$

Notice that this equation is a simplified version of the general equations given above in Sect. 5.2. In addition, it is easy to show that

$$\kappa_{12} = \frac{j\omega}{4}\int_A (\epsilon_f - \epsilon)\vec{f}_2 \cdot \vec{f}_1^* \, dx \, dy \tag{5.47}$$

A somewhat more general form of this equation is then

$$\kappa_{ij} = \frac{j\omega}{4}\int_A \Delta\epsilon(x, y, z)\vec{f}_j(x, y)\vec{f}_i^*(x, y) \, dx \, dy \tag{5.48}$$

where i, j are mode labels and $\Delta\epsilon \equiv \epsilon_f - \epsilon$. In many cases, this equation can be directly applied to determine κ_{ij}.

5.6   Summary

This chapter has introduced the formulation of coupled-mode theory and has shown its utility in describing the operating physics of several common device structures for integrated optics, i.e., a waveguide coupler and a diffraction grating (with in plane propagation). The discussion here has focused on simple problems, which involve the coupling between the two lowest order modes. In many cases, however, coupling schemes can be more elaborate, including coupling of higher order modes in devices such as mode sorters and switches. These may be approached theoretically using the same principles as discussed above. In the next chapter, a more thorough discussion of devices for simple optical couplers will be presented.

Problems

1. Using the expression for κ_{12} between two slab waveguides (5.36), calculate h needed to have $L_c = 300\,\mu$m. Assume $n_{slab} = 3.34$, $n_{cover} = 3.30$, $\lambda = 1.3\,\mu$m. First, Fix d, the slab thickness so that the V parameter is 20% below cutoff for the $m = 1$ mode. Note that this type of coupler is used for vertical coupling in III-V devices.

2. Use the effective index method to estimate h for a polymer synchronous coupler of length 3 mm.

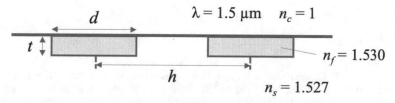

d $\lambda = 1.5\,\mu$m $n_c = 1$

t $n_f = 1.530$

h

$n_s = 1.527$

   (a) Design the height of the waveguide to be single-mode (TM-like) so that V is 20% below $V_c^{m=1}$
   (b) Design the width to be single mode using same V_c criterion (the mode is now TE-like, of course)
   (c) Find the spacing h

3. What is the maximum power transfer for a codirectional coupler if $(\beta_2 - \beta_1) \approx 3\kappa$?
4. Derive (5.36) using (5.48) and the modal fields of the slab waveguide.
5. Use (5.42) and the waveguide in Problem 1 to design the modulation depth for a grating coupler with $\kappa = 2\,\text{mm}^{-1}$.
6. A coupler made with two symmetric slab waveguides to have coupling coefficient κ. If an optical wave of power of magnitude P enters waveguide 1 at $z = 0$, try to answer the following questions.

   (a) Plot the power in waveguide 1 and 2 versus distance z in the following two cases:
      (i) $\beta_1 = \beta_2$
      (ii) $\beta_1 \gg \beta_2$
      where β_1 and β_2 are the mode propagation constants, λ is the wavelength. Use simple words to describe the differences between the two situations.

(b) For the case of two synchronous waveguides, sketch the two lowest *SYSTEM* modes. Write down the expression for each of the two propagation constants for the two modes.

7. Given two simple slab waveguides as below:

(a) Assume $d_1 = d_2 = d$. Calculate d so that each slab waveguide has a single TE mode that has a normalized frequency of 20% below the $m = 1$ mode cutoff. Assume $\lambda = 1.55\,\mu$m.

(b) If guides are spaced to have a coupling coefficient of $\kappa = 50$ cm^{-1}, what is the waveguide length L for the minimum power transfer from guide 1 to guide 2? Assume guide 1 is excited.

(c) If guide 2 now has $d_2 = 0.7d$, what is the length L for maximum transfer, assuming κ stays the same?

References

Haus, H. A. (1984). *Waves and fields in optoelectronics*. Upper Saddle River: Prentice-Hall.

Hutcheson, L. D. (1987). Integrated optical circuits and components: Design and applications. *Optical engineering* (Vol. 13, 417 p.) No individual items are abstracted in this volume. New York: Marcel Dekker, Inc.

Kogelnik, H. (1988). Theory of optical waveguides. *Guided-wave optoelectronics* (pp. 7–88). Berlin: Springer.

Marcatili, E. A. (1969). Dielectric rectangular waveguide and directional coupler for integrated optics. *Bell Labs Technical Journal, 48*(7), 2071–2102.

Marcuse, D. (1974). *Theory of dielectric optical waveguides* (Google Scholar, pp. 181–193). New York: Academic.

Marcuse, D. (1991). *Theory of dielectric optical waveguides*.

Nishihara, H., Haruna, M., & Suhara, T. (1989). *Optical integrated circuits* (Vol. 1). New York: McGraw Hill Professional.

Saleh, B. E., Teich, M. C., & Saleh, B. E. (1991). *Fundamentals of photonics* (Vol. 22). New York: Wiley.

Yariv, A. (1973). Coupled-mode theory for guided-wave optics. *IEEE Journal of Quantum Electronics, 9*(9), 919–933.

Chapter 6
Optical Couplers

Abstract The goal of this chapter is to examine in detail the practical side of integrated optical couplers. Thus, for example, these couplers are fabricated of lithium niobate via surface diffusion from a local titanium source or those fabricated of Si, including Si wires. The chapter will also include consideration of the deviation of the actual coupler geometry and materials components from that of the ideal waveguide.

6.1 Introduction

Optical couplers are one of the most important classes of integrated optical components. These devices are used in directional routing of a light signal from one waveguide to another or in dividing a signal between different lightguides. However, even more important applications rely on the fact that the amount of coupling can either be varied, as in various forms of electro-optical switches or on its sensitivity to certain property of the optical signal, e.g., its wavelength, intensity, or temporal properties. The latter function is the basis of wavelength routers or nonlinear switches.

In this chapter, we will discuss passive optical couplers. The discussion will include a consideration of both conventional and adiabatic, or spatially varying, couplers, as well as their practical implementation. This chapter also presents an excellent opportunity to consider in more detail practical examples of the coupled-mode theory in working device structures. Finally, this chapter is the first in this book to introduce the realization of a practical integrated device, namely, different forms of couplers, and thus we illustrate using a variety of couplers including coupled rings, vertical-coupled laser sources, and power dividers. Note that numerical calculations are also shown.

© Springer Nature Switzerland AG 2021
R. Osgood jr. and X. Meng, *Principles of Photonic Integrated Circuits*,
Graduate Texts in Physics,
https://doi.org/10.1007/978-3-030-65193-0_6

6.2   The Standard Two-Waveguide Directional Coupler

A backbone design step for integrated photonic circuits is the simple 2×2 coupler.
Our approach to obtaining an analytic solution for these devices follows closely that
of Nishihara et al. (1989). In essence, such a coupler consists of two waveguides—
either channel or slab—that are placed close enough so that the optical fields are
coupled from one waveguide to another by virtue of the evanescent field beyond the
confinement region. In general, the derivations assume weak coupling between the
individual waveguide modes and nearly identical waveguides. Deviations from these
assumptions will be discussed in Sect. 6.3.

The equations used for this discussion yield results identical to those in Chap. 5.
However, in this chapter, we chose an alternative approach based on using the primed
amplitudes, a_1' and a_2', that is with the rapid variation due to the optical carrier
phase factored out, since this approach is common in the literature, see Nishihara
et al. (1989), for examples. Second, this approach also develops several practical
parameters such as that for field detuning and the separation of the two system
modes.

For simplicity's sake, in our calculations, we assume that $\kappa_{1,2} = \kappa_{1,2}^* = \kappa$. If we
then factor out the rapidly varying portion of the envelope, i.e., $a_1(z) = a_1'(z)e^{-j\beta_1 z}$,
we have

$$\frac{da_1'(z)}{dz} = -j\kappa a_2'(z)e^{-j(\beta_2 - \beta_1)z} \tag{6.1}$$

$$\frac{da_2'(z)}{dz} = -j\kappa a_1'(z)e^{j(\beta_2 - \beta_1)z} \tag{6.2}$$

We now seek the solutions for these equations by finding the normal modes of
the coupled system; that is, we wish to find the system modes of the combined two-
waveguide device. The eigenmodes may be solved for the general case of input to
both waveguides by assuming that the solutions have the form

$$a_1'(z) = a_1' e^{-j\beta_c z} e^{-j\Delta(z)} \tag{6.3}$$

$$a_2'(z) = a_2' e^{-j\beta_c z} e^{+j\Delta(z)} \tag{6.4}$$

where $a_{1,2}'$ are constants and $\Delta \equiv (\beta_2 - \beta_1)/2$, $\beta_c = \sqrt{\kappa^2 + \Delta^2}$.

Insertion of these solutions into the coupled-mode equations (6.1) and (6.2) yields
the solutions for fields in this device. Note that the mode amplitude in waveguides 1
and 2 can then be written as the sum of two independent orthogonal modes, which
have even (e) or odd (o) character as indicated by their subscripts. Alternate versions
of these equations appear in the literature; see, for example, the text by Haus (1984)
or Chap. 5 of this book.

Finally, the general fields for the individual waveguide modes can thus be expressed clearly and neatly as the sum of the system modes

$$E_1(x, y, z, t) = -[a'_{1e}e^{-j\beta_e z} + a'_{1o}e^{-j\beta_o z}]f_1(x, y)e^{j\omega t} \qquad (6.5)$$

$$E_2(x, y, z, t) = -[a'_{2e}e^{-j\beta_e z} + a'_{2o}e^{-j\beta_o z}]f_2(x, y)e^{j\omega t}, \qquad (6.6)$$

which have propagation constants

$$\beta_{e,o} = \beta_m \pm \beta_c \qquad (6.7)$$

where

$$\beta_m = (\beta_1 + \beta_2)/2 \qquad (6.8)$$

The symbols $f_{1,2}$ in (6.7) and (6.8) represent the normalized field distribution in the individual guides. Notice that the expressions for $\beta_{e,o}$ are identical to the eigenvalues of β found in the formal derivations in previous chapter. Since $\beta_c \to \kappa$ as the propagation constant of the two isolated waveguides β_1 and β_2 become equal, the waveguide coupling eliminates the degeneracy in the β's for the synchronized waveguides β_e and β_o when coupling is present.

The amplitudes of the system modes, a'_{1e} and a'_{1o}, are determined by the boundary conditions. Thus for light entering waveguide 1, i.e., $a'_1(0) = 1$, $a'_2(0) = 0$, at z, the amplitudes for a'_{1e} and a'_{1o} may be found via substitution in (6.5) and (6.6) to, yielding $a'_1(z)$ and $a'_2(z)$ be

$$a'_1(z) = e^{-j\Delta z}\left(\cos\beta_c z + j\frac{\Delta}{\beta_c}\sin\beta_c z\right) \qquad (6.9)$$

$$a'_2(z) = e^{j\Delta z}\left(-\frac{j\kappa}{\beta_c}\sin\beta_c z\right) \qquad (6.10)$$

As was shown in Sect. 5.3.2, these initial conditions lead to solutions having a sinusoidal power transfer along z between each of the individual waveguides,

$$\frac{|a'_1(z)|^2}{|a'_1(0)|^2} = 1 - \frac{\kappa^2}{\beta_c^2}\sin^2\beta_c z \qquad (6.11)$$

$$\frac{|a'_2(z)|^2}{|a'_2(0)|^2} - \frac{\kappa^2}{\beta_c^2}\sin^2\beta_c z \qquad (6.12)$$

and thus κ^2/β_c^2 is the maximum fraction that can be coupled from waveguide 1 to waveguide 2. For example, if $\kappa = 0$, the optical wave sensibly remains in guide 1 ($a'_1 = 1$), for all z! Figure 6.1 shows a plot of the power in each of the two guides

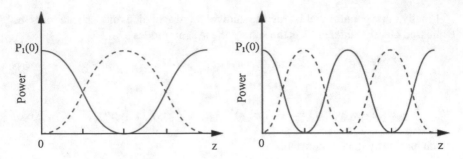

Fig. 6.1 A plot of the power in each of the two waveguides for two values of Δ

for two example values of Δ. Maximum power transfer occurs when $z_{max} \equiv L_c = \pi/(2\beta_c)$, or when $L_c = \pi/(\beta_e - \beta_o)$.

6.3   Transfer Matrix for a Coupler

The equations obtained in the preceding section suggest a simplified formulation that is useful for the overall design of a complex optical circuit, namely that of transfer matrices. The transfer matrix of a device captures the effect of an ideal device on its input waves. While this formulation is not intended to present a detailed realistic electromagnetic simulation of the actual device, it does allow for the rapid design of a large number of device elements via simple serial multiplication of the transfer matrix of each element since the formulation does realize accurately the overall properties of the device. Thus, while the limitation on transfer matrices is that such a matrix only captures the overall behavior of a specific, generally idealized geometry; it does not enable one to make a detailed exploration of the device geometry.

In this chapter, we use a typical 3 dB coupler as an example and show how to determine its transfer matrix. For a two input device, the transfer matrix allows the two inputs, a_1^i and a_2^i, to be presented in a compact, easily implemented vectorial manner and related via the matrix to the outputs, a_1^o and a_2^o.

$$\begin{bmatrix} a_1^o \\ a_2^o \end{bmatrix} = \begin{bmatrix} u_{11} & u_{12} \\ u_{21} & u_{22} \end{bmatrix} \begin{bmatrix} a_1^i \\ a_2^i \end{bmatrix} \tag{6.13}$$

The matrix elements can be obtained by manipulating the solutions obtained in the previous section for the case of arbitrary initial boundary conditions and by returning to the original variables $a_1(z)$ and $a_2(z)$. It is then found that $u_{22} = u_{11}^*, u_{21} = -u_{12}^*$, and that

$$u_{11} = \cos(\beta_c L) + j \frac{\Delta}{\beta_c} \sin(\beta_c L) \tag{6.14}$$

and

$$u_{12} = j \frac{\kappa}{\beta_c} \sin(\beta_c L) \tag{6.15}$$

for a coupler of length L. As an example, if $a_1^i = 1$ and $a_2^i = 0$, the solutions for a_1^o and a_2^o recover (6.11) and (6.12), once the transformation $a_1' \rightarrow a_1$ and $a_2' \rightarrow a_2$ is made. This matrix allows the phases and amplitudes of both outputs of the ideal directional coupler to be easily manipulated. Throughout the remainder of this chapter, transfer matrices of several other simple devices will be introduced.

6.4 More Exact Treatment of Waveguide Couplers

The traditional coupled-mode theory for optical couplers is essentially a perturbation theory. Its accuracy depends on weak waveguide coupling and near synchronization of the individual waveguide propagating modes. It is possible, however, to use more exact analyses, based on variational methods or a more exact treatment of the modes, to see that this approach fails as the validation of the above conditions begin to degrade as discussed by Hardy and Streifer (1985), Haus et al. (1987, 1989). Finally it is also possible to use numerical methods to obtain an essentially exact treatment of the coupler as long as there is also an exact understanding of the waveguide geometry and materials.

Consider, for example, two nonidentical waveguides as depicted in Fig. 6.2a. The fields of these guides have been calculated using the conventional coupled-mode analysis given above and by a more exact theory, which deals more carefully with radiative modes (Hardy and Streifer 1985). A comparison of these calculations is given in Fig. 6.2b for the two lowest order system modes. Clearly, there is an important discrepancy between the two calculations. A calculation of the system-mode propagation constants of the coupled waveguides versus thickness also shows important discrepancies. The values obtained with the more complete theory differ significantly from those obtained with traditional coupled-mode theory at smaller spacing. Note, however, that the difference in the propagation constants in both the more sophisticated and the traditional theory is extremely small except at the very smallest spacings. Radiative effects, which are not shown here, are, however, more accurately calculated by the more exact theory.

An elegant variational theoretical approach has also been applied to this same problem (Haus et al. 1987). The results of this treatment obtained similar numerical results as that by Hardy and Streifer (1985). However, the variational treatment shows clearly that the form of the traditional coupled-mode formalism can be recovered by carefully reorganizing the system modes of two waveguides. This step is done by a more complete expansion of the system modes in terms of the individual waveguide modes. Further, as mentioned earlier, Haus et al. (1987) has pointed out that the dominant effect on the modal propagation constants is a simple uniform shift in β_o

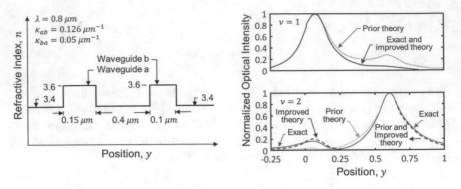

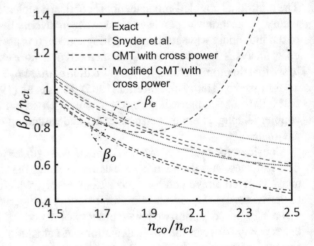

Fig. 6.2 (Left) Sketch of two nonidentical waveguides. (Right) A comparison of the fields of these guides calculated using the conventional coupled-mode analysis given above and by a more exact theory [Adapted from (Hardy and Streifer 1985)]

Fig. 6.3 The effect of uniform shift in β_o and β_e can be shown with several different versions of coupled-mode theory (CMT) [Adapted from (Haus et al. 1989)]

and β_e; this shift is accurately calculated by standard coupled-mode theory. Thus the coupling length in this case remains the same, since

$$\beta_e - \beta_o = 2\kappa_{12} \tag{6.16}$$

as in conventional theory; see Fig. 6.3 for an example. Finally, an extension of weak coupling theory has also been formulated, using the variational method, applied to vector coupled-mode theory. Again, in this case, the coupling length is that given by the first-order theory (Haus et al. 1989).

6.5 Three Examples of Actual Working Couplers

6.5.1 Lateral LiNbO₃ Couplers—A Study of the Sensitivity to Fabrication

In actuality, the fabrication of good couplers is often a difficult process. Part of the difficulty is a result of the extreme sensitivity of directional couplers to the exact value of the coupling coefficient, which makes them difficult to fabricate reproducibly by standard lithographic techniques. In fact, even slight variations in etch depth and lithographic dimensions can yield serious variation in the coupling length, since the coupling coefficient depends exponentially on waveguide separation and on the detailed profile of their evanescent fields, as given by γ. This is a particularly important issue for diffused-waveguide-based couplers since the extent of the junction is a result of both lithography and thermal diffusion.

A second issue, which was discussed in the previous section, is that the close proximity of waveguides, particularly those with weak confinement, means that each guide may perturb the mode of the other, or that inadvertently the guides coupling may cross over from the weakly coupled limit, and becomes strongly coupled or even a single waveguide. We will illustrate these considerations below, using the example of graded-index (LiNbO₃) guides. Similar studies have been done on high-index-contrast semiconductor waveguides.

Graded-index-based couplers have been examined in detail, including theoretical (WKB-based) and experimental studies for the case of Ti:LiNbO₃ systems (Noda et al. 1981). The goal in these studies was to understand clearly the dependence of coupling length on the variation of waveguide parameters, the transition region between weak and strong coupling, and the sensitivity to fabrication tolerances. In addition, the perturbation to the waveguide parameters by the presence of neighboring guide was also important.

Diffused waveguides clearly have a 3D character and thus their analysis can be complex (see Fig. 6.4). However, not surprisingly, these couplers are most sensitive to the variation in the y- (or in-plane) direction. For example, the coupling length of two weakly coupled waveguides can be written as

$$L_c = \frac{\lambda}{2(N_e - N_o)} \tag{6.17}$$

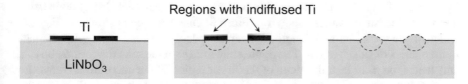

Fig. 6.4 Steps in forming Ti:LiNbO₃. Due to the three-dimensional diffusion and indiffused Ti, its complex to analyze the characters such as refractive index and loss

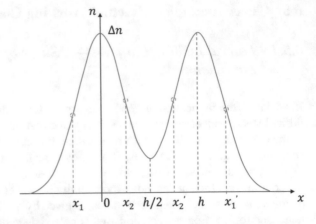

where $N_{e,o}$ are the effective indices for the odd and even system modes. However, in practice, it is found that the coupling length can be approximated by

$$L_c = \frac{\lambda}{2(N_e^y - N_o^y)} \tag{6.18}$$

i.e., using the effective indices for the y slab waveguide only.

The measured variation in the refractive index in the direction at the surface for a LiNbO$_3$ coupler is shown in Fig. 6.5, along with the physical parameters of one example of this structure. In the figure, the location of the outer turning point for the individual waveguide modes are shown in addition to the point at which light leaks across the coupling region. It is obvious from the figure that the index profile of the two guides, which are Gaussian in the isolated guide limit, significantly perturb each other on their "inner" side.

The figure also can be used to quantify the transition between weak and strong coupling. For the device to act as a weak coupler, light must be confined in the core region of either of the two waveguides. Thus, using the symbols defined in Fig. 6.5,

$$n_s + \Delta n > N_e^y, N_o^y > n_s + \Delta n_c \tag{6.19}$$

Since $N_e^y > N_o^y$, the light is no longer weakly confined when $N_e^y < n_s + \Delta n_c$. In this case, the device acts like a single waveguide.

Some of the issues associated with making such graded-index guides are illustrated by a series of calculations on coupling performance (Noda et al. 1981) done using index profiles, such as those shown in Fig. 6.5. One example is shown in Fig. 6.6, which displays the coupling length, L_c, versus spacing between waveguide centers, h, for a series of values of Δn. The figure shows the extreme sensitivity of L_c on Δn, and, hence, on n. In fact, the slope of the linear region of the curve in the figure for $\Delta n = 0.006$ indicates a value of dL_c/dh of 6 mm/µm. At short coupling lengths, the coupling becomes strong. This region is denoted by a dashed-dotted line. As

Fig. 6.6 A plot of the relation between coupling length L_c and waveguide spacing h, taking refractive-index change Δn as the varying parameter

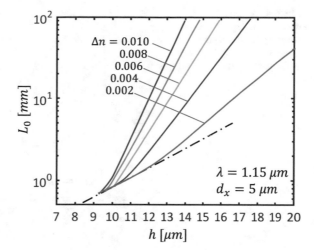

Fig. 6.7 A plot of the relation between coupling length L_c and refractive-index change Δn, taking waveguide spacing h as the varying parameter

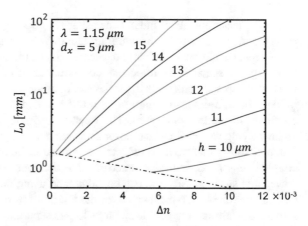

suggested by the discussion in the previous section, it is not possible to use a simple effective-index approach to calculate the coupling in this region.

A second calculation shows the dependence of coupling length on index contrast, Δn, for a series of fixed waveguide spacings. These results are given in Fig. 6.7. Again, the region of strong coupling, which below the dot-dash line, is indicated in the figure. Notice that the coupling length decreases to a very short length for small Δn. While this can be desirable for certain applications, it comes as a result of increased dimensional sensitivity. A related concern is wavelength sensitivity, which results from the variation in effective index with wavelength. These calculations, displayed in Fig. 6.8, show that the wavelength sensitivity of L_c decreases as the wavelength increases.

Although not shown here, experimental data have been compiled on Ti-diffused couplers and generally show excellent agreement with the calculated response; that

Fig. 6.8 A plot of
wavelength dependence on
coupling length, taking
refractive-index change Δn
as the parameter

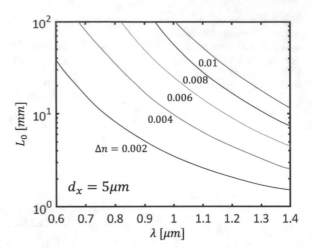

is, direct measurements of index contrast are in good agreement with those inferred
from the measurements of the coupling length versus waveguide spacing.

The results above show clearly the sensitivity of the coupling to fabrication errors
in waveguide spacing. An important consequence of this fabrication error is that it
can cause increased cross talk in the device. In this case, cross talk is calculated by the
ratio of the light-intensity output in guide 1 versus that in guide 2, where complete
crossover to guide 2 is desired. The tolerance in coupler length, ΔL, required to
attain varying values of cross talk for different coupling lengths is shown in Fig. 6.9.
The plot shows that smallest crosstalk tolerance is encountered in the case of short
coupling lengths. Once the tolerance in ΔL is obtained at a given L_c, the tolerances in
waveguide spacing, Δn, etc., can be calculated using the standard waveguide equations. For example, a typical value for the tolerance in waveguide spacing to achieve
20 dB cross talk and waveguides with $\sim$ 13 μm separation, $d_x \approx 5.0$ m, $\Delta n = 0.005$,
and $L \approx 10$ mm is 0.5% or 65 nm.

6.5.2 Lateral Si Wire Couplers

Several approaches have been developed to make a more reliable and robust coupler
technology. One obvious possibility is to make an active coupler by electro-optical
tuning or post-process trimming. Of course, this approach raises the complexity and
cost of the entire fabrication process and the final device. A second approach is to use
a vertical coupler; that is, separate the waveguide by epitaxial layers since such layers
can be grown with atomic precision and thus the coupling length can be fixed more
accurately than with lithography alone. Finally, of course, the increasing resolution

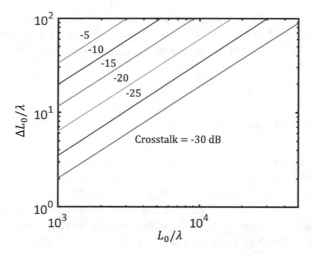

Fig. 6.9 A plot of normalized propagation constant difference vs coupling length at each value of the crosstalk

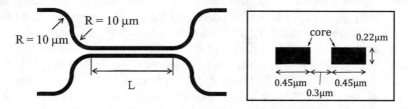

Fig. 6.10 Top and cross-sectional views of a Si wire waveguide directional coupler

and fabrication tolerance available from the patterning tools for advanced Si ICs has steadily improved lateral coupling position.

As mentioned earlier in this text, the Si wire is a basic device technology for Si photonics and thus Si have been extensively studied. Thus, one investigation examined the possibility of obtaining very compact couplers since the Si wire bending radius can be very small, i.e., a several micrometers, due to their very large refractive index contrast between the silicon core and the SiO_2 cladding, which confines the optical field within the device.

In another study (see Fig. 6.10), directional couplers were fabricated and their fundamental characteristics were measured. Thus, an SOI wafer was used, which had a 0.3 μm thick crystalline Si layer on a 1 μm buried SiO_2 layer on the Si substrate. Standard experimental planar processing was used including electron beam lithography for patterning and dry etching for pattern transfer. After patterning the waveguide, the device was clad with 1 μm of SiO_2 using chemical vapor deposition. Measurements showed that the devices had micrometer-scale coupling lengths, (see Fig. 6.10) typically ~ 10 μm, allowing them to be used as compact power dividers or combiners for optical communication signals. Because of the wavelength dependence of the waveguide materials, the measured wavelength-dependent optical outputs from the parallel and cross ports of the 800-μm-long device oscillated in magnitude with a 2.5-nm wavelength period. This behavior can be useful in designing wavelength MUX/DEMUX devices.

6.5.3 Vertical Couplers

The third coupler example is that of a basic vertical coupler using an on-chip Si wire and a III-V hetero-materials laser source. The purpose of the device is to make an electrically pumped light source on a silicon platform and to make the device more manufacturable by only patterning on the Si wafer platform. In this process, both the waveguide and the passive waveguide length are defined using the same etch step.

In addition, in this device, the laser cavity is defined in an in-plane direction solely by the length of the silicon waveguide. During fabrication, it thus does not require otherwise complex alignment in the III-V material. To summarize, this fabrication approach allows the use of CMOS fabrication tools for the Si steps and the fabrication of the optical gain step to be separately provided using the III-V materials. Bonding of the two crystal types is carried out using a low-temperature oxygen plasma step. The laser is electrically driven in the III-V materials, which in this case is AlGaInAs, and the field is coupled into Si via the coupler evanescent wave. In the example in Fang et al. (2006), the laser was CW and had an output power of 1.8 mW and a differential quantum efficiency of $\sim$13%. Thus, in this case, vertical coupling allows the buried Si waveguide device or array of devices to be excited from a surface-excited III-V slab.

6.6 Adiabatic Couplers

While directional couplers are essential for many integrated-optics applications, the "standard" design discussed above is sometimes particularly difficult to manufacture. This difficulty stems from the fact that the coupling coefficient, κ, is exponential in the waveguide separation or geometry. Thus, κ is sensitive to the effective index of the coupling layer between the two guides and, hence, to fluctuations in lithography and material composition; this fact is illustrated in Sect. 6.4 for $LiNbO_3$ guides. As a result, an alternate type of coupler has been developed, which uses a continuously controlled change in waveguide parameters to make a device, which is more robust to variations in device geometry. This device is termed an adiabatic coupler (see Fig. 6.11).

As in a standard coupler, an adiabatic coupler switches light from waveguide 1 to waveguide 2 as depicted in Fig. 6.11. While the geometry of the coupler changes in the z-direction, at any given position, the coupler has a set of local normal or system modes as shown in the inset of Fig. 6.11. These local normal modes are constituted

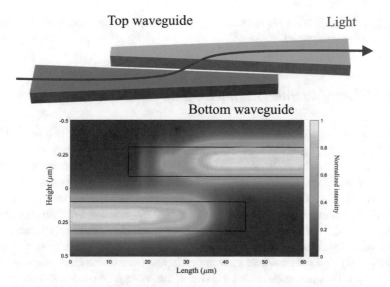

Fig. 6.11 A sketch of an adiabatic coupler

such that they would be the modes of an infinitely long ideal waveguide structure, having the same parameters as the guided-wave transition at this point. The coupler is designed so that light in one system mode remains in that mode as it propagates along the device; that is, the changes in the waveguide parameters must be sufficiently mismatched in z that coupling between system modes is prevented. Finally, adiabatic couplers designed in this way are less dimensionally sensitive than standard couplers, but adiabatic couplers come at the cost of much greater length. Note that study of these couplers also provides a gateway to understanding other forms of adiabatic devices.

An example of an adiabatic coupler is shown in Fig. 6.11, which in this case is tapered either in refractive index or in geometry. As the system modes propagate through the device, they change shape due to the local modal coupling. At the end of the device, the system modes evolve into the isolated-waveguide mode in one of the two waveguides at the end of the device. For the coupler to work successfully, no other system modes should be excited; the system modes should thus evolve adiabatically through the device. To design such a device, we wish to minimize the overlap between the system modes in any given differential element located along the device. Thus, if we divide the tapered coupler into differential lengths, we can define the overlap integral that describes coupling between two system modes,

$$c_{ij} = \int_{-\infty}^{\infty} \Psi_{i0}\Psi_{j1}\mathrm{d}x \qquad (6.20)$$

where x is the coordinate transverse to the local waveguide axis and Ψ_{i0} Ψ_{j1} are the scalar fields of system-mode i entering (subscript 0) and system-mode j leaving (subscript 1) the differential element. Thus we wish to design the device so that c_{ij} is everywhere minimized.

6.6.1   The Differential Coupled-Mode Equations

Because the geometry or effective index in such a device changes with length, it is not possible to use the standard coupled-mode equations given in the earlier section to analyze adiabatic devices. Instead, we must obtain coupled-mode equations that are differential in nature. To obtain these equations, we consider a waveguide structure, which undergoes a transition in one waveguide property, such as the waveguide width, index, thickness or, in the case of two waveguides, separation. We then consider the amplitude of the local normal modes as they propagate along this structure. As these modes travel from one differential waveguide element to the next, the modes couple weakly and thus respond to changes in response to the changing waveguide parameters. If the differential change is slight, the waveguide modes will couple only to neighboring modes. Using the electromagnetic boundary conditions for the optical fields at such a transition, it can then be shown that for a differential change in one geometric or material property (Burns and Milton 1990), designated generally by $\delta\rho$,

$$A_{j1} - A_{j0} = c_{ij} A_{i0} \tag{6.21}$$

$$A_{i1} - A_{i0} = -c_{ij} A_{j0} \tag{6.22}$$

where c_{ij} is the coupling coefficient for the transition $\delta\rho$, and where A_{j0} designates the jth local system mode at a position, 0, just before the position where the change in $\delta\rho$ occurs. The same mode after the point of differential change would be A_{j1}. Since the step is small, only a very small fraction of the original mode(s) is (are) converted by the transition.

If we define a local propagation constant β_j for the local normal mode, j, such that $A_j = |A_j| exp[j(\beta_j z + \phi_j)]$, and set

$$C_{ij} = \lim_{\delta\rho \to 0} \frac{c_{ij}}{\delta\rho} \tag{6.23}$$

then (6.21) and (6.22) can be written as

$$\frac{dA_j}{dz} = C_{ij} \frac{d\rho}{dz} A_i + j\beta_j A_j \tag{6.24}$$

$$\frac{dA_i}{dz} = -C_{ij} \frac{d\rho}{dz} A_j + j\beta_i A_i \tag{6.25}$$

These equations are similar to the usual coupled-mode equations except that they allow coupling due to changes in the device structure along z to be considered. Note that in the absence of coupling, i.e., $C_{ij} = 0$, the modes evolve purely according to their propagation phase; that is,

$$a_{i,j} \equiv A_{i,j} exp \left[-j \int_0^z \beta_{i,j} dz' \right] \tag{6.26}$$

Substitution of this equation into (6.24) and (6.25) then gives

$$\frac{da_j}{dz} = C_{ij} \frac{d\rho}{dz} a_i \exp \left[j \int_0^z (\beta_i - \beta_j) dz' \right] \tag{6.27}$$

$$\frac{da_i}{dz} = -C_{ij} \frac{d\rho}{dz} a_j \exp \left[-j \int_0^z (\beta_i - \beta_j) dz' \right] \tag{6.28}$$

These equations can be solved analytically for certain coupler geometries. These include the case where the change in the variable ρ along the waveguide axis, $d\rho/dz$, is such that

$$\frac{d\rho}{dz} = \zeta \left(\frac{\Delta \beta_{ij}}{C_{ij}} \right) \tag{6.29}$$

where ζ is a constant, and $\Delta \beta_{ij} = \beta_1 - \beta_j$, the difference in the local modal propagation constants. Note that $\Delta \beta_{ij}/C_{i,j}$ may vary with z. The general solutions to these equations when ζ is a constant are given in Burns and Milton (1990). With these solutions, it can be shown that the maximum power in mode j due to mode conversion, $P_j^{max} = (a_j a_j^*)$, from mode i with an incident power of $a_{i0} a_{i0}^*$ is given by

$$\frac{P^{max}}{P_{i0}} = \frac{4\zeta^2}{4\zeta^2 + 1} \tag{6.30}$$

for small coupling, i.e., ζ between the modes.

The mode conversion is thus controlled by the parameter ζ, and can be written using (6.29), as

$$\zeta = \frac{C_{ij}}{\Delta \beta_{ij}} \frac{d\rho}{dz} \tag{6.31}$$

and is adiabatic if $\zeta \ll 1$ (small mode conversion), and abrupt if $\zeta > 1$ (large mode conversion). As a specific example, if the index changes linearly by Δn over a device length L, then (6.31) can be written as

$$\zeta \approx \frac{C_{ij}}{\Delta \beta_{ij}} \frac{\Delta n}{L} \tag{6.32}$$

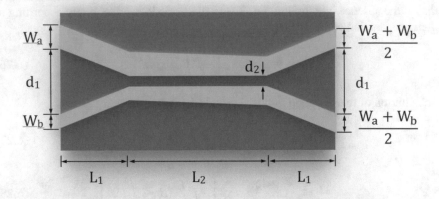

Fig. 6.12 A sketch of an adiabatic coupler

A further interesting consequence of (6.47) is that, except for mode-interference effects which oscillate, it can be shown that when the device parameters in the quantity $d\rho/dz(C_{ij}/\Delta\beta_{ij})$, that is ζ, are adjusted to be constant for a device, the resulting device has the shortest possible device length for a given allowed amount of mode conversion.

6.6.2 Adiabatic Couplers—An Example

In order to gain insight into the design of an adiabatic coupler, consider the specific geometry of the coupler shown in Fig. 6.12. In this case, the coupling region has a linear change in waveguide widths but retains a constant coupling coefficient. The complete coupler is more complex than the coupling region, since it also must incorporate input and output y-branches in order to interface with the remainder of the circuit. These features are important and have been discussed at length in Ramadan et al. (1998). However, for the sake of simplicity, only the isolated coupler will be discussed here.

Before discussing the tapered adiabatic coupler, it is important to reformulate some of the results for the fixed coupler that were obtained earlier in this chapter. Consider the system modes for this simple fixed coupler, which is weakly coupled, and shown in Fig. 6.12. The lowest order even and odd total system modes, $\Psi_{e,o}$, of this coupler with two waveguides having isolated lowest order modes Ψ_a and Ψ_b, and with waveguide a excited at the input to the device, can be written for the weakly coupling limit as (see the discussion in Sect. 6.4, for example)

$$\Psi_e = a_a'\phi_a + a_b'\phi_b \tag{6.33}$$

$$\Psi_o = -a_a'\phi_a + a_b'\phi_b \tag{6.34}$$

where a_a' and $a_b'^l$ are the amplitudes of the lowest order isolated-waveguide modes in each of the isolated guides, and $\phi_{a,b}$ are these waveguide modes containing their lateral spatial distribution and their usual z-dependent phase factor. The system modes for this case are now normalized, and thus $a_a'^2 + a_b'^2 = 1$, and locally orthogonal. In addition, there is negligible overlap between the modes of the isolated waveguides. The results in Sects. 6.1 or 6.4 can be used to show that

$$a_a' = \left[\frac{1}{2}\left(1 + \frac{\chi}{(\chi^2 + 1)^{1/2}}\right)\right]^{1/2} \tag{6.35}$$

$$a_b' = \left[\frac{1}{2}\left(1 - \frac{\chi}{(\chi^2 + 1)^{1/2}}\right)\right]^{1/2} \tag{6.36}$$

or

$$\frac{a_a'}{a_b'} = (\chi^2 + 1)^{1/2} - \chi \tag{6.37}$$

where we have explicitly chosen to write these expressions in terms of the ratio $\chi \equiv \Delta\beta/2\kappa$. Here, $\Delta\beta$ is again the mismatch in propagation constants of the lowest order modes for the isolated guides. Using the prior results in this chapter along with the quantity χ, it is readily shown that $\beta_i - \beta_j = 2\kappa(\chi^2 + 1)^{1/2}$. The local value of the parameter, $\Delta\beta/2\kappa$, thus determines the modal transfer from one waveguide to another in the device. If $\Delta\beta/2\kappa \to \infty$, the light remains in guide 1; if, however, $\Delta\beta/2\kappa$, the light is evenly divided in the two guides. This quantity, which is termed the asynchronicity parameter, is thus very important for the design of adiabatic devices.

Returning now to the tapered coupler, the local coupling coefficient c_{ij} be calculated for a linear variation of the propagation constants with length and a constant coupling coefficient $\kappa_{ij} = \kappa$ between the isolated waveguides. The essence of solving a differential coupled-mode problem is first determining the value of c_{ij}. This is typically done by using some form of model differential elements, since the system modes can be coupled (Burns and Milton 1990). However, in the case of a weakly coupled coupler, the system modes are analytic and the local coupling coefficient can be written directly using

$$c_{ij} = \int_{-\infty}^{\infty} \Psi_{i0}\Psi_{j1}dx \tag{6.38}$$

where the subscripts 0, 1 again denote light, which enters or leaves the differential element, respectively. Inserting the explicit form for Ψ in (6.33) and (6.34) in (6.38), we obtain

$$c_{ij} \approx -a_{b,1}a_{a,0} \int_{-\infty}^{\infty} \Phi_{a0}\Phi_{a1}dx + ab,0a_{a,1} \int_{-\infty}^{\infty} \Phi_{b0}\Phi_{b1}dx + a_{a,1}a_{a,1}$$
$$\int_{-\infty}^{\infty} \Phi_{a0}\Phi_{b1}dx - a_{b,0}a_b, 1) \int_{-\infty}^{\infty} \Phi_{a1}\Phi_{b0}dx \tag{6.39}$$

Using the fact that

$$\int_{-\infty}^{\infty} \Phi_{a0}\Phi_{b1}^{*}dx \cong 0 \tag{6.40}$$

$$\int_{-\infty}^{\infty} \Phi_{a1}\Phi_{b0}^{*}dx \cong 0 \tag{6.41}$$

and that

$$\int_{-\infty}^{\infty} \Phi_{a0}\Phi_{a1}^{*}dx \cong 1 \tag{6.42}$$

$$\int_{-\infty}^{\infty} \Phi_{b0}\Phi_{b1}^{*}dx \cong 1 \tag{6.43}$$

the following simple expression for c_{ij} results:

$$c_{ij} = a_{b,0}a_{a,1} - a_{b,1}a_{a,0} \tag{6.44}$$

This equation can be further simplified by substituting in (6.35) and (6.36) to yield

$$c_{ij} = \frac{f_0 - f_1}{1 + f^2} \tag{6.45}$$

where

$$f \equiv -\chi + \sqrt{1 + \chi^2} \tag{6.46}$$

In the adiabatic limit, the quantity $f_0 - f_1 \approx -(\partial f / \partial \rho)\Delta\rho$. Thus, the coupling coefficient, $C_{ij} = \lim_{\rho \to 0} c_{ij}/\Delta\rho$, can be written using (6.45) as

$$C_{ij} = \frac{1}{2(1 + \chi^2)} \frac{\partial \chi}{\partial \rho} \tag{6.47}$$

In this case, the changing waveguide variable, ρ, is the waveguide width. $\rho = W_a - W_b$, where (W_a, W_b) are the coupler waveguide widths for guides a and b. Thus $\partial\chi/\partial\rho$

is given by

$$\beta_a = \beta_a^0 + \alpha(W_a - W_a^0) \tag{6.48}$$

$$\beta_b = \beta_b^0 + \alpha(W_b - W_b^0) \tag{6.49}$$

where

$$\alpha = \frac{\beta_a^0 - \beta_b^0}{W_a^0 - W_b^0} \tag{6.50}$$

This results in $\partial \Delta\beta/\partial W = \alpha$, where (β_a, β_b) are the isolated-waveguide propagation constants. Note that the superscript 0 in the above equations corresponds to the value of the parameters at $z = 0$. Substituting Eqs. (6.49) and (6.50) into (6.48) and substituting the result of $\partial\chi/\partial\rho$ into (6.47) yields the following expression for C_{ij}:

$$C_{ij} = \frac{1}{2\Delta W^0} \frac{\chi^0}{\chi^2 + 1} \tag{6.51}$$

where $W^0 = W_a^0 - W_b^0$ and $\chi^0 = (\beta_a^0 - \beta_b^0)/2\kappa$. Thus χ^0 is the asynchronicity at the entrance to the coupler, and $\Delta W^0 = W_a^0 - W_b^0$ is the difference in waveguide widths at the input of the coupler.

Recall also that the local propagation constant difference of the system modes $\beta_i - \beta_j$ can be written as

$$\beta_i - \beta_j = 2\kappa(1 + \chi^2)^{1/2} \tag{6.52}$$

and finally the rate of change of waveguide width, $\partial W/\partial z$, for the coupler are given by simple geometry as

$$\frac{\partial W}{\partial z} = \frac{m(W_1 - W_2)}{L} \tag{6.53}$$

where L is the length of the coupler, and where $m = 1/2$ for a 3 dB coupler, and $m = 1$ for a full coupler.

Optimum adiabatic coupler design requires minimum coupling between system modes in order to minimize power loss to the other unwanted system mode. According to (6.52), this requirement is satisfied for small values of asynchronicity $\chi^0 \ll 1$ at the input of the coupler, and hence throughout the entire coupler, since by the nature of the coupler design, χ^0 is the maximum value of χ along the coupler. In this case, the coupling coefficient C_{ij} is approximately a constant given by

$$C_{ij} \simeq \frac{\chi^0}{2\Delta W} \tag{6.54}$$

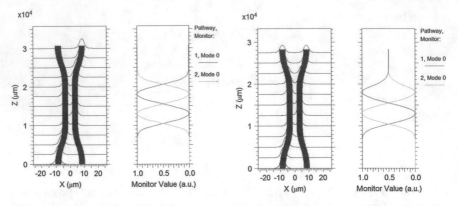

Fig. 6.13 A simulation of the fields in the waveguides of a coupler

and, furthermore, the difference in propagation constants of the system can be approximated by

$$\beta_i - \beta_j \simeq 2\kappa \tag{6.55}$$

This case corresponds to that discussed in conjunction with (6.45), and the coupled-mode equations, (6.43) and (6.44), can be directly integrated to yield the ratio of the unwanted-to-wanted system-mode power, q,

$$q \simeq \left(\frac{m\chi^0}{2\kappa L}\right)^2 \sin^2(\kappa L) \tag{6.56}$$

Excellent agreement is obtained between the values of q calculated from (6.56) and simulation done using numerical computations as illustrated in Fig. 6.13. For large values of χ^0, the assumption of uniform propagation constant difference, i.e., (6.55), is no longer valid and the oscillatory behavior in (6.56) is suppressed. However, as shown in Fig. 6.13, the simulation results continue to follow the envelope of (6.56). The above equation provides the basis for a useful design rule for the coupler length, L, which, considering the envelope of (6.56), is given by

$$L = \frac{m}{2\sqrt{q}} \frac{\chi^0}{\kappa} \tag{6.57}$$

for both types of couplers. For a given tolerance in the system-mode power ratio, the required 3 dB coupler length is found to be half that of the full coupler. However, an additional, even more stringent criterion pertains to 3 dB couplers when one considers the balance of power in the isolated waveguides. The reader is referred to Ramadan et al. (1998) for a more extensive discussion on this point.

6.7 Summary

This chapter has introduced the first of a series of passive components, i.e., the direction coupler; other important passive devices will be presented in subsequent chapters. These passive components are used in more complex devices such as filters, delay lines, etc., later in the text. In addition, many of the design concepts presented in this chapter such as adiabaticity, local normal modes, and the differential coupled-mode equations will be used frequently in the chapters which follow. For example, adiabaticity is an important component in the design of several forms of mode steering devices. The details of adiabatic couplers are discussed in this chapter.

Finally, the discussion of passive couplers in this section will be crucial to understanding many forms of coupler-based optical modulators, which will be discussed later in the text. In these devices, for example, electro-optical switching of "synchronization" of a coupler can be used to make a high-performance waveguide switch.

References

Alferness, R., Schmidt, R., & Turner, E. (1979). Characteristics of ti-diffused lithium niobate optical directional couplers. *Applied Optics, 18*(23), 4012–4016.

Burns, W. & Milton, A. (1990). Chap. 3, waveguide transitions and junctions. In T. Tamir (Ed.), *Guided wave optoelectronics*

Fang, A. W., Park, H., Cohen, O., Jones, R., Paniccia, M. J., & Bowers, J. E. (2006). Electrically pumped hybrid algainas-silicon evanescent laser. *Optics Express, 14*(20), 9203–9210.

Hardy, A., & Streifer, W. (1985). Coupled mode theory of parallel waveguides. *Journal of Lightwave Technology, 3*(5), 1135–1146.

Haus, H., Huang, W., Kawakami, S., & Whitaker, N. (1987). Coupled-mode theory of optical waveguides. *Journal of Lightwave Technology, 5*(1), 16–23.

Haus, H. A. (1984). *Waves and fields in optoelectronics*. Upper Saddle River: Prentice-Hall.

Haus, H. A., Huang, W.-P., & Snyder, A. W. (1989). Coupled-mode formulations. *Optics Letters, 14*(21), 1222–1224.

Kapon, E., Katz, J., Margalit, S., & Yariv, A. (1984a). Controlled fundamental supermode operation of phase-locked arrays of gain-guided diode lasers. *Applied Physics Letters, 45*(6), 600–602.

Kapon, E., Katz, J., & Yariv, A. (1984b). Supermode analysis of phase-locked arrays of semiconductor lasers. *Optics Letters, 9*(4), 125–127.

Louisell, W. (1955). Analysis of the single tapered mode coupler. *Bell Labs Technical Journal, 34*(4), 853–870.

März, R. (1995). *Integrated optics: Design and modeling*. Artech House on Demand.

Nishihara, H., Haruna, M., & Suhara, T. (1989). *Optical integrated circuits* (Vol. 1). New York: McGraw Hill Professional.

Noda, J., Fukuma, M., & Mikami, O. (1981). Design calculations for directional couplers fabricated by ti-diffused linbo 3 waveguides. *Applied Optics, 20*(13), 2284–2290.

Park, H., Fang, A. W., Kodama, S., & Bowers, J. E. (2005). Hybrid silicon evanescent laser fabricated with a silicon waveguide and iii–v offset quantum wells. *Optics Express, 13*(23), 9460–9464.

Ramadan, T. A., Scarmozzino, R., & Osgood, R. M. (1998). Adiabatic couplers: Design rules and optimization. *Journal of Lightwave Technology*, *16*(2), 277.

Somekh, S., Garmire, E., Yariv, A., Garvin, H., & Hunsperger, R. (1973). Channel optical waveguide directional couplers. *Applied Physics Letters*, *22*(1), 46–47.

Yamada, H., Chu, T., Ishida, S., & Arakawa, Y. (2005). Optical directional coupler based on si-wire waveguides. *IEEE Photonics Technology Letters*, *17*(3), 585–587.

Chapter 7
Passive Waveguide Components

Abstract In this chapter, we focus our discussion of waveguide circuits by considering a wide variety of passive optical guided-wave components. These devices are both important and difficult to design and fabricate and yet they play major roles in many forms of waveguide circuits. For example, these devices include tapers, Y-branches, bends, integrated mirrors, and mode conversion devices. Our intent in this chapter is to develop an understanding of the devices and relevant engineering formulae for their design.

7.1 Introduction

Most waveguide circuits rely on a variety of passive "components" such as those shown in Fig. 7.1. These devices enable connecting guided-wave links and the transitions needed to make a complete circuit. They are surprisingly difficult to analyze, as are many analog components in a microwave circuit. Generally, the most useful and general approach for analytic studies is to use a formulation based on local normal modes. This analysis shows that if the change in geometry or material is abrupt, new modes are generated. However, if the change is not abrupt, or "adiabatic," coupling between these modes does not occur; instead, the shape of each mode evolves smoothly along the propagation path, and radiation from the transition is minimized.

In this chapter, we will discuss a series of important passive devices, such as tapers, Y-branches, and bends. Our goal will be to develop both an understanding of the devices and relevant engineering formulae and techniques needed for their design. The chapter will also emphasize the loss of power and the coupling of modes in common passive transitions. Designs which control the guided modes at the output of this transition are crucial in many I/O applications. Spurious mode generation is particularly severe if the waveguides are only "quasi-" single-mode, i.e., slightly multimode. The use of such "quasi-" single-mode design is often required in real devices because of the need to minimize waveguide loss. In such quasi-single-mode

© Springer Nature Switzerland AG 2021
R. Osgood jr. and X. Meng, *Principles of Photonic Integrated Circuits*,
Graduate Texts in Physics,
https://doi.org/10.1007/978-3-030-65193-0_7

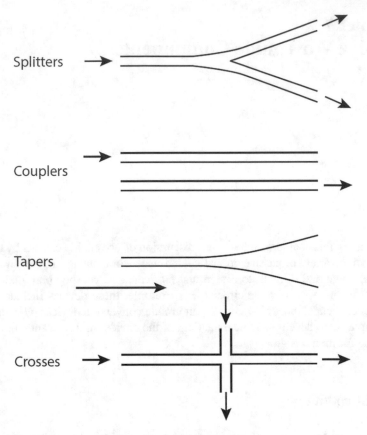

Fig. 7.1 Four of the most common passive components. These devices enable the connecting guided-wave links and transitions needed to make a complete circuit

waveguides, transitions or changes in the waveguide geometry can cause generation of higher order modes. As these modes move in and out of phase with each other along the axis of a guided-wave section, the field amplitude in the guide will change and give different lateral modal profiles.

Several approaches have been presented to analyze adiabatic transitions, including mode matching at junctions and coupling of local normal modes. In the initial set of devices discussed in this chapter, we will use each of these approaches in order to understand the operation of such components. The results of the numerical computational methods, e.g., the beam propagation technique, will also be used to provide "exact" solutions for comparison. Notice also in many places in this chapter, we have chosen to analyze structures with low-refractive index. This condition is useful for many material structures and often results in simple accurate approximate solutions for design and analysis particularly for the important materials systems, e.g., $LiNbO_3$ and SiO_2. A further advantage of examining the low-index structure is that it leads to clearer insight into the operable guided-wave physics. However, it is important to note that in many devices involving semiconductors, the refractive index contrast is high, e.g., Si wires, and in those cases it is important to use numerical methods for simulation.

7.2 Simple Transitions in Waveguide Geometry

In order to introduce the central issues of designing waveguide components, we will consider first, the simplest case: a transition between two ideal or "canonical" waveguides of different geometries, such as different width and position. In the first example, we will consider an abrupt transition between two semi-infinite, single-mode, mode-mismatched waveguides. In this case, the transition leads only to power loss. In the second example, we will consider a gradually tapered junction between two ideal waveguides, with the input guide being a narrow, single-mode structure and the output guide being a wider, possibly multimode structure. In this case, the transition should be designed to suppress coupling to other modes in the transition region. A poor design excites higher order modes or causes loss if the excited mode is not guided but radiative.

7.2.1 Reflection and Transmission at An Transition Between Two Waveguides

Consider an abrupt juncture between the two aligned, dissimilar waveguides as shown in Fig. 7.2. Light incident from the left can excite an assortment of radiative and guided modes (Marcuse 1991). The total transverse electric and magnetic fields of these guides must satisfy the boundary conditions for continuity across the interface between the two waveguides. Thus, if we assume here, for simplicity, two slab waveguides, varying in the x-direction, and having normalized incident modes E_i and H_i, the boundary conditions at the interfacial junction of the two guides cause the fields at the interface to be such that

$$E_i^n(x) + r E_i^n(x) + \int_0^\infty r_\rho(\beta) E_\rho(\beta, x) \mathrm{d}\beta = t E_t^m(x) + \int_0^\infty t_\rho(\beta) E_\rho(\beta, x) \mathrm{d}\beta$$

(7.1)

$$H_i^n(x) + r H_i^n(x) + \int_0^\infty r_\rho(\beta) H_\rho(\beta, x) \mathrm{d}\beta = t H_t^m(x) + \int_0^\infty t_\rho(\beta) H_\rho(\beta, x) \mathrm{d}\beta$$

(7.2)

where $i, r,$, and t denote the incident, reflected, and transmitted normalized guided modes, and ρ denotes the radiative modes. The quantities r and t denote the complex reflectivity and transmissivity of the incident fields, respectively. Note that similar mode matching analysis will also be used in Sects. 7.3.1 and 7.4.

Using the modal orthogonality condition

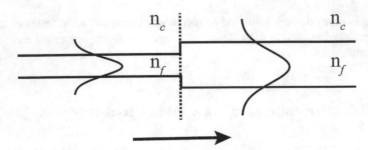

Fig. 7.2 A plot showing an abrupt juncture between the two aligned, dissimilar waveguides. The material quantities are those defined earlier in the book

$$\frac{1}{2}\int_{-\infty}^{\infty} E_n(x)E_m(x)\mathrm{d}x = \frac{\omega\mu_0}{\beta_n}\delta_{nm}$$

(7.3)

where δ_{nm} is the Kronecker-delta symbol, it is possible to solve these two equations for t and r, under the condition that we may neglect the radiative modes for small waveguide mismatch. The solutions are then (Pollock 1995)

$$t = \left(\frac{\beta_i\beta_t}{\beta_i + \beta_t}\right)\frac{1}{\omega\mu_0}\int_{-\infty}^{\infty}\mathrm{d}x\, E_i^n(x,0)\cdot E_t^m(x,0)$$

(7.4)

and

$$r = \frac{\beta_i - \beta_t}{\beta_i + \beta_t}$$

(7.5)

Note that these equations reduce to the usual Fresnel equations as the field becomes increasingly more plane-wave-like. It is important to emphasize that the radiative fields have been neglected in this particular case.

The first equation shows that, neglecting radiation, the complex transmissivity is given by the modal overlap of the modes in the two waveguides multiplied by the complex Fresnel coefficient. Since the reflected light involves mode coupling between identical modes, i.e., those in the same guides, the modal overlap is unity and only the Fresnel reflection coefficient is important.

These simple expressions are important for a number of practical applications, including the frequent practical situation of coupling between a fiber and an integrated waveguide. A good example of such a practical application is the case of coupling between an idealized Gaussian fiber mode with a diffused rectangular guide having a Gaussian distribution in x and y (Burns and Hocker 1977). In that case, the expression for the transmission reduces to

$$t = T\frac{4}{\left(\frac{w_x}{a} + \frac{a}{w_x}\right)\left(\frac{w_y}{a} + \frac{a}{w_y}\right)}$$

(7.6)

where w_x and w_y are the half-widths of the rectangular waveguide mode and a is the mode radius of the circular Gaussian mode. This quantity on the right-hand side reaches a maximum when $a = (w_x w_y)^{1/2}$. T is the usual Fresnel transmissivity (see above)

$$T = \frac{N_f N_g}{N_f + N_g} \qquad (7.7)$$

where $N_{f,g}$ is the effective index for the fiber and the guide.

7.2.2 Modal Coupling in Waveguide Transitions

In the discussion above, the transition region is a clearly defined plane. If guided modes are excited in the second waveguide of the transition, they propagate in a well-defined way because the waveguide (if canonical) is uniform. This situation is not true in a tapered transition, in which modal coupling can occur throughout the transition region. In fact, we have earlier seen an example of such distributed or differential mode coupling in the case of the adiabatic coupler discussed in Chap. 6.

From a more general perspective, in this section, we introduce the central problem of following the conversion of modes in a transition region and then designing the transition to reduce undesired mode conversion to an acceptable level. Consider now the design of a tapered or horn waveguide junction, such as that shown in Fig. 7.3. Tapers such as this are extremely important for proper matching of waveguide modes between a fiber and the optical input waveguide for a chip as well for changing the mode size for two "on-chip" devices.

The calculation of mode conversion in a tapered device is most conveniently done using the differential coupled-mode formulation, i.e, as shown in (6.43) and (6.44). The crucial quantity that is needed for this calculation is the coupling coefficient between the local system modes in the taper, c_{ij}. To obtain this coupling coefficient, it is first necessary to compute a general c_{ij} for a channel waveguide with a differential step discontinuity. This is done in the same manner as for the step discontinuity in the coupler, discussed in Chap. 6, or in the different waveguides in (7.1) and (7.2), namely, matching of the electromagnetic field at the discontinuity. In fact, the solution for a taper is outlined in Burns and Milton (1990). In the case of a symmetrically tapered waveguide of width w, the differential shown in Fig. 7.4 can then be used in

Fig. 7.3 A plot showing a tapered or horn waveguide juncture between the two aligned, dissimilar waveguides whose width are W_0 and W, respectively

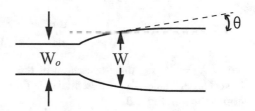

Fig. 7.4 A plot showing a
symmetrically tapered
waveguide of width w and
differential element δw

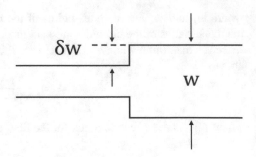

conjunction with the modal wavefunction to obtain a specific value of c_{ij}, or for the
case of the differential coupled-mode equations, the form of c_{ij} per differential, that
is C_{ij}. Formally, this reduces to solving equations which are similar to those given
in (7.1) and (7.2). For the case of a wide, well-confined waveguide of width w, i.e.,
well removed from cutoff, the result is

$$c_{ij} = -\frac{3}{4}\frac{\delta w}{w} \tag{7.8}$$

or

$$C_{ij} = -\frac{3}{4w} \tag{7.9}$$

where again, a well-confined mode is assumed.

Earlier in Chap. 6, we found that an analytic solution to the differential coupled-
mode equations may be found if the differential element in the varying waveguide
parameter, i.e., the width of the taper, w, in our case, is given by

$$\frac{dw}{dz} = \zeta\frac{\Delta\beta_{ij}}{C_{ij}} \tag{7.10}$$

where ζ is a constant. This equation may be solved to determine the shape needed
for this analytic solution (Burns and Milton 1990).

The difference in propagation constants for the first two propagating modes, β_0
and β_2 in a planar waveguide may be obtained by simplification of the dispersion
relation for a slab waveguide of thickness d (see Appendix A); such a calculation
shows that $\Delta\beta_{02} \approx (2\pi)^2/\beta_0 d^2$, and where in our problem $d = w$ (Burns and Milton
1990). Then, (7.10) may be rewritten as

$$\frac{dw}{dz} = -\zeta\frac{8\pi}{3}\frac{\lambda_g}{w} \tag{7.11}$$

where $\lambda_g = 2\pi/\beta_0$, the wavelength of the phase front in the taper or horn for the
lowest order mode. This differential equation may be integrated to yield the parabolic
horn shown in Fig. 7.3,

Fig. 7.5 A calculation of mode conversion versus length. Notice that as the taper length increases, the power remains essentially the same in a single mode

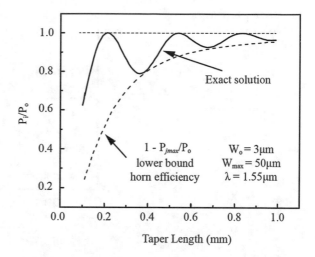

$$w = (w_0^2 + 2\alpha\lambda_g z)^{\frac{1}{2}} \qquad (7.12)$$

where $\alpha = -8\pi\zeta/3$, and w_0 equals the width of the input waveguide. Thus, the design of the taper involves determining a set of criteria, which "fixes" α for the particular application of interest. For the case of a transition between a thin, single-mode waveguide and a wider, somewhat multimode waveguide, and these design criteria yield, for example, the acceptable upper limit on the mode conversion for a specific device design.

Recall that in Chap. 6, a general expression for the maximum value of mode conversion in (7.10) was obtained when solving the differential coupled-mode equations under the condition of constant ζ. In particular, for small ζ, the maximum converted power from the input power, to power in the jth mode is

$$\frac{P_j^{max}}{P_{i0}} = \frac{4\zeta^2}{4\zeta^2 + 1} = \frac{(3\alpha/4\pi)^2}{(3\alpha/4\pi)^2 + 1} \qquad (7.13)$$

Because of the symmetry of the parabolic taper, the dominant power transfer is from the E_{11}^x into the E_{31}^x mode; transfer from an odd to even mode would give a zero overlap integral. The parabolic shape is thus the most efficient since it maintains constant coupling through the length of the horn. For a low value of mode conversion, $3\alpha/4\pi \ll 1$. Thus once α (or ζ) is set, the specific horn shape is given by (7.12). As an example, a horn having $\alpha = 1$ will have a maximum mode conversion from $i \to j$ of only $\sim 6\%$.

As in the case of the tapered coupler, it is possible to compute the actual value of the conversion by referring back to the differential coupled-mode equations. A calculation of mode conversion versus length is shown in Fig. 7.5. Notice that as the taper length increases, the power remains essentially the same in a single mode.

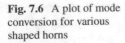

Fig. 7.6 A plot of mode
conversion for various
shaped horns

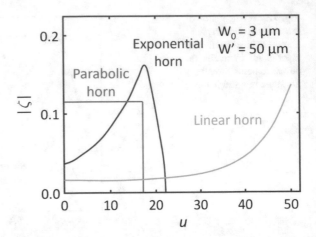

The oscillations seen in the exact solution to the problem result from mode beating
between the i and j system modes; that is, the quantity (Burns and Milton 1990)

$$u = \int_0^z \Delta\beta_{ij}dz' \tag{7.14}$$

passes through successive values of π.

A plot of the mode conversion ζ for various shaped horns is shown in Fig. 7.6.
The horizontal axis is u, which is defined in (7.14). The parabolic shape has the
lowest mode conversion for the per device length. This result agrees with the earlier
statement in Chap. 6 that mode conversion is minimized when the parameter ζ is
constant.

7.3 Y-Branches

Y-branches are crucial elements in many integrated optical devices, including Mach–
Zehnder interferometers, power dividers, and light routing junctions. The "classic"
shape for this device is shown in Fig. 7.7. There are several different design issues
for Y-branches, and these cause performance trade offs, including coupling between
the exit waveguides in the junction region for small splitting angle, θ; radiative loss
at large θ; and device compactness or overall size, which also requires small θ. The
dependence of the modal behavior of the Y-branch on θ leads to the use of Y-branches
with relatively large θ being used as power dividers, while those with small θ are
used for mode routers. The next section will consider importance of modal overlap
in devices for each of these functions.

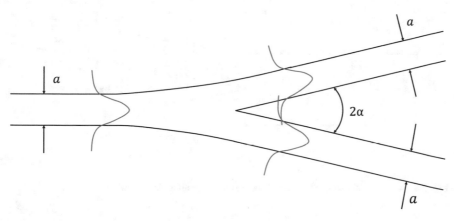

Fig. 7.7 A sketch of a "classic" shape Y-branch

7.3.1 Mode Overlap to Analyze Y-Branches

One of the first concerns in designing a satisfactory Y-branch is the degree of mode matching between the mode of the input guide and those of the two output waveguides. This matching controls the excitation of the local modes of the transition. To realize efficient mode matching, we must find the overlap integral between the incident mode and the lossy radiative and the undesired higher order modes at the junction. In performing this procedure, it is generally found that the mode overlap is controlled by the tilt of wavefronts. Thus, large angles reduce this overlap, with the unmatched field being lost into the substrate.

The procedure for mode matching of Y-branches involves a straightforward matching of the scalar fields at $z = 0$, using the field spatial distribution. Several different approximations to the overlap integral have been described in the literature for a variety of Y-branch structures. In all cases, at small angles, the mode matching in the Y- branch is good and thus the loss is low while at larger angles, the loss increases.

Consider now the mode matching at the initial junction of a simple "model" Y-branch shown in Fig. 7.8 (Kuznetsov 1985). This particular Y-branch is selected for discussion here because it has a simple analytic solution; having an analytic solution makes it possible to understand clearly how the device impacts the interrelation of field and geometry. Since the guides have a constant geometry in z away from the transition, it is only necessary to do the mode matching between ideal guides at the transition. This procedure allows a quantitative measure of how well a single mode in the input waveguide drives the output modes in the two exit guides.

To calculate the mode matching in the Y-junction in Fig. 7.8, we can again apply (7.4) to obtain the overlap integral of this junction. Recall that this overlap integral gives the complex transmission of the device. Note that it is assumed in this calculation that only the lowest order TE mode is present in the slab waveguide. This procedure allows us to determine the modal field in the input guide using the

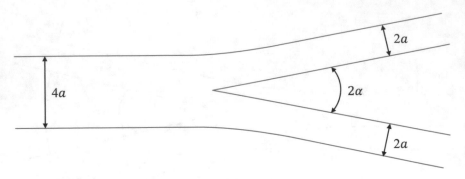

Fig. 7.8 A sketch of a simple "model" Y-branch with an analytical solution

well-known slab waveguide modes given earlier. Recall that the field distribution is characterized by the transverse wavevector, k_t, and the evanescent decay constant, γ, which is determined from the following transcendental equation

$$k_t = \frac{1}{2a} \tan^{-1}\left(\frac{k_t}{\gamma}\right) \tag{7.15}$$

and the relations

$$\gamma^2 = \beta^2 - n_s^2 k^2 \tag{7.16}$$

and

$$\beta^2 = n_f^2 k^2 - k_t^2 \tag{7.17}$$

and where the quantity $2a$ related to the waveguide width is defined in Fig. 7.8.

Inserting the expressions for the modal field in (7.4) then gives an analytic, but still daunting, expression for the complex transmissivity:

$$t = \frac{e^{j\nu a}}{\left(1 + \frac{1}{2k_t a}\right)}\left[\frac{1}{2}e^{-j2\gamma a}\,\text{sinc}(\nu - 2\gamma)a + \frac{1}{2}e^{j2\gamma a} \cdot \text{sinc}(\nu + 2\gamma)a\right.$$

$$\left. + \text{sinc}(\nu a) + \cos^2(2\gamma a)\frac{e^{j\nu_0 a}}{(2k_t - j\nu)a}\right] \tag{7.18}$$

where $\nu = \beta \sin \alpha$, and α is one half of the splitting angle.

A plot of this function is provided in Fig. 7.9. Notice as the branching angle in the Y increases, the transmission of the Y-branch decreases. This decrease is due to the increased coupling of the incident light into unguided or radiative modes. The weak oscillations (solid line) are caused by the interference of the radiative light with the guided light in the waveguide. Notice that the "simplified" analytic expression works extremely well for small angles. At larger angles, radiation from the tilted waveguides becomes increasingly important, and this regime is not handled well

Fig. 7.9 A plot of the relation between transmission and α from (7.18). As the branch angle increases, the transmission decreases

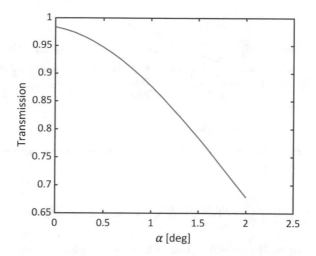

by the simple mode-overlap integral. However, several more exact approaches have been made toward solving for the radiative loss in wider-angle Y-branches.

One approach, which is known from comparison with accurate computation to be reliable, is the "volume current method" (VCM), which considers radiation from the changing waveguide structure (Kuznetsov and Haus 1983). Along the propagation distance, this method obtains the radiation field in the far field via the polarization current in the waveguides. The radiation modes then act back on the Y-branch by beating with the guided modes. The method uses the vector potential for radiation field

$$\vec{A}_r(\vec{r}) = \frac{\mu_0}{4\pi} \frac{e^{-jk_1 r}}{r} \int \int_\nu \partial \nu' J(\vec{r}') e^{jk_1 \hat{r} \cdot \vec{r}'} \tag{7.19}$$

and inserts into this expression the polarization current from the waveguide modes. The radiative power is then obtained from the form of the Poynting vector containing the vector potential.

Application of this approach to the Y-branch yields the solid line shown in Fig. 7.9. The weak oscillations shown in the solid-line plot are caused by interference of the radiative modes with light in the waveguide. As seen in the figure, the power loss computed simply from the quantity $|t|^2$, obtained from (7.18), agrees well with the VCM method at small branch angles, but diverges and is lower at large angles. This more rapid drop off in Y-branch transmission in the VCM calculation is due to increasing coupling to the radiative modes in the spreading waveguides, that is, as they decouple during the power dividing process. The figure also plots an "exact" numerical computation, showing that the VCM picture accurately reflects the actual loss in the Y.

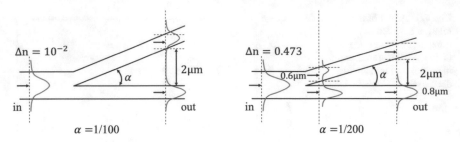

Fig. 7.10 A sketch of an "mode splitter" or "mode router" which routes selectively a certain input mode into a given output port

7.3.2 Guided-Mode Coupling in Y-Branches: Mode Splitters

In the above section, two factors were considered in the design of Y-branches: mode matching at the input and loss due to radiation from radiative modes in the diverging guided-wave structure. In addition, it is also important to consider coupling between guided modes after the transition region of the Y-branch. In particular, after the entrance to the Y-branch is excited by the input mode, modes traveling along the exit waveguides can couple with each other. This coupling can be useful or deleterious depending on the application of the Y-branch.

An extremely important application of Y-branches, or other 1-to-N splitters, which are designed to use mode coupling, is to route selectively a certain input mode into a given output port. These are termed "mode splitters" or "mode routers." An example of this device is shown in Fig. 7.10. This device is an asymmetric Y-branch, designed to separate one system mode from another, adiabatically. This capability, in turn, allows selective routing of one mode of an input multimode wave to a specific output waveguide arm. Such devices have been used , for example, in conjunction with other passive elements to separate one polarization or one wavelength from another. In these devices, mode steering is an important function, while mode conversion is not desired.

Mode routers require that the Y-branch to be asymmetric either in real or effective index or in geometry. To analyze mode coupling, consider, for simplicity, a Y- branch which has a symmetric separation geometry for the two waveguide arms with different effective indices. Such a device has a changing geometry in z and thus mode coupling must be analyzed using the differential coupled-mode equations introduced in Chap. 5 and applied subsequently to treating mode coupling in tapers and adiabatic couplers. To solve these equations, we must determine the differential coupling coefficients for the system modes i and j of the waveguide branch as the waveguides separate (Burns and Milton 1990). An appropriate differential, which when used in a series of increments would lead to a symmetric Y-branch, is a "step" in the waveguide spacing or separation, δw, shown in Fig. 7.11. The mode coupling coefficient is then found through a procedure similar to that discussed earlier in this chapter, and in detail in Chap. 6 for the case of the adiabatic coupler, namely, boundary-value

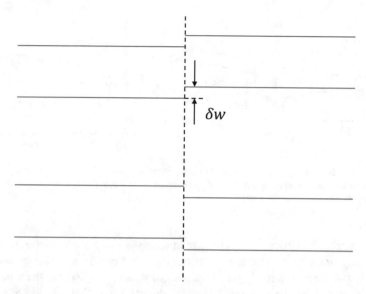

Fig. 7.11 A sketch of a differential "step" in the waveguide spacing or separation

matching at the discontinuity. Again, recall that the equations for the system modes in a canonical or ideal coupler are expressed in terms of the asynchronicity parameter χ, see the discussion following (6.53). The modal coupling coefficient is then found to be

$$C_{ij} = \frac{\gamma \chi}{2(\chi^2 + 1)} \tag{7.20}$$

$$c_{ij} \equiv C_{ij}\delta w = \frac{\gamma \chi}{2(\chi^2 + 1)}\delta w \tag{7.21}$$

where γ is defined by the fact that the local coupling between the waveguides separated at their edges by w is given by $\kappa = const \times \exp(-\gamma w)$.

Once the expression for c_{ij} is found, the differential coupled-mode equations can then be solved to determine the extent to which modal power transfer occurs in the Y-branch (Burns and Milton 1990). For example, in a mode splitter (see the following paragraph), the output to the Y-branch junction is often designed such that when it is excited with the symmetric, lowest order mode, light will be transferred only to the guide with the highest effective index of refraction. This index is determined either by the geometric or/and materials parameters of the waveguides. In this case, modal separation will occur with low modal cross talk if mode conversion, that is c_{ij} remains small throughout the transition.

A computation of c_{ij} for a Y-branch shows that there are two operational regimes for this device: (1) power splitter or divider and (2) mode splitter. Each of these operational regimes is dominant at different values of angular separation and waveguide asymmetry, a parameter termed χ. Basically, the difference between these regimes is

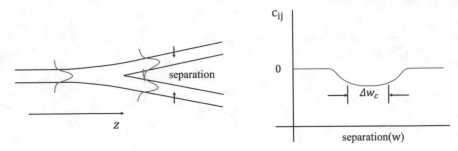

Fig. 7.12 A sketch of a 1×2 splitter(left); the corresponding coupling coefficient c_{ij}(right). Significant coupling only occurs in the range of separation $w < \delta w_c$. The corresponding δz_c can be determined as shown in the left figure

that in one case, the system modes remain in phase to allow mode conversion during the splitting process; in the other, the modes are sufficiently asynchronous that no conversion of the two excited system modes occur, and the power in the input mode is divided according to a mode matching formulae such as that given in Sect. 7.3.1. Note, of course, that significant conversion into radiative modes can still occur in this regime.

Typically, mode conversion occurs dominantly in one spatial region of the device (see Fig. 7.12). The factors controlling the transfer between the two modes are both the magnitude of the peak c_{ij} and its spatial overlap. These quantities depend on the separating waveguide slope and the separation, and their effective index (or real index) asymmetry as shown in Fig. 7.13. Thus, the localization of the mode conversion allows us to apply a semiquantitative criterion for use when mode coupling may be neglected. If δz_c is the longitudinal distance, over which conversion occurs, i.e., where c_{ij} is largest, then when

$$(\beta_{i0} - \beta_{j1})\delta z_c \ll \pi/2 \tag{7.22}$$

that is, the modes remain in phase, mode conversion can occur, and the power in the input guide of the Y is split between the two output guides. If, on the other hand,

$$(\beta_{i0} - \beta_{j1})\delta z_c \gg \pi/2 \tag{7.23}$$

then the modes will remain out of phase and thus coupling between the modes will not occur. The modes then evolve naturally to permit mode splitting or separation. Note that δz_c and δw_c (see Fig. 7.12) are related by the geometry of the Y-branch.

To obtain a more specific formula for the case of Y-branches, consider a symmetric Y-branch, which has output waveguides that separate $w(z)$ linearly with z (the region of width δw_c over which c_{ij} is large, is related to δz_c by $\delta w_c = \theta \delta z_c$). Then, using the expression for c_{ij} (7.21) and the values for $\delta \beta_{ij}$ in terms of $\delta \beta$, it can be shown (see Appendix B) that

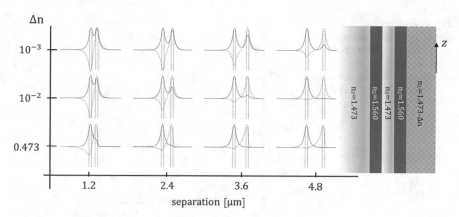

Fig. 7.13 A plot of modes with different waveguide separation and their effective index

$$\frac{\delta\beta}{\theta\gamma} > 0.43 \text{ for a mode splitter (no mode conversion)} \tag{7.24}$$

$$\frac{\delta\beta}{\theta\gamma} < 0.43 \text{ for a power splitter (mode conversion)} \tag{7.25}$$

and where $\delta\beta$ is understood to be $\neq 0$. Here, $\delta\beta$ is the asynchronicity of the equivalent isolated output waveguides, θ is the angular separation, and γ is again obtained from the local waveguide coupling coefficient $\kappa = const \cdot \exp -\gamma w$; each of the quantities is important. In these equations

$$\theta\gamma \sim \delta z_\gamma \tag{7.26}$$

where δz_γ is the distance along for separation by approximately one γ.

The relation $\delta\beta/\theta\gamma > 0.43$ is, thus, the criterion for an adiabatic splitter. If we make use of the simple asymmetric structure such as that shown in Fig. 7.10, and use the appropriate coupled-mode analysis, a representative plot of mode conversion versus $\delta\beta/\theta\gamma$ can be obtained. This plot is shown in Fig. 7.14. Note the rapid drop off in converted mode amplitude above a value of $\delta\beta/\theta\gamma \approx 0.1$; thus the criteria given in (7.24) and (7.25) hold. These equations can then be used in the design of a linear variation of θ distance separating asymmetric Y-branches for mode splitting or power dividing by using them to find, say, θ if γ and $\delta\beta$ are known, etc. Analytic solutions for shaped Y-branches are given in Burns and Milton (1990).

7.3.3 Advanced Y-Branch Design

The guided-wave physics discussed above suggest several approaches for improving Y-branches. The first improvement is to adjust the geometry of the Y transition to

Fig. 7.14 A representative plot of mode conversion versus $\delta\beta/\theta\gamma$, demonstrating that when $\delta\beta/\theta\gamma > 0.43$, the separating waveguide acts as a mode splitter; <0.43, as a power splitter

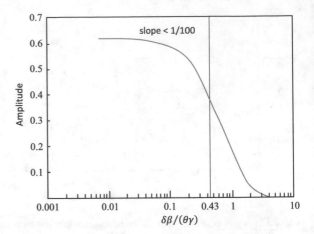

allow the mode from the input waveguide to broaden before coupling into the output arms. This process is done simply by using the tapered geometry shown in Fig. 7.7. An even more effective transition can be realized by adiabatically introducing the dividing region through the use of index tapering in the central region of the Y-branch (see Fig. 7.15). This tapered region allows a natural evolution of the symmetric system mode in the transition region and thus suppresses formation of other higher order system modes, even for relatively large splitting angles. Index tapering also allows more efficient use of the splitting region by inserting it into the taper.

A second important improvement in design is the reduction of coupling to radiative modes in the splitting region. This coupling to unguided modes can be problematic on the basis of simple loss considerations or/and because it leads to unexpected variation in the optical output with device dimensions. One approach to reducing loss has been to decrease the guide index in the transition region (see Fig. 7.16). In a physical sense, in the ray optics picture, this low-index region allows a greater modal angle for light entering from the input guide, and hence reduces the phase front tilt for light entering the branching region. A second approach uses coupling to the radiation modes to occur but then adjusts the device dimensions to increase coupling of these modes back into the device (Johnson and Leonberger 1983). In effect, the approach uses the oscillations seen earlier in Fig. 7.9 to best advantage. This approach is identical to the "coherent coupling" approach which will be described in Sect. 7.4 for waveguide bends. Simulations by these same authors have shown that the approach can be used to reduce loss by $\sim 0.5 - 1$ dB in small angle LiNbO$_3$ Y-branches. The same approach can be used to reduce loss for two closely spaced Y-branches in series.

Finally, one approach, shown in Fig. 7.17, avoids modal splitting, and instead uses two parallel linear couplers to split off light from a central feeder guide. The device, thus, eliminates tilted-mode-front mismatch in the apex region of the Y, and hence reduces radiation loss. The structure is in essence a form of the coupled parallel waveguide arrays discussed in the previous chapter. From the viewpoint of fabrication, the device trades off the difficulty of precisely fabricating the apex in

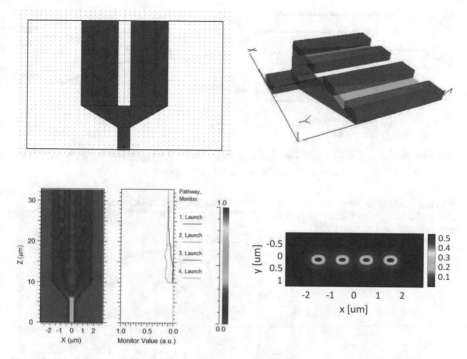

Fig. 7.15 A sketch of a Y-branch with adiabatically introduced dividing region through the use of index tapering in the central region

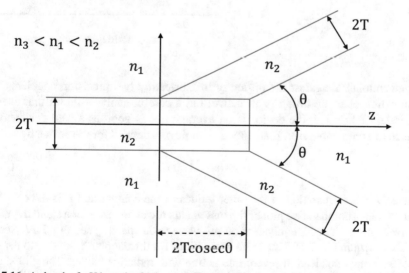

Fig. 7.16 A sketch of a Y-branch with lower index placed in the transition region to reduce the loss

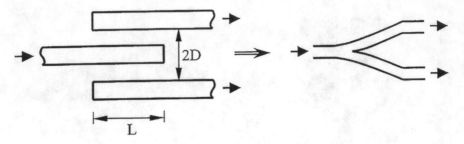

Fig. 7.17 The sketch of a triple guide coupler which avoids modal splitting, and instead uses two parallel linear couplers to split off light from a central feeder guide

Fig. 7.18 A comparative plot of the loss versus θ that can be made for a standard Y and a "triple guide" Y

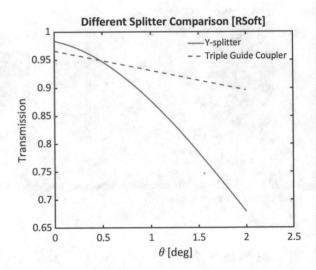

a conventional Y against the difficulty in fabricating two precision couplers. The elimination of mode tilting is attractive, but a careful analysis shows that even if accurately fabricated, the device is on average not as good as a well-designed Y-branch of comparable size. If an effective angle α is defined for the device by setting

$$\alpha = \tan^{-1}(D/L) \tag{7.27}$$

where D is the separation of the outer guides of the coupler and L is the coupling length, then a comparative plot of the loss versus α can be made for a standard Y and a "triple guide" Y, with equivalent output waveguide spacing; see Fig. 7.17. Such a plot is shown in Fig. 7.18. The large oscillations for the coupler Y are due to the fact that its abrupt ends lead to severe interaction with radiative modes. The plot shows that the coupler device is only as good as the standard Y for certain geometries and, in general, it has a higher loss than the standard Y.

7.3.4   Power Loss Due to Coherent Beam Combination and Splitting in Single-Mode Waveguides

Y-branches are often used as phase-sensitive combining elements in optical Mach–Zehnder interferometers; see, for example, their use in modulators and switches in Chaps. 12 and 13, respectively. This application shows that in at least some applications, modal coherence must be considered in the design of Y-branches. Further, these coherent phenomena, when viewed from the point of view of mode splitting, lead to some initially unexpected phenomena for optical loss in the branches.

Consider first the two lowest order modes incident on a symmetric Y-branch, depicted in Fig. 7.18. If the modes arrive in phase, the lowest order mode splits into two equal in-phase components, while the antisymmetric mode splits into an in-phase and an out-of-phase mode. Superposition shows that this leads to one excited waveguide and one unexcited waveguide.

If the waveguide is now reversed so as to be used as a beam combiner, the same combinations of system modes are also important. Consider first exciting only one arm of the beam combiner. If the guide is a single-mode guide, feeding one of the input arms of the Y-branch with the lowest order mode having power P excites the transition with both the symmetric and antisymmetric waveguide modes each having one half of the power, or $P/2$. Since the antisymmetric mode does not propagate, the power is automatically cut in half in such a combiner. This 3 dB power loss is thus inherent for a device having single-mode waveguides. A similar inherent loss occurs when combining with $N \times 1$ combiners.

On the other hand, if both arms of the combiner are excited, the output then depends on the phase of the two excited modes. Using the arguments presented in the previous paragraph, feeding each arm with an in-phase lowest order mode will excite one in-phase system mode in the converging arms of the Y-branch and one out-of-phase mode. These two out-of-phase modes will, in turn, have a 180° difference in phase from each other and cancel. However, the in-phase system modes will be added coherently and the output power will be the sum of the two modes, or $2P$. If the two modes are out of phase at the input to the combiner, then the out-of-phase system mode will be excited, and after it enters the single-mode output guide it will be lost to radiation.

7.3.5   Example: Asymmetric Y-Branch for Mode Sorting

In efforts to realize high data-rate data transmission systems, various orthogonal (non-coupling) parameters are used to provide additional channels and, thus, bandwidth; examples of these parameters include wavelength, spatial modes, and polarization. Recently Y-branch-like devices, fabricated in Si from SOI have been utilized to sort and manipulate modal pathways for this application. Thus an asymmetric Y-junction, in which each arm supports a mode with a different wavevector, has been designed

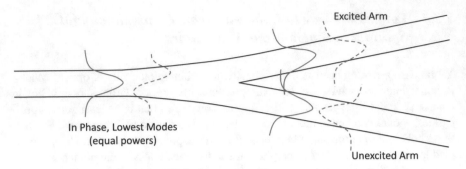

Fig. 7.19 A sketch of an asymmetric y-branch which can be used for mode sorting

and demonstrated to act as a mode sorter (Riesen and Love 2012; Burns and Milton 1975). In such a mode sorter, the arm supporting the fundamental mode with the smaller (larger) effective index, n_{eff}, adiabatically excites the first odd (even) mode of the Y-junction stem, and vice versa as shown in Fig. 7.19. This functionality enables a pair of asymmetric Y-junctions to be used as a mux/demux in a multimode link. The criterion for mode sorting, which has been derived from coupled-mode theory, defines a mode conversion factor (MCF) based on Y-junction properties:

$$MCF = |\beta_A - \beta_B|/\theta\gamma_{AB} \qquad (7.28)$$

where θ is the angle between the Y-junction arms, $\beta_{A(B)}$ is the wavevector of the fundamental mode supported by Arm-A (Arm-B), $\gamma_{AB} = 0.5[(\beta_A + \beta_B)^2 - (2kn)^2]^{1/2}$ is related to the evanescent decay constant of the two modes, k is the free-space wavevector, and n is the cladding index of refraction (Riesen and Love 2012; Burns and Milton 1975). As described earlier in this chapter, it can be shown that for MCF >0.43 (MCF <0.43), an asymmetric Y-junction acts as a mode sorter (power divider) (Riesen and Love 2012; Burns and Milton 1975).

These devices have been recently fabricated and tested. They are shown to work as designed to selectively address individual modes in a multimode Si waveguide. The frequency response of the mux/demux pair depends upon the length of the multimode section as well as the Y-junction angle. The measured cross talk can be as low as -30 dB, < -9 dB over the C band, with insertion loss < 1.5 dB (Driscoll et al. 2013). The measured outcoupled power for both Arm-A and Arm-B excitation as a function of wavelength shows that the power exiting the cross port (P_x) is substantially lower than that at the through port (P_t), indicating that the device serves as a mode mux/demux. The device operates equally well for both modes of the multimode link since P_x and P_t are nearly identical for both inputs.

In addition, the insertion loss for this device was <1.5 dB over the C band. This loss is higher than the 0.2 dB estimated from 3D finite-difference time domain calculations mostly due to optical scattering from fabrication imperfections, especially at the sharp branch point, and propagation loss. Further reduction in loss is possible through an optimized design, including a reduction in the critical dimension of the branch point

(Zhang et al. 2013). The cross talk, defined as P_x/P_t, was < -9 dB over the C band, with a minimum of -30 dB near $\lambda = 1580$ nm. Finally an important aspect of this device as well as its more advanced follow on devices is that they are fabricated using Si processing tools and thus the fabrication quality is excellent.

7.4 Abrupt Angle Bends

An abrupt fixed-angle bend has similar properties to a Y-branch; in fact, it can be viewed as having similar mode-matching properties as one half of a Y-branch. Again, this device, shown in Fig. 7.20, may be analyzed by the two methods mentioned in Sect. 7.3.1, i.e., mode matching at the junction and the VCM approach. As before, for a judicious choice of geometry, the mode matching treatment gives an analytic formula for the transmission, t, of the bend:

$$t = \frac{e^{jva}}{\left(1+\frac{1}{\gamma a}\right)} \left\{ \operatorname{sinc}(va) + \tfrac{1}{2}\operatorname{sinc}(v+2\kappa_t)a + \tfrac{1}{2}\operatorname{sinc}(v-2\kappa_t)a + \right.$$
$$\left. \cos^2(\kappa_t a) + \left[\frac{e^{-jva}}{(2\gamma+jv)a} + \frac{e^{jva}}{(2\gamma-jv)a}\right] \right\}$$

(7.29)

where $\kappa_t \equiv k_t$ is the transverse wavevector, γ is the modal evanescent decay constant outside of the waveguide, and $v = \beta \sin \alpha$, where β is the propagation constant in the waveguides; these definitions are essentially the same as those given earlier for the case of the Y-branch. Note that in this case, the widths of the input and output arms of the bend are the same, i.e., $2a$. The plot of t versus bend angle, α, in Fig. 7.21, shows that the transmission decreases with increasing bend angle due to the mode mismatch of the tilted waves at the junction, just as seen earlier in a Y-branch.

The bend performance can also be analyzed by the volume current method (VCM). This method can give an accurate estimate of radiation loss in the structure, as

Fig. 7.20 A sketch of an abrupt fixed-angle bend

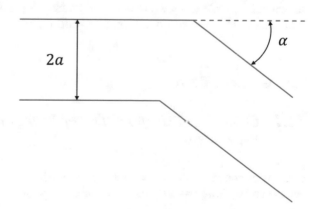

Fig. 7.21 A plot of t versus
bend angle, α, showing that
the transmission decreases
with increasing bend angle

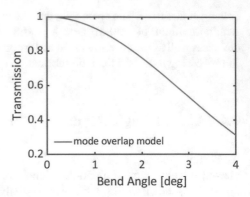

discussed earlier for Y-branches. In particular, the use of sharp bends leads to radiation
into unguided modes. This radiation can couple back into the waveguides and lead
to oscillation in a plot of the loss versus waveguide parameters due to interference
effects.

The coupling between guided and unguided modes can be manipulated to greatly
reduce loss in the bends. This approach, called "coherent coupling," was first
described by Taylor (1974). The approach has been experimentally demonstrated
in a series of abrupt LiNbO$_3$ bends and has been shown to yield significant improvement in the overall loss of a circular curve consisting of a series of small angle
abrupt bends (Johnson and Leonberger 1983). The bend loss is due to constructive
interference between the many radiative modes and one guided mode. Constructive
interference occurs when the length, L between successive bends (see Fig. 7.22) is
such that

$$L = \frac{(2m+1)}{2\Delta N}\lambda \quad m = 0, 1, 2\ldots, \tag{7.30}$$

where ΔN is the difference between the effective index of the guided mode and
the weighted effective index of the excited unguided modes. The oscillations in
transmission versus L for $\lambda = 0.63\,\mu\text{m}$ measured for the simple test bend structure
shown in Fig. 7.22 are shown in the same figure. More complicated structures have
been made, with loss reduction from 50 to 6dB reported, again for $\lambda = 0.63\,\mu\text{m}$.

7.5 Circular Bends

7.5.1 Conformal Transformation of Bends into Straight
Waveguides

Circular or continuous bends are a major, recurring design element in PICs. These
bends can be viewed as a series of the bent waveguides discussed in the preceding

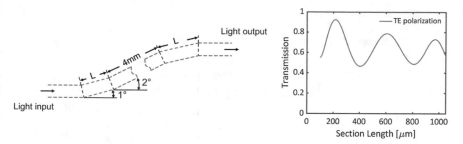

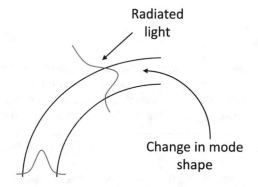

Fig. 7.22 Left: A sketch of a bend consisting of a progressive series of small angle abrupt bends. Right: A plot of transmission versus L for $\lambda = 0.63\,\mu\text{m}$ for the bend showing on the left

Fig. 7.23 A sketch of a simple circular bend. Note that the bend causes shifting of a guided-wave mode toward the outside wall of the guide

section. Such a curved waveguide leads to a continuous series of infinitesimal mode mismatches along the guide, thereby causing loss in the guide. The bends also cause shifting of a guided-wave mode toward the outside wall of the guide, which enables coupling to radiative modes, as depicted in Fig. 7.23. As mentioned in Chap. 2, this radiative loss from a curved waveguide can be serious if the radius of the bend is small and the index contrast is low. In fact, the dimensions of many PICs are often controlled by the acceptable level of loss in small-radius bends.

There are a number of approaches to solving for bend loss and for the mode shape, but use of a conformal transformation gives the clearest results (Heiblum and Harris 1975; Smit et al. 1993). This conformal transformation can be illustrated through a calculation for the TE mode of a curved 2D waveguide, such as that sketched in Fig. 7.24.

The transformation starts by considering the Helmholtz equation in cylindrical coordinates:

$$[\nabla^2_{r,\phi} + k^2(r, \phi)]\Psi(r, \phi) = 0 \tag{7.31}$$

where

$$\nabla^2_{r,\phi} = \frac{1}{r}\frac{\partial}{\partial r}\left(r\frac{\partial}{\partial r}\right) + \frac{1}{r^2}\frac{\partial^2}{\partial\phi^2} \tag{7.32}$$

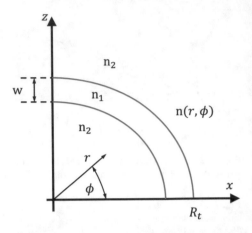

Fig. 7.24 A sketch of a conformal transformation for the TE mode of a curved 2D waveguide

and $\Psi(r, \phi)$ denotes the scalar-field of a waveguide mode. Note for simplicity, here we consider only uniform bends with no structural variation along ϕ; thus the index can only be a constant or a function of r, $n(r)$. In this case, the modal field can be separated into its angular and radius-dependent terms:

$$\Psi(r, \phi) \equiv U_\gamma(r) e^{\gamma_\phi \phi} \tag{7.33}$$

where γ_ϕ is a (complex) propagation constant along the angular direction specified by ϕ. As for all propagation constants, γ_ϕ may have a real or an imaginary component:

$$\gamma_\phi = \alpha_\phi + j\beta_\phi \tag{7.34}$$

Thus in this case, the modal phase is constant for specific values of ϕ, as opposed to specific values of z in a straight waveguide. In this expression, α_ϕ is the angular loss coefficient and β_ϕ is the angular propagation constant; both are expressed in units of (radians)$^{-1}$. Thus, the unit loss in dB/rad for a 90-degree bend with an angular loss coefficient of α_ϕ is $-20 \log(e^{-\alpha_\phi/2})$.

It is possible to transform this bent waveguide into an equivalent straight waveguide by adopting a conformal transformation of the lateral direction of the waveguide; that is, a transformation, which maps r onto u,

$$r = R_t \exp \frac{u}{R_t} \tag{7.35}$$

where R_t is an arbitrary reference radius, which is generally the radius, R, of the bend. This transformation will result in a new equivalent-index profile and simplified Helmholtz equation. Inserting this variable transformation for r into (7.30) yields

$$\left[\frac{\partial^2}{\partial u^2} + (k^2 n_t^2(u) - \gamma_t^2) \right] U_t(u) = 0 \tag{7.36}$$

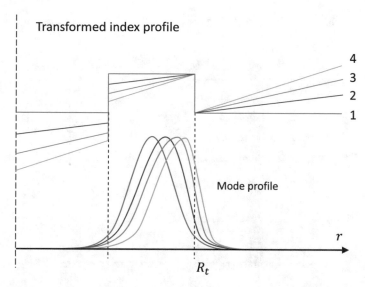

Fig. 7.25 A plot of a transformed index profile and the mode profile for a uniform bend of radius R

where

$$n_t(u) = n(r(u)) \cdot e^{u/R_t} \tag{7.37}$$

and

$$\gamma_t = \gamma_\phi / R_t \tag{7.38}$$

With (7.35), the wave equation for a straight waveguide is recovered with a new effective index $n(u)$, in the transformed Cartesian coordinates (u, φ). The propagation coordinate is thus now φ, and the transformed lateral direction is u. This equation can be solved to obtain both the transformed propagation constant, γ_t (or transformed effective index), and transformed modal field, U_t. From the definition of γ_ϕ and from (7.37), it follows that $\beta_\phi = \beta_t R_t$ and $\alpha_\phi = \alpha_t R_t$.

There are several advantages of using the above conformal transformation. The first is that after transformation, analytic or numerical calculations for linear guides can be readily performed on the transformed structure. In addition, the waveguide physics are more easily and intuitively understood using the transformed structure. Figure 7.25 shows a transformed index profile and the mode profile for a uniform bend of radius R.

For large bends, such that

$$\frac{u}{R} \ll 1 \tag{7.39}$$

the transformed index may be simplified. Then the exponential in (7.36) can be approximated with a Taylor's series to yield

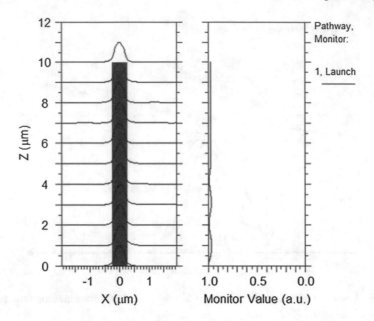

Fig. 7.26 A plot of a beam propagation numerical simulation for the equivalent straight waveguide, showing this outward shifting of the modal profile as z increases

$$n_t(u) = n(r)\left(1 + \frac{u}{R}\right) \tag{7.40}$$

and where the approximation,

$$u = R_t \ln(r/R) \approx -R + r \tag{7.41}$$

has also been used. This transformation index profile then becomes the original profile, $n(r)$, multiplied by a linearly increasing index profile in going across the waveguide (see Fig. 7.25).

In order to find the modal shape, it is first necessary to find the first transformed modal shape, $U_t(u)$, by solving for the equivalent straight waveguide. Then, using the explicit r-dependent form of u in the expression for U_t, i.e.,

$$U = U_t(u(r)) \tag{7.42}$$

the actual mode shape can be obtained. For example, when $u/R \ll 1$, application of (7.40) shows that the modal profile is simply translated linearly toward the outside of the waveguide. A beam propagation numerical calculation for the equivalent straight, given in Fig. 7.26, shows this outward shifting of the modal profile as z increases.

Referring again to Fig. 7.23, note that for a sufficiently tight radius, the field at the inner edge of the waveguide becomes negligible, and this side of the waveguide does not contribute to the modal lateral confinement. The guided mode is then classified as a whispering gallery mode. Waveguides have the lowest loss when they are operated so as to have whispering gallery modes, since optical scattering from one of the waveguide walls is eliminated. Thus, a guide of width w and radius R operates in a whispering galley mode when

$$w > \frac{\delta n}{n + \delta n} R \tag{7.43}$$

where δn is the index contrast in the untransformed waveguide.

7.5.2 Improved Design of Curved-to-Straight Waveguide Junctions

The tilting of the transformed index in curved bends can cause difficulties in the design of waveguide devices. In particular, the mode shifts away from the center of the waveguide toward the outer wall when a mode from a straight guide enters a bent guide; this point can be seen in Fig. 7.26. Hence, the mode overlap between the two single waveguides is not unity and, as a result, the transition will lead to loss at the transition. In addition, if, as is common in many integrated systems, quasi-single-mode waveguides are used in the optical circuit (see Chap. 3), then a curved-to-straight waveguide transition can trigger excitation of the higher modes of the waveguide. This excitation results from the fact that there is overlap between the single-mode input and the higher modes of the guide due to the lateral shift in the bend. One solution to this problem is to offset the center of the two waveguides at the beginning of the transition, thus maximizing the overlap for the two regions. This approach is shown in Fig. 7.27. Another approach to countering the mode shifting problem is to "push" the mode back with a region of higher effective index, e.g., a heavier Ti:doped region (see Fig. 7.26) or a more deeply etched region. This approach is called a "Crown" waveguide.

Fig. 7.27 A sketch of an improved design of curved-to-straight waveguide junction which offsets the center of the two waveguides at the beginning of the transition to maximizing the overlap for the two regions

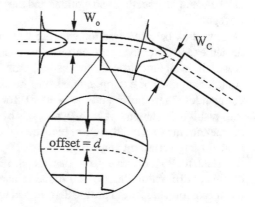

7.5.3   An Explicit Equation for Bend Loss

In many instances, the design of waveguide bends is done with simple loss calculations using approximate formulae. This approach is useful for low-index contrast waveguides. For example, an explicit solution to the Helmholtz equation for a slab waveguide containing a bend of radius R has been solved by Marcuse (1971). This solution is obtained for a slab waveguide of width w, surrounded by an index n_s. The formula gives the linear loss coefficient (i.e., loss/length) of the modal field in the circular bend as

$$\alpha \simeq \frac{2\gamma^2 k_t^2 \exp(\gamma w)}{(N_{eff}^2 - n_s^2)k^2\beta(2 + \gamma w)} \exp\left(\frac{-2\gamma^3}{3\beta^2}R\right) \tag{7.44}$$

for $\alpha R \ll R$ and $(N_{\text{eff}} - n_s) \ll n_s$, i.e., low-index contrast. A normalized loss curve, given, for example, in Marcatili (1969), can be constructed for various curve radii R versus the waveguide normalized frequency. Note that low-loss bends require large R and the highest possible Δn.

7.6   90° Waveguide Crossovers

Waveguide crossovers are critical for reducing overall device "footprints," since they are often required for folded device configurations. For example, such crossovers have been successfully used in GaAs spiral waveguides to allow the exiting radial guide to cross through the bounding azimuthal guides. Similar 90° waveguide crossovers have been used to fold the structure of a long short-pulse optical delay line in GaAs (Hu et al. 1998); in this case, folding provides efficient use of the chip area by shortening the length of the device. More recently, it has been found that in the case of Si photonics switching arrays that crossover loss is the *major* factor in scaling up the number of switching nodes (Rumley et al. 2015; Stern et al. 2015). There are generally two important performance factors for a crossover: the cross talk between intersecting waveguides and the optical attenuation in each intersecting waveguide.

Cross talk between two intersecting waveguides at a crossover has been investigated experimentally and theoretically. These investigations have shown that the cross talk is significant only for a small crossing angle and decreases very quickly at larger angles. For example, one set of measurements of intersecting Ti: LiNbO3 waveguides showed cross talk $< - 35$ dB for a crossing angle $>5°$ (Bogert 1987). This result is in agreement with an analysis by Agrawal et al. (1987), which showed the maximum cross talk between two equal width, intersecting waveguides is negligible for large crossing angles.

In fact, for 90° crossovers, the loss in such a crossover can be treated approximately by simple diffraction theory. The mode in one of the crossing waveguides far from the intersection has a field profile determined by its isolated cross-sectional structure.

As the guided mode reaches the intersection, it enters a short length of an infinitely wide slab, and diffraction occurs. This near-field diffracted light is collected by the intersecting waveguide, causing loss and cross talk. Since the length of the slab region is short, i.e., usually equal to the waveguide width, the diffractive loss is small. For example, the diffractive losses in the GaAs spiral waveguides mentioned above were as low as 0.012 dB for a crossover formed by 2 µm waveguides (Hayes and Yap 1993). Optical simulations have been performed on 90° step-index waveguide crossovers to estimate their loss. The simulation shows, for example, that for a ~5 µm waveguide, the diffractive loss was < 0.1 dB per crossover (Hu et al. 1998).

One potential practical problem with crossovers is patterning imperfections, e.g., the rounding at the corners, where two waveguides intersect. These imperfections can occur either during lithographic patterning or pattern transfer, e.g., etching. For example, rounded corners effectively make the diffraction section at a crossover longer and wider, causing additional diffraction of the light and thus an increased loss at the crossover. However, simulations have shown that for a short diffraction length, the total scattering loss remains small; e.g., with length equal to twice the waveguide width, the scattering is ~0.1 dB (Hu et al. 1998).

More recently the growing interest in complex Si photonics systems has put high demand on finding extremely low loss solutions for waveguide crossings (Rumley et al. 2015). And in fact, the continued improvements in Si patterning techniques, e.g., lithography and etching as well as in simulation tools have greatly reduced crossing loss. For example, Ma et al. (2013) has demonstrated experimentally compact, spectrally broadband Si crossings for use at 1550 and 1310 nm. Specifically CMOS-compatible processes based on 248 nm optical lithography and single etch step were used to fabricated the structures. Their characterization measurements used reliable multiple-die measurement to obtain transmission insertion losses of 0.028±0.009 dB at 1550 nm and 0.017±0.005 dB at 1310 nm, with cross talk lower than −37 dB.

7.7 Turning Mirrors

Another approach to folding an optical path on a PIC chip is the use of miniature flat turning mirrors. These mirrors are mounted in the end of waveguides and tilted at angles beyond the critical angle; as a result, the ideal mirror reflectivity for plane waves should be 100%. Turning mirrors are particularly important for use with the weakly confining waveguides that are needed to mode match with input/output optical-fiber connectors. For such weakly confining waveguides, it would otherwise be necessary to use a relatively large radius of curvature bends to steer modes in the optical current.

Turning mirrors have been investigated using a variety of numerical techniques (Chung and Dagli 1991, 1995). These studies show that losses have three origins: mirror displacement, angular misalignment, and surface roughness. Not surprisingly, loss is also a function of the optical polarization, since the polarization determines in part the optical phase at the air-semiconductor interface.

Fig. 7.28 A plot of the
computed transmission
versus rotation angle of the
turning mirror for the
TM-like waveguide mode for
different waveguide width W

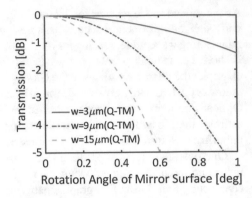

In addition, the effect of displacement error has also been computed for a rib waveguide with a TE mode (corresponding to the TM mode of the slab used in the effective index model). The results show that output intensity peaks at a displacement, which is $0.1\,\mu$m before the ideal mirror position. This displacement is due to the fact that for small beams, diffractive effects cause the beam phase front to be distorted from that of an ideal plane wave. This displacement is smaller for the TM mode than for the TE mode: 0.1 versus $0.02\,\mu$m, respectively. Note that for the TE mode, the mirrors must be positioned accurately to avoid a significant loss penalty, particularly for a small diameter beam. Thus a displacement error of $0.2\,\mu$m can lead to 5% reflection loss for the TE-like mode at $1.55\,\mu$m.

Figure 7.28 shows the computed transmission for an error in the rotation angle of the turning mirror for the TM-like waveguide mode. Although not shown in the figure, the angular sensitivity of the TM and TE modes is approximately equal. Not surprisingly, the sensitivity to mirror rotation is greatest for the wider waveguides, due to the more plane-wave-like nature of the guided wave in a wide waveguide. The investigation that led to Fig. 7.28 suggested that, in general, the error due to angular misalignment was a less severe constraint than that due to mirror displacement.

Loss due to mirror roughness has been examined (Chung and Dagli 1995) by assuming either random or sinusoidal mirror roughness in the mirror surface, e.g., $d(z) = d_0 + d_m \cdot random(z)$, where $random(z)$ is obtained from a random number generator. Regarding random roughness, the results of a calculation which examined transmission versus d_m (the random roughness amplitude) showed that the loss was much more pronounced for the TE-like mode compared to the TM-like mode. This result follows from the larger reflectivity across all angles for the TE versus the reflectivity of the TM mode. In addition, it was found that the loss was approximately independent of waveguide width. In contrast, when the roughness was examined under the assumptions of sinusoidal roughness, i.e., $d(z) = d_0 + \sqrt{2}d_{\mathrm{rms}}(\sin 2\pi z/\Gamma)$, where Γ is the spatial period, the conclusions were different. In this case, it was found that the loss increase was sensitive to waveguide width, although it remained insensitive to mode polarization. Specifically, it is found that turning mirror loss is most pronounced for roughness, having a period between $0.1-2.0$ of the waveguide width.

Above this period, the loss drops off exponentially with spatial period. These results are in accord with the phase distortion of the reflected wave. In addition, loss increases nonlinearly with the amplitude of the spatial period.

The above loss phenomena are important in the fabrication of turning mirrors. Typically, these mirrors are etched by use of some form of ion etching in conjunction with a robust surface mask. The loss mechanism, which is of highest concern, in etched surfaces is that of the surface roughness. Generally, it is possible to characterize the etching by decomposing the surface roughness into its dominant sinusoidal component via Fourier analysis; in one case, for example, reactive ion etching (RIE) etching was shown to yield a Fourier component with $\sim 5 - 30\,\mu$m period. In practice, mirrors with ~ 0.8 dB and 1 dB loss have been fabricated for TE- and TM-like polarized modes. In general, mirrors are etched more deeply than the feeder waveguides to enable high reflection over the entire vertical extent of the structure.

7.8 Transfer Matrices

In Chap. 6, we introduced the concept of transfer matrices for the specific case of couplers. Such transfer matrices make design of many standard elements of an integrated optical system relatively easy. The final section in this chapter provides the transfer matrices of the most common passive elements. These elements, which relate output to the input fields, are listed in Table 7.1. The use of these elements requires some explanation. Specifically, in the case of a single port input or output, it is necessary to fix the nonexistent port at a value of 0. Thus for a Y-branch 3 dB splitter with unity input, the splitter output will be written as

$$
\begin{bmatrix} a_1^o \\ a_2^o \end{bmatrix} = \frac{1}{\sqrt{2}} \begin{pmatrix} 1 & 0 \\ 1 & 0 \end{pmatrix} \begin{pmatrix} 1 \\ 0 \end{pmatrix}
\tag{7.45}
$$

or

$$
= \frac{1}{\sqrt{2}} \begin{pmatrix} 1 \\ 1 \end{pmatrix}
$$

An excellent example of the use of component transfer matrices to obtain the matrix for a more complex device is provided by the example of a simple Mach–Zehnder. For such a device, we multiply the matrices starting with the input of the device. For a Mach–Zehnder, shown in Fig. 7.29, the transfer matrix is

$$
\frac{1}{\sqrt{2}} \begin{pmatrix} 1 & 1 \\ 0 & 0 \end{pmatrix} \begin{pmatrix} \exp(i\Delta\phi/2) & 0 \\ 1 & \exp(-i\Delta\phi/2) \end{pmatrix} \frac{1}{\sqrt{2}} \begin{pmatrix} 1 & 0 \\ 1 & 0 \end{pmatrix}
\tag{7.46}
$$

or $\cos(\Delta\phi/2)$. The text by März (1995) provides a more extensive discussion of these matrices

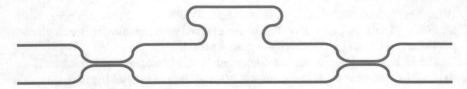

Fig. 7.29 A sketch of a Mach–Zehnder interferometer

Table 7.1 Transfer matrices for some of the multiport devices in this chapter

Component	Matrices	Description
Lossy waveguide	$\sqrt{\ell}$	power loss, ℓ
Phase shifter	$\begin{pmatrix} \exp(i\,\Delta\phi/2) & 0 \\ 1 & \exp(-i\,\Delta\phi/2) \end{pmatrix}$	phase shift, ϕ, in a waveguide
Crossing	$\begin{pmatrix} 0 & i \\ i & 0 \end{pmatrix}$	exchanges input and output ports
Splitter	$\dfrac{1}{\sqrt{x+1}} \begin{pmatrix} \sqrt{x} & 0 \\ 1 & 0 \end{pmatrix}$	$1 : x$ power splitter
Lossless combiner	$\dfrac{1}{\sqrt{x+1}} \begin{pmatrix} \sqrt{x} & 1 \\ 0 & 0 \end{pmatrix}$	$x : 1$ combiner
Lossy combiner	$\dfrac{1}{\sqrt{x+1}} \begin{pmatrix} \sqrt{x} & 1 \\ 1 & -\sqrt{x} \end{pmatrix}$	

7.9 Conclusion

This chapter has examined the analysis and design of many of the basic passive components that are present in PICs. Their performance in terms of loss, mode conversion, and cross talk is at the very heart of designing high-performance photonic circuits. Thus, careful analysis of these components is an essential step in building the more complex PICs described in succeeding chapters. Use of the transfer matrices included at the end of this chapter simplify analysis of more complicated devices by straightforward multiplication of a chain of matrices in the simple components comprising the longer device. However, an accurate prediction of performance must consider the nonideal performance of each component in the circuit; this performance can only be obtained with an accurate numerical simulation package.

Appendix A

To calculate $\Delta\beta_{ij}$ between the ith and jth TM modes for a slab waveguide of width, d, first write the usual waveguide expression for the propagation coefficient of mode m,

$$k_{ym}^2 + \beta_m^2 = k^2 n_f^2 \tag{7.47}$$

But if mode m is far from cutoff,

$$k_{ym} \cong (m+1)\pi/d_{em} \tag{7.48}$$

where d_{em} is the effective waveguide width for mode m.

Then, using a Taylor series expansion,

$$\beta_m \cong kn_f - \frac{(m+1)^2\pi^2}{2kn_f d_{em}^2} \tag{7.49}$$

Notice that d_{em} is also mode dependent. This can be simplified by noting that the effective width of a waveguide is just approximately the actual width d for high index contrast or to the next level of approximation

$$d_{em} \cong d_{eo} \tag{7.50}$$

$$\cong d + \frac{2}{k}\left(\frac{n_s}{n_f}\right)^2 \left(\frac{1}{n_f^2 - n_s^2}\right)^{1/2} \tag{7.51}$$

for a TM mode and where d_{eo} is the effective width of the lowest order mode.

Then,

$$\beta_0 - \beta_m \equiv \Delta\beta_{0m} = \frac{\pi^2(m^2 + 2m)}{d^2(\beta_0 + \beta_m)} \tag{7.52}$$

so that $\Delta\beta_{02} \approx (4\pi^2)/(d^2\beta_0) = (2\pi\lambda_g)/d^2$, where $\lambda_g \equiv 2\pi/\beta_0$ and β_0 is the propagation constant of the lowest-order mode. Note also that away from cutoff, $\beta_0 + \beta_m \approx \beta_0$. This derivation will also recur in Chap. 9, Imaging Devices. For a channel waveguide, $\Delta\beta_{02} \approx \Delta\beta_{13}$ (Burns and Milton 1990).

Problems

1. (a) Calculate the lowest-order mode profile for a simple slab waveguide with the following parameters:
   $n_f = 3.4$
   $n_s = n_c = 3.38$
   $w = 4\,\mu m$
   $\lambda = 1.5\,\mu m$

(b) Calculate the displacement of the peak position if the waveguides has a 1 cm radius bend in it. Use a simple tilted-axis approach to calculate.

(c) Estimate the loss in this waveguide using (7.44).

2. Design a parabolic horn waveguide such that the power loss is 20%. Start with the waveguide in *Problem.* 1 and expand the lowest order mode by a factor of 1.5. (Note: you need to describe the shape in width and calculate the length L of the horn waveguide.)

3. Prove (7.43) with sketches, normalized quantities and appropriate assumptions.

$$w > \frac{\delta n}{n_s + \delta n} R$$

where δn is the index contrast, n_s is the substract refractive index and R and w are radius and width of the waveguide.

Hint: you could use transformation method or ray picture to prove this.

4. Calculate the transmission loss for the lowest mode of a GaAs waveguide butted up against a 6-μm-single-mode-fiber. The GaAs waveguide is square with a 2 μm $\times$ 2 μm Gaussian shape mode. Use the following parameters:

$$\lambda = 1.5 \,\mu\text{m}; \; n_{GaAs} = 3.4; \; n_{SiO_2} = 1.5$$

5. An excited waveguide feeds a second slab waveguide with a 1-meter radius of curvature, but otherwise is the same as the straight waveguide, which feeds it.

(a) Qualitatively, sketch the mode amplitude versus transverse coordinate for *BOTH* waveguides on the *SAME* graph; label all features on the curves. Explicitly state which modal phenomena happen near the trasition region between the two guides.

(b) Sketch the geometry of a common waveguide design change to the above layout that is used to make higher-performance straight-bend connections.

6. Explain using words how you would design a low-loss waveguide mode expander, that is to expand from width w to $5w$. You should supplement your words with simple equations.

References

Agrawal, N., McCaughan, L., & Seshadri, S. (1987). A multiple scattering interaction analysis of intersecting waveguides. *Journal of Applied Physics, 62*(6), 2187–2193.

Benson, T. (1984). Etched-wall bent-guide structure for integrated optics in the iii–v semiconductors. *Journal of Lightwave Technology, 2*(1), 31–34.

Bogaerts, W., Dumon, P., Van Thourhout, D., & Baets, R. (2007). Low-loss, low-cross-talk crossings for silicon-on-insulator nanophotonic waveguides. *Optics Letters, 32*(19), 2801–2803.

Bogert, G. (1987). Ti: Linbo3 intersecting waveguides. *Electronics Letters, 23*(2), 72–73.

Burns, W., & Milton, A. (1975). Mode conversion in planar-dielectric separating waveguides. *IEEE Journal of Quantum Electronics, 11*(1), 32–39.

Burns, W., & Milton, A. (1990). Chap. 3,"waveguide transitions and junctions". In T. Tamir (Ed.), *Guided wave optoelectronics*.

Burns, W. K., & Hocker, G. (1977). End fire coupling between optical fibers and diffused channel waveguides. *Applied Optics, 16*(8), 2048–2050.

Chen, H., & Poon, A. W. (2006). Low-loss multimode-interference-based crossings for silicon wire waveguides. *IEEE Photonics Technology Letters, 18*(21), 2260–2262.

Chu, F., & Liu, P.-L. (1991). Low-loss coherent-coupling y branches. *Optics Letters, 16*(5), 309–311.

Chung, Y., & Dagli, N. (1991). Analysis of integrated optical corner reflectors using a finite-difference beam propagation method. *IEEE Photonics Technology Letters, 3*(2), 150–152.

Chung, Y., & Dagli, N. (1995). Experimental and theoretical study of turning mirrors and beam splitters with optimized waveguide structures. *Optical and Quantum Electronics, 27*(5), 395–403.

Driscoll, J. B., Grote, R. R., Souhan, B., Dadap, J. I., Lu, M., & Osgood, R. M. (2013). Asymmetric y junctions in silicon waveguides for on-chip mode-division multiplexing. *Optics Letters, 38*(11), 1854–1856.

Hanaizumi, O., Miyagi, M., & Kawakami, S. (1985). Wide y-junctions with low losses in three-dimensional dielectric optical waveguides. *IEEE Journal of Quantum Electronics, 21*(2), 168–173.

Hayes, R. R., & Yap, D. (1993). Gaas spiral optical waveguides for delay-line applications. *Journal of Lightwave Technology, 11*(3), 523–528.

Heiblum, M., & Harris, J. (1975). Analysis of curved optical waveguides by conformal transformation. *IEEE Journal of Quantum Electronics, 11*(2), 75–83.

Hu, M., Huang, J., Scarmozzino, R., Levy, M., & Osgood, R. (1997). A low-loss and compact waveguide y-branch using refractive-index tapering. *IEEE Photonics Technology Letters, 9*(2), 203–205.

Hu, M., Huang, Z. J., Hall, K., Scarmozzino, R., & Osgood, R. M, Jr. (1998). An integrated two-stage cascaded mach-zehnder device in gaas. *Journal of Lightwave Technology, 16*(8), 1447.

Izutsu, M., Nakai, Y., & Sueta, T. (1982). Operation mechanism of the single-mode optical-waveguide y junction. *Optics Letters, 7*(3), 136–138.

Johnson, L., & Leonberger, F. (1983). Low-loss linbo 3 waveguide bends with coherent coupling. *Optics Letters, 8*(2), 111–113.

Kuznetsov, M. (1985). Radiation loss in dielectric waveguide y-branch structures. *Journal of Lightwave Technology, 3*(3), 674–677.

Kuznetsov, M., & Haus, H. (1983). Radiation loss in dielectric waveguide structures by the volume current method. *IEEE Journal of Quantum Electronics, 19*(10), 1505–1514.

Ma, Y., Zhang, Y., Yang, S., Novack, A., Ding, R., Lim, A. E.-J., et al. (2013). Ultralow loss single layer submicron silicon waveguide crossing for soi optical interconnect. *Optics Express, 21*(24), 29374–29382.

Marcatili, E. (1969). Bends in optical dielectric guides. *Bell Labs Technical Journal, 48*(7), 2103–2132.

Marcuse, D. (1970). Radiation losses of tapered dielectric slab waveguides. *Bell Labs Technical Journal, 49*(2), 273–290.

Marcuse, D. (1971). Bending losses of the asymmetric slab waveguide. *Bell Labs Technical Journal, 50*(8), 2551–2563.

Marcuse, D. (1991). *Theory of dielectric optical waveguides* (2nd ed.). San Diego: Academic.

März, R. (1995). *Integrated optics: Design and modeling*. Artech House on Demand.

Pollock, C. R. (1995). *Fundamentals of optoelectronics*.

Riesen, N., & Love, J. D. (2012). Design of mode-sorting asymmetric y-junctions. *Applied Optics, 51*(15), 2778–2783.

Rumley, S., Nikolova, D., Hendry, R., Li, Q., Calhoun, D., & Bergman, K. (2015). Silicon photonics for exascale systems. *Journal of Lightwave Technology, 33*(3), 547–562.

Smit, M. K., Pennings, E. C., & Blok, H. (1993). A normalized approach to the design of low-loss optical waveguide bends. *Journal of Lightwave Technology, 11*(11), 1737–1742.

Stern, B., Zhu, X., Chen, C. P., Tzuang, L. D., Cardenas, J., Bergman, K., et al. (2015). On-chip mode-division multiplexing switch. *Optica*, *2*(6), 530–535.

Taylor, H. F. (1974). Power loss at directional change in dielectric waveguides. *Applied Optics*, *13*(3), 642–647.

Zhang, Y., Yang, S., Lim, A. E.-J., Lo, G.-Q., Galland, C., Baehr-Jones, T., et al. (2013). A compact and low loss y-junction for submicron silicon waveguide. *Optics Express*, *21*(1), 1310–1316.

Chapter 8
Components for Polarization Control

Abstract Controlling the state of polarization for a propagating mode on an integrated optical chip is a particularly challenging aspect of integrated optical engineering. A significant aspect of that challenge is the need for fabrication with several different forms of materials. In fact in many cases the use of a hybrid approach is required. In this chapter, a wide variety of approaches from simple metal over layers to complex layered structure, and carefully fabricated optical nanostructures are examined for their use in polarization control.

8.1 Introduction

Controlling the state of polarization for guided light in a PIC is one of the most challenging aspects of integrated optical design. In some cases, for example, polarization control is essential for the operation of certain devices within the optical "circuit." An excellent example is the need for a well-defined, specific polarization states in order to obtain optical isolation using a nonreciprocal optical element. In other cases, the need for polarization control is purely to ensure a predictable or balanced performance in the optical circuits. For example, in many cases, fibers employed for communication or sensing applications do not maintain a constant polarization. Thus the input of an optical chip used to receive the output from the fiber may have an uncertain polarization state. This lack of polarization control can then pose a serious problem for signal detection and amplification since many photodetector and optical amplifiers based on semiconductors are polarization sensitive, i.e., they have different responses for different polarization of the light. In such cases, a polarization diversity operation/network may be an effective solution. For this application, devices for polarization manipulation are required for functions such as polarization of the input optical signals, conversion and rotation of the polarization states, and splitting/combination of optical signals with different degree of polarization.

© Springer Nature Switzerland AG 2021

R. Osgood jr. and X. Meng, *Principles of Photonic Integrated Circuits*,
Graduate Texts in Physics,
https://doi.org/10.1007/978-3-030-65193-0_8

153

In this chapter, we will discuss several different components for polarization control. These components include polarizing elements, polarization rotators, and polarization splitters. Unlike most of the components discussed up to this point, devices for polarization include not only integrated devices, but also may require the employment of hybrid integration techniques such as those used for vertical polarizing plates. In practice, such hybrid integration is difficult and expensive, as the cost advantages of full integration such as planar processing are lost. However, hybrid technology often offers specialized advantages such as the more rapid implementation of certain new or advanced materials systems or easing the addressing of optical requirement for an off-chip application.

8.2   Polarizing Elements

The most fundamental polarizing element is a simple polarizer, which selects from the two possible orthogonal polarization states. A number of different realizations of such a polarizing element have been reported in the literature; however, the main approaches to achieve the polarization control can be divided into three classes: plates of metallic thin films or other absorbing media inserted in a slot normal to a waveguide axis; birefringent or absorbing thin films deposited on a waveguide surface; and polarization routers which guide each of the two states into different waveguides. These three classes will be described in the following subsections. As will be shown below, the best polarizing devices have extinction ratios of $\sim$30 dB, with insertion losses of $\sim$1 dB.

8.2.1   Thin Polarization Plates

An important commonly used approach to polarizing a guided mode is the insertion of a simple polarizing plate in a direction normal to the guided-wave axis. These thin-film polarizers have been used to change the polarization state and to control the polarization in isolators and circulators. As will be mentioned below, this approach using insertion of a plate is best considered to be a micro-optics or hybrid optical approach, although the dimensions are close to those seen in many integrated optics devices. This approach is particularly attractive for waveguides with low-index materials such as silica and large beam spot sizes $\sim$8 μm. These devices are more problematical for materials with high indices (such as semiconductors) and/or devices with small beam sizes because reflection and/or diffraction loss are typically unacceptable.

There are several optical considerations, which must be borne in mind, in selecting polarization plates. Typically, these plates are inserted into a circuit by cutting a slot into a waveguide structure, generally using a thin diamond saw. The slot and the plate introduce four facets in the waveguide which back-reflect, causing a loss, R:

$$R = \left(\frac{n_f - 1}{n_f + 1}\right)^2 \left(\frac{n_p - 1}{n_p + 1}\right)^2$$

where n_f is the usual index of refraction of the waveguide and n_p is the index of the plate. If index matching fluid can be used, the losses are substantially reduced.

In addition, because typically there is no provision for index guiding in the plate, diffraction occurs after light leaves the input facet and propagates freely in the plate. This diffracting beam expands and can then lead to coupling loss at the facet of the output waveguide. If the waveguide is, say, rectangular in cross section with dimensions d and w, the coupling efficiency over a slot of length L will be approximately

$$\eta \approx \frac{d^2 w^2}{1.4 \lambda^2 L^2} \tag{8.1}$$

This loss can be substantial even for relatively modest values of L. For example, a GaAs ridge waveguide with $d = 6\,\mu m$ and $w = 5\,\mu m$ will produce a 10.7% loss for $L = 50\,\mu m$ (Sieger and Mizaikoff 2016).

The most obvious choice of material for a polarizer plate would be Polaroid-like polymers, see Fig. 8.1, which have in fact been used for imaging devices as will be discussed in Chap. 9. Unfortunately, these plates are generally relatively thick, and thus lead to high diffractive losses. One example of a thinner polarization plate is that made of lamipol. Lamipol is a film polarizer which uses an alternating series of thin metal and SiO_2 layers and is sliced into a plate in a plane perpendicular to the plane of the SiO_2/Al layers (Sato et al. 1993). It thus resembles a miniature wire polarizer such as that used for commercial infrared polarizers. Lamipol slices with thicknesses between ~10 and ~30 μm have been fabricated (see Fig. 8.1), and an extinction ratio of >40 dB for a 20 μm plate has been reported (Okuno et al. 1994).

Fig. 8.1 A sketch of a slice of lamipol showing the materials in the lamipol

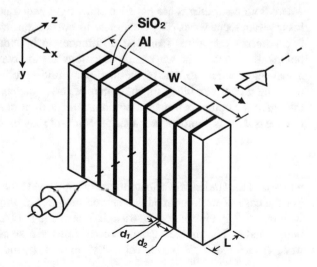

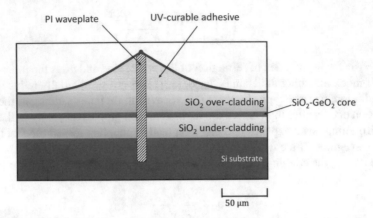

Fig. 8.2 A cross-sectional view of a TE/TM polarization mode converter with polymide half-waveplate

8.2.2 Thin Absorbing Film on a Buried Waveguide

The simplest approach to making a waveguide polarizer is to deposit a metal film on the "cover" surface of a waveguide, as shown in Fig. 8.2. Polarization of the guided wave is then achieved through the strong differences in the absorption loss in the metal for TE and TM components. This differential ratio of absorption in the metal film follows from the fact that metal films respond with both real and imaginary parts of the dielectric constant at optical frequencies:

$$\epsilon_{metal} = \epsilon_r + j\epsilon_i \qquad (8.2)$$

where both components are negative since light penetrates into the metal only as a lossy, evanescent wave. The difference in absorption in a waveguide between the two polarizations arises from the differences in boundary conditions of the two polarization states. The waveguide equations for the two cases can be obtained in a manner identical to that of the slab-waveguide problem in Chap. 3, by simply inserting a complex index for the upper or lower dielectric layer. These equations then give the complex propagation constant for light in a metal-clad waveguide of thickness d. This complex propagation constant may be written as

$$k_z = \beta - j\alpha \qquad (8.3)$$

where α is the guided-wave absorption constant due to the presence of the metal film. For the case of well-confined waveguide modes, i.e., those formed in a waveguide with reasonably high-index contrast, it is possible to obtain approximate values for both β and α. The simplest example considers the case of a metal film with a thickness, d, such that $d \gg 1/\alpha$. Specifically, for $\beta \gg \alpha$, and a relatively thick, strongly confined waveguide $1/\gamma \ll d$, the propagation constants for the two polarizations may be written as

$$\beta_e \approx kn_f - \frac{1}{2n_f k}\left(\frac{(m+1)\pi}{d}\right)^2 \tag{8.4}$$

$$\beta_m \approx kn_f - \frac{1}{2n_f k}\left(\frac{(m+1)\pi}{d}\right)^2 \tag{8.5}$$

where the arctangent terms, such as those found in the normalized dispersion curves, have been approximated as zero and is not seen in (8.4) and (8.5). The corresponding absorption coefficients are then

$$\alpha_e \approx \frac{-1}{\beta_E}\frac{[(m+1)\pi]^2}{d^3\,Re(\gamma_m)}\eta_i \tag{8.6}$$

and

$$\alpha_m \approx \frac{-1}{\beta_m}\frac{[(m+1)\pi]^2}{d^3\,Re(\gamma_m)}\left(\frac{\epsilon_r}{n_f^2}\left(\eta_i + \frac{\epsilon_i}{\epsilon_r}\eta_r\right)\right) \tag{8.7}$$

where

$$Re(\gamma) \approx k\sqrt{n_f^2 - \epsilon_r} \tag{8.8}$$

$$\eta_r + j\eta_i = \sqrt{1 + j\frac{\epsilon_i}{\epsilon_r - n_f^2}} \tag{8.9}$$

and m is an integer.

Expressions containing higher order approximations for β and α may be found in (Nishihara et al. 1989). Actual computations of metal-clad waveguide loss are given in Fig. 8.3 for the case of a waveguide covered with an aluminum film and a wavelength of 1.55 μm. Notice in the figure that TE light has a much smaller loss than TM, namely, 0.6 dB/cm versus 36 dB/cm. Note that, even for the TE case, the absorption loss in the metal-clad waveguide is not negligible. However, the use of a thin dielectric buffer layer between the metal and guiding layers can further alleviate the loss (see below). Note that in order to obtain the loss in dB/cm from $\alpha(\mu m^{-1})$, it is necessary to multiply α by 87, 000.

As suggested above, the eigenvalue equations for TE and TM show that the relative absorption constants for the two polarizations depend on the ratio of the dielectric constants of the metal and guiding layers. Aluminum is one of the most practical cladding materials, because it has a high ratio of the imaginary to real component of the dielectric constant and it is used in standard Si fabrication. Also, for the same reason, waveguides with lower index materials have a higher ratio of the two dielectric coefficients.

Fig. 8.3 A plot of
computations of metal-clad
waveguide loss

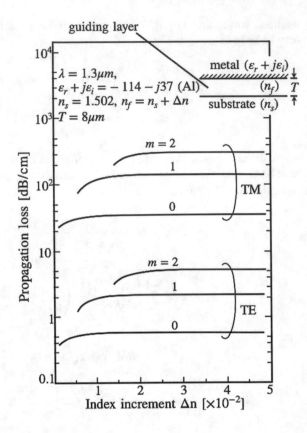

The effects of the metal overlayer can be enhanced or attenuated by interposing
a dielectric between the overlayer and waveguide. For example, a dielectric layer
such as SiO_2 or polymer is often interposed between the electrode and substrate in
semiconductor or $LiNbO_3$ electro-optical devices so as to reduce absorption loss in
the metal. In fact, if the thickness of this dielectric layer is properly adjusted, the
selective loss of the TM mode can be greatly increased (Čtyroký and Henning 1986)
so as to improve polarization selectivity. Finally, adjustment of the dielectric layer
also changes the shape and the nature of the waveguide mode.

The effects of a thin metal/dielectric/waveguide/substrate structure can be understood by treating the optical response of this stack as that of a four-level dielectric
waveguide. Such an analysis shows that in addition to the usual dielectric slab-
waveguide modes obtained without the metal film, addition of the film allows the
formation of a surface plasmon mode, sometimes designated as a TM_{-1} mode. This
mode is important for a small thickness of the dielectric spacer layer. As the thickness of this spacer layer, t_{sl}, is increased, two effects are noticed. First, the loss of
the TM_{-1} mode decreases monatomically. However, in the case of other modes, as
t_{sl} increases, the propagation loss first increases until a characteristic distance t_p is
reached after which the modal loss then decreases. This value of $t_{sl} \approx t_p$ is approxi-

Fig. 8.4 Plots of effects of a buffer layer on the TM mode in a metal-clad optical waveguide using Ti-diffused LiNbO₃ C-plate

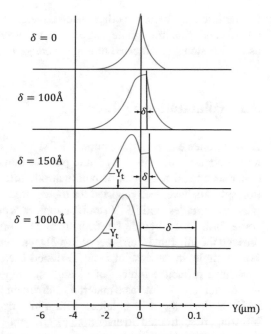

mately the same for all of the first few modes beyond TM_{-1}. Figure 8.4 shows this behavior for a characteristic slab-waveguide stack for both the TM_{-1} and the TM modes.

Second, the shape of the modes evolves and changes character as t_{sl} increases. The variation in modal shapes is shown for TM_{-1} and TM_0 waveguides in Fig. 8.4 as t_{sl} increases. Notice that as the transverse modal shapes evolve, they change into the next-higher slab-waveguide modes. When $t_{sl} \approx t_p$, this evolution is essentially complete. An approximate value for t_p can be obtained (Nishihara et al. 1989) for the case of a relatively low-loss metal film, i.e., $\epsilon_{metal} \approx \epsilon_r$, and for a waveguide away from cutoff, i.e., $N_{eff} \approx n_f$. In this case,

$$t_p \approx \frac{k}{\sqrt{n_f^2 - n_{sl}}} \tanh^{-1} \left(\frac{n_{sl}^2}{|\epsilon_r|} \sqrt{\frac{n_f^2 - \epsilon_r}{n_f^2 - n_b^2}} \right) \tag{8.10}$$

Finally, there is a second approach to realizing a waveguide polarizer via the use of a waveguide overlayer. This method uses birefringence cladding materials to provide a mode-sensitive effective index to waveguide structure. For a guiding layer with refractive index of n_f, a birefringent cladding film with indices n_c^e and n_c^m in the cladding are chosen such that $n_c^e < n_f < n_c^m$, or vice versa. In this way, one mode will propagate within waveguide while the other mode will leak into cladding. Typically, birefringent materials such as calcite (CaCO₃) and polyimide have been used to demonstrate the polarization functionality. Birefringence in overlayers has been artificially created in other materials, e.g., disordered superlattices on semiconductors and even through the use of standard waveguide fabrication techniques such as Ti diffusion and annealed proton exchange in LiNbO$_3$.

8.3   Polarization Splitters

As mentioned earlier, a waveguide TE/TM splitter or router is a basic element in photonic integrated circuits that require polarization control, e.g., polarization-diversity coherent optical receivers, or polarization shift keying. Several types of polarization splitters have been reported for the common material systems such as LiNbO$_3$, glass waveguides, and III–V semiconductors. These devices require the presence of some form of waveguide birefringence. This birefringence can be obtained either through the inherent birefringence in a 3D waveguide or through material properties. For example, in the case of z-cut Ti-doped LiNbO$_3$, the two waveguide transverse axes have refractive indices of n_e and n_o, respectively. While III–V semiconductors do not have an intrinsic material birefringence, polarization-dependent behavior in III–V devices can be obtained with a birefringent passive overlayer, waveguide birefringence, strained layers, or the electro-optic effect. For glass waveguides, birefringence has been typically obtained through the addition of a thin high-index layer such as Si$_3$N$_4$, TiO$_2$, and Si.

In addition to thin film materials methods, there are two general categories of waveguide polarization routers or sorters. One type uses asymmetric Y-branches to do mode sorting by choosing the index of one of the two output waveguides so as to guide either the TE or the TM mode in that output arm. The other method uses interference within either a Mach–Zehnder interferometer or a directional coupler in order to accomplish polarization-sensitive manipulation of the phases of the modes in each arm. The phase of these modes subsequently allows "polarization-tagged" modal steering, typically in a symmetric Y-branch. This latter device type was the first used in polarization routing; however, it is the more wavelength- and dimensional-sensitive of the two types. In this section, we will describe several examples of each of these two types.

Figure 8.5 shows the basic structure used to accomplish mode sorting in an asymmetric Y-branch. The device consists of single-mode input and output waveguides joined through a dual-mode transition region located just prior to the split in the Y-branch. As described in Chap. 7, an asymmetric Y-branch, if designed as a mode sorter, will cause the fundamental mode to be routed into the guide with the higher β and the first higher order mode into the arm with the lower β. This mode sorting occurs efficiently when the guide is designed such that

$$\frac{\Delta\beta}{\theta\gamma} > 0.43 \tag{8.11}$$

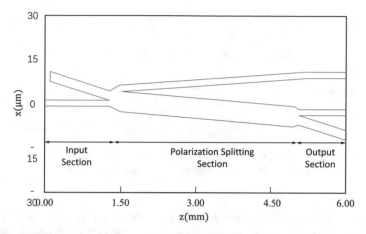

Fig. 8.5 A schematic of an asymmetric Y-branch, which acts as a mode evolution type polarization splitter on InGaAsP/InP

where the symbols are those defined for the same inequality in Chap. 7, i.e., the adiabatic criteria discussed earlier. When used as a mode sorter, the asymmetric Y in Fig. 8.5 is designed such that the propagation constant β for the fundamental mode is associated with a different output waveguide for each of the lowest order TE and the TM modes. Further, the dual-mode waveguide transition region is designed so that only the fundamental modes of each polarization are excited. This requirement is essential to keep the device polarization crosstalk low.

Several different versions of such a device have been successfully demonstrated. For example, one device (Goto and Yip 1989) used a branching waveguide in LiNbO$_3$, in which one arm had its index formed by proton exchange, while the other was formed by Ti in diffusion. The proton-exchange process created a strong waveguide for the fundamental TM mode, while the Ti-diffused region was used to form the waveguide for the TE mode; see the discussion in Chap. 2. In this device, extinction ratios of 20 dB at $\lambda = 0.633\,\mu$m were achieved for each polarization. A second device design (van der Tol and Laarhuis 1991), also in LiNbO$_3$, used the fact that the ordinary and extraordinary indices in z-LiNbO$_3$ exhibits a reversal in relative magnitude as the Ti concentration is increased. Thus a careful choice of the waveguide doping, as controlled by the thickness and width of the deposited Ti metal strip prior to diffusion in each of the arms, can cause the higher β to be in opposite arms for each of the polarization modes. As described above, this allows separation of the two polarizations from an input with arbitrary polarization. Simulations showed that this device should have $\sim$30 dB extinction ratio for $\lambda = 1.55\,\mu$m light and a waveguide separation angle of 0.1°. A related device design, using ridge waveguide birefringence, has been described and demonstrated in InGaAsP/InP (Van der Tol et al. 1993) and shown to have a 20 dB splitting ratio at $\lambda = 1.55\,\mu$m. This device was $\sim$3 mm in length.

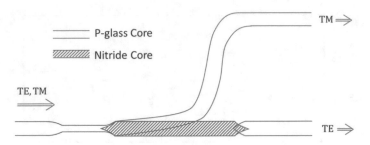

Fig. 8.6 A schematic of an adiabatic-tapered asymmetric Y-branch, which is designed to "sort" the lowest order TM and TE modes from a mixed-polarization input signal

One of the highest performance polarization splitters has been made using mode sorting in a silica-on-silicon structure (Shani et al. 1991). This device, depicted in Fig. 8.6, used an adiabatic-tapered asymmetric Y-branch to "sort" the lowest order TM and TE modes from a mixed-polarization input signal. The design used the strong waveguide birefringence obtained with a Si_3N_4 layer loaded on a p-glass silica waveguide. In this structure, a correct choice of the nitride-film thickness caused its effective index for the mode to be larger than that for the p-glass waveguide, while the reverse was true for the TM mode. Thus the signal is routed through the nitride-covered arm, while the TM mode propagates through the pure P-glass arm. Finally, as shown in Fig. 8.6, after separation, the nitride film is tapered again so as to allow rejoining the purely P-glass waveguide structure. In this glass waveguide, extinction ratios of $20 - 34$ dB were achieved with an insertion loss of ~ 1.5 dB at $\lambda = 1.55 \, \mu$m.

Silica-on-silicon technology allows easy and loss-free integration of multiple structures. Hence, it is also possible to fabricate a series of two asymmetric Y-branch mode sorters; this integrated mode sorter device, shown also in Fig. 8.6, allowed even higher extinction ratios to be obtained, namely, -35 to -45 dB, than for a single device. These extinction ratios indicate that higher performance has thus been achieved in glass waveguides than in the $LiNbO_3$ devices discussed earlier. However, this performance comes at a price since it requires that the device lengths for glass waveguides be long, i.e., >14 mm. Thus the length scale of the integration is very large.

It is also possible to make an active variant of the mode-sorting device. In one example, made in AlGaAs/GaAs, the electro-optical effect was used to change the phase in the arms of a Mach–Zehnder. Note that while the principle of electro-optical phase retardation is described later in Chap. 12, we include a brief discussion of electro-optical devices in this chapter in order to provide a complete discussion.

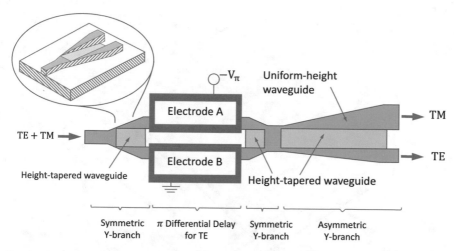

Fig. 8.7 Top view of a Mach–Zehnder TM/TE splitter using height-tapered Y-branches. The insert shows a height-tapered symmetric Y-branch used as the input of the interferometer [Adapted from (Hu et al. 1997)]

Thus, depending on the applied voltage and the polarization state, the output of the Mach–Zehnder would generate the lowest symmetric and antisymmetric waveguide modes at the entrance of an asymmetric Y-branch (see Fig. 8.7) (Hu et al. 1997). The Y-branches in this device used tapering of the waveguide thickness, and, hence, refractive index, to ensure adiabatic transition regions. The electrode configuration and the interferometer path lengths were chosen such that for zero applied voltage, TM light was routed through one waveguide while, when the voltage was switched on, only light was passed by the other waveguide. The device had a 20 dB extinction ratio and could be used either in 1.3 and 1.55 μm bands, after readjustment of the voltage.

The second type of polarization router relies on the use of polarization-dependent modal interference. An example of an early device of this type is shown in Fig. 8.8. In this structure, a single-mode input waveguide, containing a combination of TE- and TM-polarized light, feeds a dual-mode mixing region. When this region is excited by the input waveguide, equal intensities of the first symmetric and asymmetric modes of the wider waveguide are created for each of the TE and TM waves, separately. This mode mixture propagates through the wider waveguide. The length of the guide and the modal birefringence are designed such that at the end of the device, the TE modes have phases such that they will couple out on one of the output waveguides, while the TM modes will have the relative phases to couple out on the other waveguide.

Several devices of those described above have been made in LiNbO$_3$ that utilize polarization-sensitive directional coupling. One of the first of these devices, shown in Fig. 8.9, was designed using waveguide birefringence so as to have a length which was exactly designed to form the cross and bar state for each of the two polarizations, using the interference approach just described. This passive device was designed for 0.78 μm; it was fabricated and found to have a ~12 dB splitting ratio for both

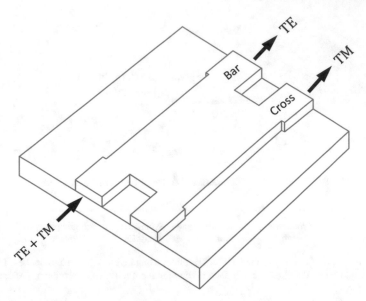

Fig. 8.8 A schematic of an early polarization router which relies on the use of polarization-dependent modal interference

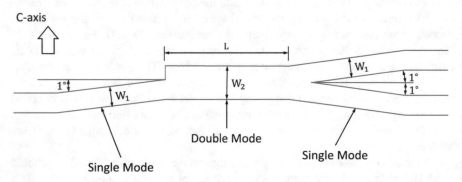

Fig. 8.9 A Schematic of a TE–TM mode splitter fabricated by diffusing titanium into lithium niobate

polarizations at a wavelength of 0.78 μm (Yap et al. 1984). The performance of the device was limited by the difficulty of optimizing the design for both TE and TM splitting; it also requires high fabrication tolerance. A related version of this device has been discussed which utilizes the "birefringence" in the coupling coefficients in a passive Ti: LiNbO$_3$ waveguide directional coupler (Alferness and Buhl 1984). This "birefringence" would enable the coupler to be made such that $\kappa_{TE}L = \pi$ while $\kappa_{TM}L = \pi/2$, and hence, allowing splitting of the polarizations.

A second design (Maruyama et al. 1995) used material birefringence in the waveguide, obtained with the use of the two main different doping techniques, i.e., Ti diffusion or annealed proton exchange (APE) to fabricate the waveguides on two

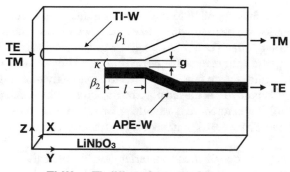

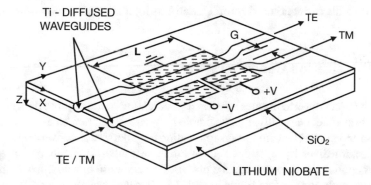

Fig. 8.10 A Schematic of a TE–TM mode splitter using material birefringence in the waveguide, obtained with the use of two different doping techniques [Adapted from (Maruyama et al. 1995)]

Fig. 8.11 A schematic of an active version of the interference-based asymmetric coupler splitter

different arms of the coupler. Specifically, in this case, the input waveguide of the waveguide directional coupler and one of the coupler waveguide arms was formed by Ti-diffusion, while the other output arm was formed by APE (see Fig. 8.10). The device operated as a result of the fact that only a TE-like mode can propagate in the APE waveguide (see Chap. 2) and thus only this mode can couple into the APE-formed output waveguide (see Fig. 8.10). Note that the TM mode remained in the original input waveguide. With this device, an extinction ratio of $> 12\,dB$, with an insertion loss of $2.4\,dB$, was demonstrated for a coupler length of $< 1mm$ and at $\lambda = 0.813\,\mu m$.

Active versions of the interference-based asymmetric coupler splitters have also been demonstrated. An example by (Mikami 1980) is shown in Fig. 8.11. In this device, with the voltage on, the electro-optical effect is used to make the two parallel waveguides highly asymmetric for the TM light entering at the input. Thus transfer to the adjacent guide does not occur. On the other hand, the index is not changed by the applied voltage and thus coupling to the adjacent guide does occur. More complex electrode geometries can be used which make the device less sensitive to the precise voltage. For example, one device has used a Mach–Zehnder interferometer with one arm covered by a split reversed $-\Delta\beta$ electrode (Alferness and Buhl 1984). In addition, the waveguides were fabricated so as to be strongly birefringent to TE and TM polarization. This device achieved polarization cross talk of each polarization, at 1.33 µm of ~30 dB for extinction of a single polarization and a cross talk of −24 dB when both TE and TM were simultaneously extracted. The device was ~1 cm in length.

Finally, recently, more sophisticated interference-like optical structures have been designed and tested for polarization routing. These have used phase manipulation involving the multimode imaging devices discussed in Chap. 9. One design, for example, used a polarization-sensitive interferometer coupled in and out by 3 dB multimode imaging devices (Soldano et al. 1994). These devices are based on the fact that the input phase of a 2 × 2 MMI imaging device (see Chap. 9) depends on the phase of the input beams. This device was made in InGaAsP/InP and was measured to have ~19 and 15 dB extinction ratios for TE and TM light at 1.5 µm, respectively.

8.4 Polarization Converters

A polarization converter rotates any input polarization to the specific desired output polarization state. Generally, it is desirable to have 90° rotation so as to convert TE modes to TM modes. One important application for polarization conversion devices is in polarization diversity receivers, such as might be used in a coherent optical system. Another application is to average the polarization response of a complex symmetric waveguide path, such as that found in a PHASAR-based wavelength router, over the two, TE and TM, polarization directions.

Both active and passive polarization converters have been reported. For example, active devices for polarization mode conversion have been described based on electro-optical, acousto-optical, and even magneto-optical tuning. However, active devices are generally, by their very nature, more complicated and require external power and stability control. Thus, a passive polarization converter is preferable, especially when, as in most cases, a fixed degree of conversion is needed.

Thus far in this chapter, it has been sufficient to specify the polarization state as either TE or TM, or more exactly, as TE- or TM-like. With polarization converters or even more important, with polarization controllers, it is sometimes useful to adopt a more complex notation. In this case, the complex normalized amplitudes are specified by a Jones vector (Saleh and Teich 2007)

$$\begin{bmatrix} \alpha_{TE} \\ \alpha_{TM} \end{bmatrix} = \begin{bmatrix} \cos\theta \\ \sin\theta e^{j\phi} \end{bmatrix} \tag{8.12}$$

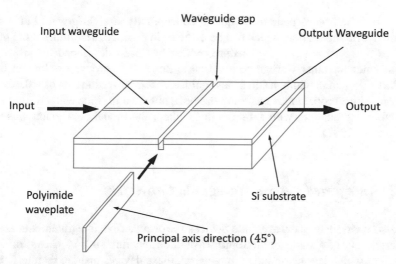

Fig. 8.12 A schematic of a passive, plate-type converter with an inserted half-wave plate made of a thin polyimide sheet

which specifies both the relative phase, ϕ, of the two components and the angle of the polarization vector, θ. For example, right circularly polarized light is given by $\phi = \pi/2$ and $\theta = \pi/2$.

8.4.1 Passive, Plate-Type Converters Placed Normal to the Waveguide Axis

One conceptually simple passive, micro-optics solution to polarization conversion is to insert a half-wave plate into a slot within the waveguide circuit. This mounting method is similar or even identical to that used for the polarization plates mentioned earlier in this chapter. The inserted half-wave plates are made of polyimide sheets, which can be as thin as $10 - 15\,\mu\text{m}$ (see Fig. 8.12). These sheets enable one component to be converted into another, say TE into TM without attenuation. This device thus allows polarizations to be interchanged. Thus if inserted into the middle of a symmetric device, such as a bend, each of the two original components will sample identical path lengths.

In this approach, a short path length, within the plate and its slot, is essential to reduce the diffraction losses as the waveguide mode propagates across the free-space region within the slot and the plate. Several methods have been made to reduce the diffraction losses. One approach uses a relatively thick half-wave plate coupled with an expanded waveguide core to reduce diffractive loss. The second uses a thin plate to reduce waveguide diffraction, in conjunction with a standard waveguide core.

Use of a half-wave plate technique is conceptually straightforward but it does rely on hybrid integration, which can be difficult in practice. Mounting of the plates requires a precision sawing procedure, which has been developed for silica, but typically not for other waveguide material systems. The groove or slot can be as small as ∼18 μm in width in silica waveguides. A total insertion loss of <0.26 dB at 1.55 μm with a proper index matching has been obtained in silica with no significant extinction degradation, with a 14 μm thick plate; the modal conversion was 99% (Inoue et al. 1997).

8.4.2 Integrated Passive Waveguide Converters

It is also possible to use waveguide devices to form passive integrated polarization converters out of a standard isotropic waveguide medium such as GaAs, InP, and Si each of which has a high index. In general, these devices operate by interference between hybrid modes (containing both TE and TM components) of waveguides, which are designed to have the desired hybrid-modal structure.

The first general class of devices uses asymmetric waveguide structures to form hybrid modes. These hybrid modes can then interfere to cause an overall polarization rotation. Consider three kinds of such asymmetric structures: asymmetric-loaded waveguides, tilted-loaded rib waveguides and angle-sidewall waveguides; these are investigated in (Sun et al. 2015; Heidrich et al. 1992; El-Refaei et al. 2004). All three generally have a longitudinally periodic structure and multiple sections. The transverse waveguide asymmetry between any two adjacent sections is designed to be in opposite direction. Phase-matching and hence efficient polarization rotation occurs when the interference length (or period) is such that the two hybrid modes have π phase difference. This interference length, L_π, is determined by the propagation constants β_1 and β_2 of the two fundamental hybrid modes:

$$L_\pi = \frac{\pi}{\beta_1 - \beta_2} \tag{8.13}$$

After one L_π length, the lateral asymmetry is reversed to allow a second addition of the hybrid modes. After several such periods, "pure" TE and "pure" TM modes are then obtained.

The most readily understandable device of this type is that having a single section. Consider the case of an angle-sidewall device with a 45° sidewall. It is possible, by a careful use of crystallographic and anisotropic etching, to make such a passive polarization converter design (Tzolov and Fontaine 1996) which has sufficiently hybrid modes that it does not retain longitudinal periodicity. This device has recently been demonstrated in GaAs (Huang et al. 2000). This design uses the structure of an asymmetric, angle-sidewall waveguide such that its optical axes are rotated 45° with respect to that of a comparable symmetric step waveguide. Thus the modes of the angle-sidewall waveguide are sufficiently hybrid that only one section is needed

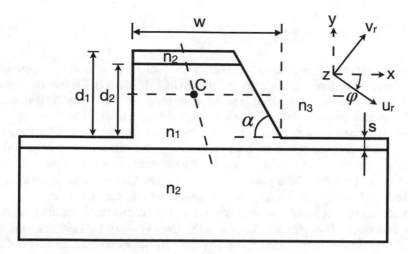

Fig. 8.13 A sketch of an asymmetric, angle-sidewall waveguide (Huang et al. 2000)

for polarization rotation; its rotation per unit length is thus larger than that used in the multiple-section designs.

The asymmetric waveguide structure of this single section device is shown in Fig. 8.13. The optical axis of the device is at a 45° angle with respect to the fixed coordinate system of the device (x, y in Fig. 8.13); thus the TE (x axis) and TM (y axis) components of each of two zero-order modes will have comparable field amplitudes and very similar field distributions. When linearly polarized, say, TE polarized light is launched into this longitudinally invariant, single-mode waveguide, only the two zero-order hybrid modes are excited and at each half beat-length, a 90° polarization rotation is achieved; for this example, after a single section, the output would then be TM polarized. The beat-length L_π is determined from the propagation constants β_1 and β_2 of the two zero-order hybrid modes:

$$L_\pi = \pi/(\beta_1 - \beta_2) = \frac{\lambda}{2\Delta N_{\text{eff}}} \tag{8.14}$$

where ΔN_{eff} is the difference between the effective modal index of these two hybrid modes and λ is the operating wavelength. In this 90° polarization rotation case, the converted power into TM-polarized light can be expressed as a function of the interaction length, z, for TE input light:

$$P_{TM}(z) = P_{TE}(0) \sin^2\left(\frac{z}{L_\pi}\right) \tag{8.15}$$

In addition, the percent conversion, η, is given by

$$\eta = \frac{P_{TM}(z)}{P_{TM}(z) + P_{TE}(z)} = \sin^2\left(\frac{\pi \cdot z}{2L_\pi}\right) \qquad (8.16)$$

where $P_{TM}(z)$ and $P_{TE}(z)$ are the power in TM and TE polarization, respectively. Equation 8.16 assumes that there is no differential loss between TE and TM modes. Also as seen from (8.16), any output polarization state can be obtained by controlling the interaction length, z, between the two fundamental modes.

As mentioned earlier in this section, multiple sections are needed for the more general case when the modes are not hybrid enough to allow a rotation of 90° in one period. Of course, in this case, the device length increases because more sections are needed to achieve 100% polarization conversion. Multiple section polarization converters also suffer period-to-period coupling losses at each junction.

A second type is based on the conversion of hybrid supermodes and can also be done in a single short section. The device (Mertens et al. 1995, 1998) uses the fact that in strip-loaded waveguides, the TM_{11} and te_{21} modes degenerate in β and can thus couple for certain waveguide dimensions. This coupling allows the resulting set of coupled modes to be excited by an input TM wave. The excited modes may then interfere at specific lengths, again given by $L_\pi = \pi/(\beta_2 - \beta_1)$, so as to yield a rotated polarization in the form of an output wave. This device works well in practice; 93% conversion has been obtained from a 250 μm-long InP device (Mertens et al. 1998). However, these devices are not single-mode devices, since they use conversion between the TM_{11} and TE_{21} modes. Thus, for TM to TE conversion, the output is in the first-order mode. In order to obtain the single-mode output required for practical applications, either a mode converter must be used or a monomode waveguide at the output must be employed to collect a half of the first-order output TE mode. The latter approach would result in a minimum of a 3 dB insertion loss. The design also converts a single-mode TE input to TM polarization.

8.4.3 Active Polarization Converters

A more complex but active approach for polarization rotation has been demonstrated using an electro-optical device. A sketch of this device is given in Fig. 8.14. The device is a Ti-diffused waveguide on x-cut LiNbO$_3$, oriented along the z-direction. The waveguide is covered with an interdigitated waveguide, with the field oriented along the waveguide, but reversed over a distance of length Λ. This field reversal field produces a dielectric periodicity similar to that described earlier for the multisector polarization rotator.

Because of the z-orientation of the applied electric field, $E^{(0)}$ the TM and TE modes are coupled via the off-diagonal element of the electro-optical tensor, r_{51}. The coupling coefficient in this case (see Chap. 12) is given by

$$\kappa = \frac{\Gamma\pi}{\lambda}n_s^3 r_{51}\frac{V^{(0)}}{d} \qquad (8.17)$$

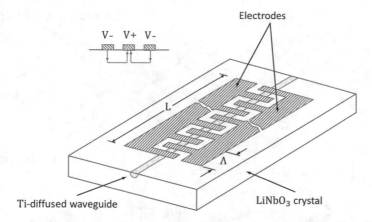

Fig. 8.14 A schematic of an active EO polarization rotator in which any output polarization state can be obtained by controlling the interaction length, z, between the two fundamental modes

where d is the electrode spacing, $V^{(o)}$ the applied voltage, n_s is the zero-bias substrate refractive index, and Γ the overlap factor as given in Chaps. 12 and 13. The interdigitated electrode grating also provides the phase matching for the TM and TE modes which have different effective indices:

$$\frac{2\pi}{\lambda}(N_{\text{eff}}^{TM} - N_{\text{eff}}^{TE}) - \frac{2\pi}{\Lambda} = 0 \qquad (8.18)$$

An active polarization controller allows input light to be changed from an input of arbitrary polarization to an output of a second arbitrary polarization at the output. This capability is important in fiber-optic communication systems, which are not completely polarization insensitive or which require a specific polarization, e.g., coherent communication systems.

This device has been made in LiNbO$_3$, so as to allow electro-optic phase control and rotation of the polarization. The device is generally oriented in the z-direction of the LiNbO$_3$ slab so as to minimize any wavelength shifts. A sketch of this device is given in Fig. 8.15.

The performance of this device can be analyzed by multiplying the Jones matrices of each of the three segments shown in the figure together. A detailed discussion of this approach is provided in (Alferness and Buhl 1981); here we will only provide a brief qualitative description based on this more complete discussion in these references. Specifically, the device uses an input phase shifter which is adjusted so that the relative phase, $\Delta\phi_i$, between the two polarizations is $\pm\pi/2$ at the entrance to the mode converter. This phase adjustment allows the TE–TM converter to achieve a

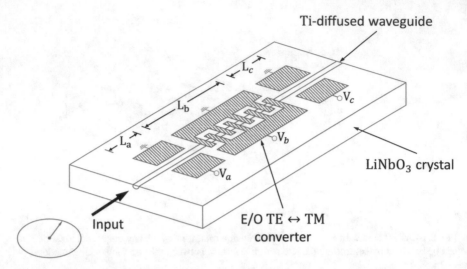

Fig. 8.15 The schematic of active polarization controller which allows input light to be changed from an input of arbitrary polarization to an output of a second arbitrary polarization at the output [Adapted from (Alferness and Buhl 1981)]

rotation of the input polarization that is linearly controlled by the coupling coefficient, κ. i.e., $\theta_o - \theta_i = \pm \kappa L_2$ and hence the electro-optical voltage. The relative phase of the output, $\Delta\phi$, is then controlled by the output phase shifter. Thus if the phase shift imparted by the output phase shifter is $\Delta\phi_s$, the output phase is given by $\phi_0 = \Delta\phi_s \pm \pi/2$ when $\Delta\phi_i = \pm\pi/2$.

The device has several useful modes of operation. For example, if the first phase shifter is adjusted so that $\Delta\phi_i = -\pi/2$ and the second shifter is adjusted to set $\Delta\phi_0 = 0$, the device then acts as a polarization rotator, with the rotation controlled by the voltage through κ. The device can also be used in a feedback-controlled mode in conjunction with a polarization sensor. Operated in this form, the device can keep the polarization fixed in an optical system.

8.5   Polarization Devices for Si Photonics

As it is clear from the earlier discussion in this chapter, maintaining or controlling polarization is key to high-performance photonics systems and thus it is not surprising that the increasing role of Si photonics in optical subsystems and systems has also generated much interest in novel methods for polarization control via integrated Si waveguide devices. Our intent in this section is to present several of these devices and examine their performance parameters and operational methods. Some recent polarizer research has been nicely summarized in an extensive article by (Dai et al. 2012) (see in addition (Wang and Dai 2008; Dai 2012)). Note that the reader should keep in mind that the polarization properties of pure Si waveguide devices, which are attributable to silicon's high-index contrast can also be found also in earlier devices based on GaAs or other III–V materials.

One interesting approach to polarization in an integrated Si photonics which makes use of the light bending radius that is enabled by Si's high-index contrast. This approach is shown in (Paredes et al. 2016); it uses a simple series of waveguide bends to achieve high loss for TM light and thus making an integrated TE-pass polarizer. The device was measured to have a 20 dB TM/TE extinction ratio over a 50 nm spectral bandwidth with a 1–2 dB insertion loss.

A second device type uses asymmetrical-coupling based on polarization beam splitting (PBS) (Dai et al. 2012). One version of this device consists of a combination of SOI nanowires to form the asymmetric coupler. It allows separation of TE- and TM-polarized light within a very short propagation length. Using this approach, a short (25 μm long) PBS was designed based on silicon-on-insulator nanowires. Numerical simulations showed that the PBS had good fabrication tolerance and a broadband (50 nm) response and an extinction ratio of 15 dB. The device was designed for phase matching of any TM polarization light to dominate within a short length determined by the strength of the coupling region. For the TE polarized wave, the phase-matching condition was not fulfilled and hence the coupler was always in the bar state. Polarization management devices have also been demonstrated on the silicon nitride (Si_3N_4) on silicon-on-insulator (SOI) platform (Sacher et al. 2014). SiO_2-clad silicon nitride (Si_3N_4) waveguides which are an alternative Si photonics materials platform are sometimes preferred for polarization control since they are insensitive to thermal drift (i.e., they have low thermo-optical effects) and they have low waveguide loss (due to the low-index contrast of the nitride, which reduces sidewall-roughness. In addition, Si_3N_4 waveguides can be readily integrated onto SOI photonic platforms or vice versa to achieve CMOS-compatible integration.

Polarization control in these devices includes a broadband polarization rotator-splitter using a $TM_0 - TE_1$ mode converter in a composite Si_3N_4-silicon waveguide (Sacher et al. 2014). In one device, the measured polarization cross talk, insertion loss, and polarization-dependent loss were less than $\sim$19 dB, 1.5 dB, and 1.0 dB, respectively, over an 80 nm bandwidth. These same materials' platform enabled a polarization controller composed of polarization-rotator-splitters, multimode interference couplers, and thin film heaters to be fabricated and tested. In particular the polarization rotator-splitter (PRS) and a polarization controller were designed for the Si_3N_4-on-SOI integrated photonics platform. These polarization devices are of the type needed for control of polarization in polarization diversity operation and in polarization demultiplexers and multiplexers.

8.6   Conclusion

This chapter has described devices to be used for manipulation of polarization in integrated photonic circuits. The devices use a wide range of materials and approaches, including the recent emphasis on Si devices. Historically, however, the most advanced devices are in the area of LiNbO$_3$ for switch arrays. The approaches use passive micro-optical insertion of half wave plates and integrated devices as well as active polarization rotators and routers. While polarization control is crucial in many areas of integrated optics, research continues on methods to improve device performance.

References

Albrecht, P., Hamacher, M., Heidrich, H., Hoffmann, D., Nolting, H., & Weinert, C. (1990). Te/tm mode splitters on ingaasp/inp. *IEEE Photonics Technology Letters*, 2(2), 114–115.

Alferness, R., & Buhl, L. (1981). Waveguide electro-optic polarization transformer. *Applied Physics Letters*, 38(9), 655–657.

Alferness, R.C., & Buhl, L. (1984). Low-cross-talk waveguide polarization multi-plexer/demultiplexer for $\lambda = 1.32~\mu$m. *Optics Letters*, 9(4), 140–142.

Čtyroký, J., & Henning, H.-J. (1986). Thin-film polariser for ti: Linbo3 waveguides at $\lambda = 1.3~\mu$m. *Electronics Letters*, 22(14), 756–757.

Dai, D. (2012). Silicon polarization beam splitter based on an asymmetrical evanescent coupling system with three optical waveguides. *Journal of Lightwave Technology*, 30(20), 3281–3287.

Dai, D., Bauters, J., & Bowers, J.E. (2012). Passive technologies for future large-scale photonic integrated circuits on silicon: polarization handling, light non-reciprocity and loss reduction. *Light: Science & Applications*, 1(3), e1.

El-Refaei, H., Yevick, D., & Jones, T. (2004). Slanted-rib waveguide ingaasp-inp polarization converters. *Journal of Lightwave Technology*, 22(5), 1352.

Glance, B. (1987). Polarization independent coherent optical receiver. *Journal of Lightwave Technology*, 5(2), 274–276.

Goto, N., & Yip, G. L. (1989). A te-tm mode splitter in linbo/sub 3/by proton exchange and ti diffusion. *Journal of lightwave technology*, 7(10), 1567–1574.

Hayakawa, T., Asakawa, S., & Kokubun, Y. (1997). Arrow-b type polarization splitter with asymmetric y-branch fabricated by a self-alignment process. *Journal of lightwave technology*, 15(7), 1165–1170.

Heidrich, H., Albrecht, P., Hamacher, M., Nolting, H.-P., Schroeter-Janssen, H., & Weinert, C. (1992). Passive mode converter with a periodically tilted inp/gainasp rib waveguide. *IEEE Photonics Technology Letters*, 4(1), 34–36.

Hu, M., Huang, J., Scarmozzino, R., Levy, M., & Osgood, R. (1997). Tunable mach-zehnder polarization splitter using height-tapered y-branches. *IEEE Photonics Technology Letters*, 9(6), 773–775.

Huang, Z., Scarmozzino, R., Nagy, G., Steel, J., & Osgood, R. (2000). Realization of a compact and single-mode optical passive polarization converter. *IEEE Photonics Technology Letters*, 12(3), 317–319.

Inoue, Y., Ohmori, Y., Kawachi, M., Ando, S., Sawada, T., & Takahashi, H. (1994). Polarization mode converter with polyimide half waveplate in silica-based planar lightwave circuits. *IEEE Photonics Technology Letters*, 6(5), 626–628.

Inoue, Y., Takahashi, H., Ando, S., Sawada, T., Himeno, A., & Kawachi, M. (1997). Elimination of polarization sensitivity in silica-based wavelength division multiplexer using a polyimide half waveplate. *Journal of Lightwave Technology*, 15(10), 1947–1957.

Maruyama, H., Haruna, M., & Nishihara, H. (1995). Te-tm mode splitter using directional coupling between heterogeneous waveguides in linbo/sub 3. *Journal of Lightwave Technology*, 13(7), 1550–1554.

Masuda, M., & Koyama, J. (1977). Effects of a buffer layer on tm modes in a metal-clad optical waveguide using ti-diffused linbo 3 c-plate. *Applied Optics*, 16(11), 2994–3000.

Mertens, K., Opitz, B., Hovel, R., Heime, K., & Schmitt, H. (1998). First realized polarization converter based on hybrid supermodes. *IEEE Photonics Technology Letters, 10*(3), 388–390.

Mertens, K., Scholl, B., & Schmitt, H. J. (1995). New highly efficient polarization converters based on hybrid supermodes. *Journal of Lightwave Technology, 13*(10), 2087–2092.

Mikami, O. (1980). Linbo3 coupled-waveguided te/tm mode splitter. *Applied Physics Letters, 36*(7), 491–493.

Nishihara, H., Haruna, M., & Suhara, T. (1989). *Optical integrated circuits* (Vol. 1). New York: McGraw Hill.

Okuno, M., Sugita, A., Jinguji, K., & Kawachi, M. (1994). Birefringence control of silica waveguides on si and its application to a polarization-beam splitter/switch. *Journal of Lightwave Technology, 12*(4), 625–633.

Paredes, B., Zafar, H., Dahlem, M.S., & Khilo, A. (2016). Silicon photonic te polarizer using adiabatic waveguide bends. In *2016 21st OptoElectronics and Communications Conference (OECC) held jointly with 2016 international conference on Photonics in Switching (PS)*, pp. 1–3. Piscataway: IEEE.

Sacher, W. D., Huang, Y., Ding, L., Barwicz, T., Mikkelsen, J. C., Taylor, B. J., et al. (2014). Polarization rotator-splitters and controllers in a si 3 n 4-on-soi integrated photonics platform. *Optics Express, 22*(9), 11167–11174.

Saleh, B.E. & Teich, M.C. (2007). *Fundamentals of photonics* (pp. 260–269). Hoboken: Wiey.

Sato, T., Baba, K., Hirozawa, T., Shiraishi, K., & Kawakami, S. (1993). Fabrication techniques and characteristics of al-sio/sub 2/laminated optical polarizers. *IEEE Journal of Quantum Electronics, 29*(1), 175–181.

Shani, Y., Alferness, R., Koch, T., Koren, U., Oron, M., Miller, B., et al. (1991). Polarization rotation in asymmetric periodic loaded rib waveguides. *Applied Physics Letters, 59*(11), 1278–1280.

Sieger, M., & Mizaikoff, B. (2016). Optimizing the design of gaas/algaas thin-film waveguides for integrated mid-infrared sensors. *Photonics Research, 4*(3), 106–110.

Soldano, L., De Vreede, A., Smit, M., Verbeek, B., Metaal, E., & Green, F. (1994). Mach-zehnder interferometer polarization splitter in ingaasp/inp. *IEEE Photonics Technology Letters, 6*(3), 402–405.

Suematsu, Y., Hakuta, M., Furuya, K., Chiba, K., & Hasumi, R. (1972). Fundamental transverse electric field (te0) mode selection for thin-film asymmetric light guides. *Applied Physics Letters, 21*(6), 291–293.

Sugimoto, N., Terui, H., Tate, A., Katoh, Y., Yamada, Y., Sugita, A., et al. (1996). A hybrid integrated waveguide isolator on a silica-based planar lightwave circuit. *Journal of Lightwave Technology, 14*(11), 2537–2546.

Sun, Y., Xiong, Y., & Winnie, N. Y. (2015). Compact soi polarization rotator using asymmetric periodic loaded waveguides. *IEEE Photonics Journal, 8*(1), 1–8.

Suzuki, Y., Iwamura, H., and Mikami, O. (1990). Te/tm mode selective channel waveguides in gaas/alas superlattice fabricated by sio 2 cap disordering. *IEICE Transactions (1976–1990), 73*(1), 83–87.

Takahashi, H., Hibino, Y., & Nishi, I. (1992). Polarization-insensitive arrayed-waveguide grating wavelength multiplexer on silicon. *Optics Letters, 17*(7), 499–501.

Tzolov, V. P., & Fontaine, M. (1996). A passive polarization converter free of longitudinally-periodic structure. *Optics Communications, 127*(1–3), 7–13.

Uehara, S., Izawa, T., & Nakagome, H. (1974). Optical waveguiding polarizer. *Applied Optics, 13*(8), 1753–1754.

Van Der Tol, J., Hakimzadeh, F., Pedersen, J., Li, D., & Van Brug, H. (1995). A new short and low-loss passive polarization converter on inp. *IEEE Photonics Technology Letters, 7*(1), 32–34.

Van der Tol, J., Pedersen, J., Metaal, E., Oei, Y., Van Brug, H., & Moerman, I. (1993). Mode evolution type polarization splitter on ingaasp/inp. *IEEE Photonics Technology Letters, 5*(12), 1412–1414.

van der Tol, J. J., & Laarhuis, J. H. (1991). A polarization splitter on linbo/sub 3/using only titanium diffusion. *Journal of Lightwave Technology, 9*(7), 879–886.

Veselka, J., & Bogert, G. (1987). Low-insertion-loss channel waveguides in linbo3 fabricated by proton exchange. *Electronics Letters*, *23*(6), 265–266.

Wang, Z., & Dai, D. (2008). Ultrasmall si-nanowire-based polarization rotator. *JOSA B*, *25*(5), 747–753.

Wei, P.-K., & Wang, W.-S. (1994). A te-tm mode splitter on lithium niobate using ti, ni, and mgo diffusions. *IEEE Photonics Technology Letters*, *6*(2), 245–248.

Yap, D., Johnson, L., & Pratt, G, Jr. (1984). Passive ti: Linbo3 channel waveguide te-tm mode splitter. *Applied Physics Letters*, *44*(6), 583–585.

Chapter 9
Imaging Devices

Abstract This chapter provides a detailed discussion of two important devices for using the reconstruction of a number of spatial modes to carry out spatial multiplexing or demultiplexing. The small number of modes involved means that the device is relatively efficient. The two devices considered in this chapter are the multimode interference device and the star coupler.

9.1 Introduction

This chapter includes discussion of two important imaging devices: multimode interference devices and star couplers. Imaging, in this case, refers to reconstruction of an object in the two-dimensional plane of a guided-wave slab by using guided-wave modes. Multimode interference devices operate by the interference of a discrete number of guided modes while the star coupler operates by the excitation of a continuum of "free-space" modes, with the image formed in the far field. Further, both devices enable $1 \times N$ splitting or $N \times M$ interconnection, although in the case of the multimode interference devices, these numbers, N and M, are small—say, $1 - 16$—while for star couplers, their magnitudes maybe 100s! In addition to power division, these devices may be used for a variety of other functions; these functions are extensive and in many cases subtle. These functions, which rely on their efficient operation and easy design, will be explained in this chapter and other places throughout the text.

9.2 Multimode Interference Devices (MMIs)

9.2.1 Overview

Modal effects and manipulation have been shown to be an effective tool for lateral imaging in waveguides. The devices, which use multimode interference (MMI), can

© Springer Nature Switzerland AG 2021 177
R. Osgood jr. and X. Meng, *Principles of Photonic Integrated Circuits*,
Graduate Texts in Physics,
https://doi.org/10.1007/978-3-030-65193-0_9

Fig. 9.1 A numerical simulation of the imaging process of multimode waveguide

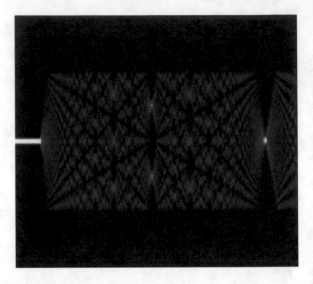

be used to achieve optical routing and coupling functions. MMI-based devices have the advantages of large bandwidth, small device dimension, polarization insensitivity, and low optical loss. Their useful properties have led to the demonstration of a wide variety of advanced applications such as coherent receivers, variable power splitters, WDM routers, etc. The application to wavelength routers, which is of particular importance, will be discussed in Sect. 9.3. A general sketch of the multimode waveguide and its imaging process is shown in Fig. 9.1.

9.2.2   Self-imaging Principle and Mode Propagation Analysis

The self-imaging properties of waveguide structures were first described in detail by Bryngdahl (1973), Ulrich (1975). Modal imaging or "self-imaging" results from the repeating of the input field along the propagation direction in the multimode waveguide. Typically these MMI devices have a 2D character: the imaging guide has a single mode in the vertical direction and multiple modes in the horizontal direction. Thus, imaging involves only the lateral modes of the waveguide.

The most common approach to analyze a multimode waveguide uses a guided-mode-propagation analysis (MPA) of a 2D step-index multimode waveguide (Ulrich and Kamiya 1978). To use this approach, the dispersion relation of the MMI waveguide must be determined. To obtain the dispersion relation, we first assume that the physical width of the guide is W, and that the waveguide has a high-index contrast. Such a waveguide will have n lateral modes with mode numbers $m = 0, 1, ..., (n - 1)$ as depicted in Fig. 9.2:

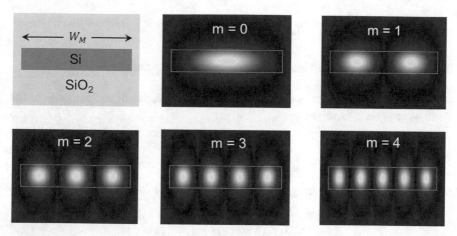

Fig. 9.2 A multimode waveguide with five lateral modes

$$k_{ym}^2 + k_z^2 = k_0^2 n_f^2 \tag{9.1}$$

or

$$k_{ym}^2 + \beta_m^2 = k_0^2 n_f^2 \tag{9.2}$$

where $k_0 = 2\pi/\lambda_0$ and $k_{ym} = (m+1)\pi/W_{em}$. Further, β_m and W_{em} are the propagation constant and the effective mode width, respectively, for mode m. Since the waveguide is assorted to have high contrast, $W_{em} \approx W$, or to a somewhat better approximation:

$$W_{em} \approx W_{e0} = W + \frac{\lambda_0}{\pi}\left(\frac{1}{n_f^2 - n_c^2}\right)^{\frac{1}{2}} \tag{9.3}$$

for TE modes away from cutoff. In what follows, we will adopt the shorthand notation of $W_{e0} = W_e$. For highly multimode waveguides and for the case $k_{ym}^2 \ll k_0^2 n_f^2$, i.e., when the only lower order modes are excited, (9.2) can be approximated using a Taylor's series as

$$\beta_m \approx k_0 n_f - \frac{(m+1)^2 \pi^2}{2k_0 n_f W_e^2} \tag{9.4}$$

The propagation constant and, hence, the phase shift of mode m can be conveniently referenced to that of $m = 0$; thus,

$$\beta_0 - \beta_m \approx \frac{m(m+2)\pi}{3L_\pi} \tag{9.5}$$

where

$$L_\pi = \frac{\pi}{\beta_0 - \beta_1} = \frac{4n_f W_e^2}{3\lambda_0} \tag{9.6}$$

Thus, L_π is the modal beat length of the two lowest order modes.

At the input we assume only the guided, i.e., no radiative or leaky, modes of the wide waveguide are excited

$$\Psi(y, 0) = \sum_m c_m \Psi_m(y) \tag{9.7}$$

where the overlap factor, c_m, gives the amplitude for the modes which are excited:

$$c_m = \int \Psi(y, 0) \Psi_m(y) / \left(\int \Psi_m^2(y) dy \right)^{\frac{1}{2}} \tag{9.8}$$

Equation 9.8 is based on modal orthogonality.

Then, referring the phase to that of the $m = 0$ mode,

$$\Psi(y, z) = \sum_{m=0}^{n-1} c_m \Psi_m(y) \exp(j(\beta_0 - \beta_m)z) \tag{9.9}$$

and/or at $z = L$:

$$\Psi(y, z) = \sum_{m=0}^{n-1} c_m \Psi_m(y) \exp\left(j \left(\frac{m(m+2)\pi}{3L_\pi} \right) L \right). \tag{9.10}$$

It will be seen below that, at certain intervals of the distances to L_π, the output field $\Psi(y, z)$ will be a reproduction or self-image of the input field $\Psi(y, 0)$. The quality of images is determined by c_m, the modal coefficients that determine modal weighting, and the phase factor of that same mode.

9.2.3 Imaging Modalities: General, Restricted, and Overlap Imaging

In this section, we examine conditions for forming images of the lateral modes in an MMI device or waveguide region with the application being, in general, for power splitting. There are several operational mechanisms for image formation that depend on how the modes in the MMI region are excited by the input field and how the excited modes interfere at the output end of the device. If all modes are excited, the image mechanism is termed "general interference." If only certain modes are excited, the mechanism is termed as "restricted interference." Both types of interference result in uniform power splitting. A third interference mechanism, called "overlap imaging," may be used to obtain nonuniform power splitting.

General Interference

A single image in an imaging waveguide having parameters L_π and W_e will occur at a length L, such that

$$\exp\left(j\left(\frac{m(m+2)\pi}{3L_\pi}\right)L\right) = (-1)^m \tag{9.11}$$

irrespective of the values of c_n, i.e., the relative modal content excited by the input waveguide. At this axial distance (z), all of the original modes will be in phase again, and the image of the input waveguide will be reconstituted. The length, L, for imaging in this case is

$$L = p(3L_\pi) \tag{9.12}$$

where $p = 0, 1, 2\ldots$, and its value labels the image periodicity along z.

In addition, further examination of the equation for modal fields (9.9) shows that at distances:

$$L = \frac{p}{2}(3L_\pi) \tag{9.13}$$

where $p = 1, 3, 5, \ldots$, there is a double image described by

$$\Psi\left(y, p\frac{3}{2}L_\pi\right) = \frac{1+(-j)^p}{2}\Psi(y, 0) + \frac{1-(-j)^p}{2}\Psi(-y, 0) \tag{9.14}$$

thus having periodic images at $3L/2, 9L/2, 15L/2\ldots$. Such a scheme allows the formation of a x^2-fold splitter.

In general, N-fold multiple images can also be found. This form of imaging can be shown clearly by realizing that the modal field is approximately sinusoidal in a highly confined waveguide, i.e.,

$$\Psi_m(y) \simeq \sin(k_{ym}y) \tag{9.15}$$

Then, if

$$L = \frac{p}{N}(3L_\pi) \tag{9.16}$$

where the integers p and N are such that $p > 0$ and $N > 1$, and p and N do not have a common divisor,

$$\Psi(y, L) = c \sum_{q=0}^{N-1} \Psi_{in}(y - y_q) \exp(j\phi_q) \tag{9.17}$$

where q identifies each of N images along y,

$$y_q = p(2q - N)W_e/N \tag{9.18}$$

$$\phi_q = p(2q - N)q\pi/N \tag{9.19}$$

and

$$|c| = \frac{1}{\sqrt{N}} \tag{9.20}$$

Thus, at $z = L$, N images are formed with amplitude $1/\sqrt{N}$ and phase ϕ_q. In this case, as shown by (9.18), the N images are not equally spaced, but are periodically distributed laterally across the waveguide in uniform fractions of units of W_e.

Consider an $N \times N$ MMI coupler with $p = 1$, so as to have the shortest length of the imaging region (Bachmann et al. 1994). In the device, waveguides of finite width are used as inputs and outputs. These waveguide dimensions impose the spatial restrictions on the light distribution. In particular, for properly designed MMI couplers, different images of the input field at the output usually should not overlap. For a given MMI effective width W_e and a chosen number of images, N, the input waveguide can be placed at an arbitrary position which is defined by an additional free parameter α and $0 < \alpha < W_e/N$. Figure 9.3 shows the situation for N even and N odd, respectively. Inputs and outputs are numbered with indices i and j, respectively. The resulting phase for imaging input i to output j can be given now in a very compact and direct form: when $i + j$ is even,

$$\phi_{ij} = \pi + \phi_{N-(j-i)/2} = \phi_0 + \pi + (\pi/4N)(j - 1)(2N + j - i) \tag{9.21}$$

and when $1 + j$ is odd,

$$\phi_{ij} = \pi + \phi_{N-(j-i)/2} = \phi_0 + (\pi/4N)(i + j - 1)(2N + j - i + 1) \tag{9.22}$$

where ϕ_0 is a constant phase, explicitly given by

$$\phi_0 = -\beta_0 L_\pi/N - \pi/N - (\pi/4)(N - 1) \tag{9.23}$$

Thus, all the output-phase relationships for a practical $N \times N$ MMI coupler can be determined by Eqs. 9.21, 9.22, and 9.23. These relationships permit the straightforward application of MMI couplers in the more complex integrated optical devices

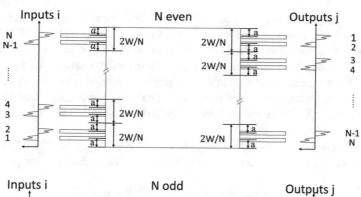

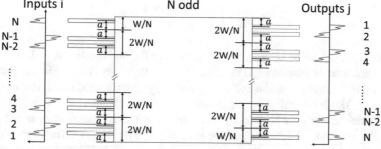

Fig. 9.3 $N \times N$ MMI couplers with even N(top) and odd N(bottom) [Adapted from (Bachmann et al. 1994)]

such as generalized Mach–Zehnder switches and phased arrays for wavelength (de)multiplexing, which will be discussed in Sect. 9.2.5.

Restricted Interference

Thus far, in this discussion, no restrictions have been placed on the modal excitation; in fact, by selecting the modes to be excited leads to restrictive or selective imaging as explained earlier. In practice, this method is achieved, for example, by shortening the imaging distance, and hence the device length.

One possible condition for restricted interference is paired interference in the MMI. In this case, selective mode excitation of certain pairs of waveguides is used to obtain a desirable interference criterion. Then, it can be shown that the phase factor

$$\exp\left(j\left(\frac{m(m+2)\pi}{3L_\pi}\right)L\right)$$

is periodic in $L = pL_\pi$ instead of $3pL_\pi$, if $c_m = 0$ for $m = 2, 5, 8, \ldots$, that is, if only waveguides 0 and 1, and 3 and 4, etc. are excited. When only these particular modes are excited, single images of the input field $\Psi(y, 0)$ are obtained at $L = pL_\pi$. The device length is then three times shorter than that for a general interference device. Similarly, as in the case of the general interference device, N-fold images are found at $L = (p/N)L_\pi$. Thus, twofold images will occur at $pL_\pi/2$ for $p = 1, 3, 5, \ldots$

The selected set of modes needed for paired imaging at $L = pL_\pi$ can be excited by launching a symmetric even input field $\Psi(y, 0)$ at $y = \pm W_e/6$ (see Fig. 9.2). By inspection, this choice of launching position eliminates the excitation of the $m = 2, 5, 8, \ldots$ mode, since the spatial overlap of these modes with the input waveguides is zero. However, the number of input waveguides is limited to two in this case. Such short imaging devices have been made with InP waveguides, e.g., one device was 107 µm long, with 0.9 dB excess loss and -28 dB cross talk, and was designed as a 3 dB coupler (Spiekman et al. 1994).

A second restricted interference condition can also be obtained using symmetric interference that is exciting only even *symmetric* modes. In this case, the phase factor

$$\exp\left(j\left(\frac{m(m+2)\pi}{3L_\pi}\right)L\right)$$

is periodic with $L = p(3L_\pi/4)$, when $c_m = 0$ for the odd modes, $m = 1, 3, 5, \ldots$

In symmetric excitation of MMIs, the input waveguide is always in the center of the multimode waveguide and excited by only a symmetric mode of the input guide. In this case, single images of the input field $\Psi(y, 0)$ are obtained at $L = p((3L_\pi)/4)$, a fact, which can readily be seen by inserting sample even values of m in the phase factor above. Thus, N-fold images are at $L = (p/N)((3L_\pi)/4)$. The device length is, in principle, four times shorter than for a general interference device. N-fold output images of the input field $\Psi(y, 0)$ are symmetrically located transverse to the multimode waveguide axis, that is, the y-axis, with equal spacing W_e/N.

Table 9.1 summarizes the various properties for selective excitation.

Overlap Imaging

In both cases, i.e., general and restricted interference, each of the N-fold output images will have equal output intensity. However, it is possible to achieve nonuniform power splitting by selecting a specific set of positions for access waveguides such that images of unequal intensities are formed (Bachmann et al. 1995). In essence, this approach makes use of the phase of the images and achieves different image inten-

Table 9.1 Summary of the characteristics of the general-, paired-, and symmetric-interference multimode imaging schemes (after Soldano and Pennings 1995)

Interference mechanism	General	Paired	Symmetric
Inputs × Outputs	$N \times N$	$2 \times N$	$1 \times N$
First single image distance	$3L_\pi$	L_π	$3L_\pi/4$
First N-fold image distance	$(3L_\pi)/N$	L_π/N	$(3L_\pi)/4N$
Excitation requirement	None	$c_m = 0$ for $m = 2, 5, 8, \ldots$	$c_m = 0$ for $m = 1, 3, 5, \ldots$
Inputs locations	Arbitrary	$y = \pm W_e/6$	$y = 0$

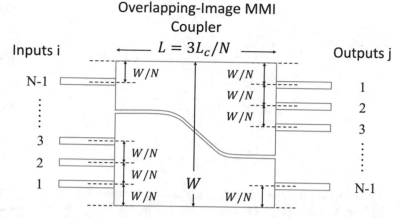

Fig. 9.4 An overlap-imaging MMI coupler

sities by causing spatial overlap of two images with different phases; this approach is thus called overlap imaging. To operate a MMI device, it is necessary to make use of the relative phase in each of the images, which are overlapped. While the output-phase relationship for the general interference case is given in (9.21), the discussion of the phase in images has not been otherwise addressed in this chapter.

The concept of overlap imaging can be illustrated for the case of general interference in the generic MMI device shown in Fig. 9.4. When $\alpha = 0$ or $\alpha = W/N$, output images merge together in pairs, thus an overlap-imaging MMI coupler is formed and the number of images at the output is reduced. For practical, i.e., the short version of the MMI coupler shown in Fig. 9.4, the possible input and output waveguide positions are

$$y_i^{in} = i\frac{W}{N} \tag{9.24}$$

where $i = 0, 1, 2, 3, \ldots, N - 1, N,$

$$y_j^{out} = W - j\frac{W}{N} \tag{9.25}$$

where $j = 0, 1, 2, 3, \ldots, N - 1, N,$ and $i + j$ is an even integer. Channels at the edge of the coupler are denoted by numbers in parenthesis. Formulae for phases and intensities are obtained using (9.26). This equation shows the results obtained from the general interference results by considering the interference between the two outputs that are merged:

$$r_{ij}^2 = \frac{4}{N}\cos^2\left[(N-1)i\frac{\pi}{2N} - b\frac{\pi}{2}\right] \tag{9.26}$$

for

$$\cos\left[(N-j)i\frac{\pi}{2N} - b\frac{\pi}{2}\right] > 0 \qquad (9.27)$$

$$\Phi_{ij} = -(i^2+j^2)\frac{\pi}{4N} + j\frac{\pi}{2} + b\frac{\pi}{2} \qquad (9.28)$$

and for

$$\cos\left[(N-j)i\frac{\pi}{2N} - b\frac{\pi}{2}\right] < 0 \qquad (9.29)$$

$$\Phi_{ij} = -(i^2+j^2)\frac{\pi}{4N} + j\frac{\pi}{2} + b\frac{\pi}{2} + \pi \qquad (9.30)$$

where Φ_{ij} and r_{ij} are the phase and intensity at output channels j for input channel i, respectively, and $b = 0$ for symmetric input, whereas $b = 1$ for antisymmetric input. Output channels j with odd values of $i + j$ always have zero light intensities. Uniform as well as nonuniform power splitting is then possible. For example, nonuniform power splitting into two output channels can have the following possible splitting ratios: $100 : 0, 85 : 15, 72 : 28$, and $50 : 50$.

For overlap imaging, when $y_i^{in} = k(W/2)$, or $y_i^{in} = k(W/3)$, i.e., the input position is at either the center or one- or two-thirds of the MMI width, the output intensity is uniform. These two cases are exactly the same as the paired and symmetric-interference cases discussed earlier. Note, however, by considering them as special cases for general interference achieved through overlap imaging, the output-phase relationship can be easily derived. Moreover, overlap imaging also gives the output results for both the symmetric and antisymmetric input fields, which will be useful for mode-filtering application.

9.2.4 Properties of MMIs

Resolution of an MMI

The spatial resolution of MMIs is determined by several factors. First, if the number of modes is m, the spatial resolution ρ is, by simple Fourier analysis, approximately equal to

$$\rho \simeq \frac{W_e}{m} \qquad (9.31)$$

since the high spatial frequency component is set approximately by the width of nodes in the highest order lateral mode. This equation shows that the higher the mode number, for a fixed W_e, the better the resolution. Thus, since the higher the index contrast, the larger the number of modes supported in the imaging region, and high-index contrast gives high resolution. In addition, the weighing of the amplitude

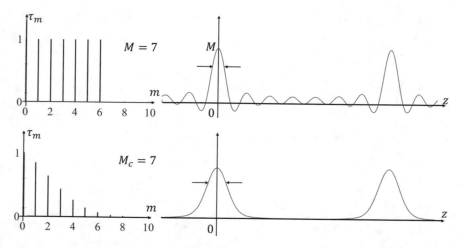

Fig. 9.5 The weighing of the amplitude of each of the spatial harmonics can affect the spatial resolution of MMI

of each of the spatial harmonics, i.e., the modal amplitudes, such as shown in Fig. 9.5, can also affect the spatial resolution.

Tolerance

If w_0 is the focused image width, then by analogy with Gaussian optics (Saleh and Teich 1991), the Rayleigh range of an image is

$$\delta L \simeq \frac{\pi n_f w_0^2}{4\lambda} \tag{9.32}$$

But since $w_0 \propto \rho$, it follows that $w_0 \sim W_e/m$. Thus, generally, the wider the access waveguide, the better the tolerance in length. Also, from the expression for the characteristic image length,

$$L_\pi = 4n_f W_e^2/3\lambda_0 \tag{9.33}$$

the tolerances related to fabrication parameters can be obtained by simple differentiation of (9.16),

$$\frac{\delta L}{L} = 2\frac{\delta W_e}{W_e} \simeq \frac{\delta\lambda}{\lambda} \simeq \frac{\delta n_f}{n_f} \tag{9.34}$$

For example, for a 2×2 MMI coupler with $L \sim 0.25\,\text{cm}$, $\delta L \sim 15\,\mu\text{m}$ (from 9.34), and $W_e \sim 8\,\mu\text{m}$, $\delta W_e \sim 0.25\,\mu\text{m}$ (Pennings et al. 1991). Thus, in this case, the device width and its tolerance are the major fabrication challenges in MMI devices.

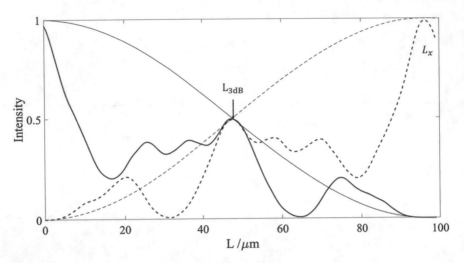

Fig. 9.6 The output intensities in the two output waveguides as a function of length. In this example, a 2 × 2 MMI coupler is used and compared to a 2 × 2 directional coupler

Balance and Phase Balance in MMI Splitters

One of the most important device applications of multimode imaging devices is their use as $1 \times N$ splitters and, in particular, for 1×2 splitting for use in Mach–Zehnder interferometers, etc. In this application, MMI performance is relatively insensitive to dimensional errors since the power in each arm is equally affected by a length or width error, that is, the relative power in each of the arms is balanced. A similar behavior is true for the relative phase balance of each arm. A good comparison of the insensitivity of MMI to dimensional tolerance can be had by comparing their length tolerance with those of a 3 dB directional coupler. In particular, in Fig. 9.6, the output intensities in the two output waveguides for a typical 2 × 2 3 dB MMI coupler and a typical 2 × 2 3 dB directional coupler are plotted as a function of the length of their coupling region (Weinert and Agrawal 1995). At a length of L_3 dB, the two coupler intensity curves versus length intersect with each other with the maximum but opposite slope in this plot, whereas the intensity curves for each of the MMI arms have well-defined overlapping maxima. For 3 dB coupling applications, this behavior makes MMI couplers more fabrication tolerant than directional couplers.

Bandwidth

Equation 9.34 shows for a wavelength deviation of $\delta\lambda_0$ from the design wavelength λ_0 causes an image to be focused at $L + \delta L$ instead of L, the original imaging length. This defocusing sets the wavelength bandwidth of the device for a given level of acceptable loss. To determine this bandwidth, $\Delta\lambda$, we again assume Gaussian beam-like behavior of the image, having a waist, w_0, at the focal point. The relation between transmission loss, T, at length L due to a wavelength deviation of $\delta\lambda$ from the design wavelength can be obtained by calculating the overlap integral of the image spot at length L with that at $L + \delta L$ (Besse et al. 1994). This calculation yields the following equation:

$$T = \sqrt{1 + z^2} / \sqrt{1 + 5z^2 + 4z^4} \tag{9.35}$$

where

$$z \approx \frac{2\lambda\delta L}{\pi n w_0^2} \tag{9.36}$$

Using (9.34), this expression can be recast in terms of the wavelength bandwidth for an acceptable loss, $\Delta\lambda$.

$$2|\Delta\lambda| \simeq Z \cdot \frac{\pi}{4} \cdot aN \cdot \frac{w_0^2}{W^2} \cdot \lambda \tag{9.37}$$

In (9.37), N is the number of images, a is a integer whose value is between 1 and 4, depending on the different imaging mechanisms, W is the MMI width, w_0 is the input-field width. Thus, the optical bandwidth is related to the main design parameters through a loss factor.

9.2.5 Materials

Soldano and Pennings (1995) have discussed and investigated the important advantages of MMI devices. Others, such as Bryngdahl (1973), Ulrich (1975), have shown how relatively complex functionalities can also be obtained for certain materials. In many ways, these advantages derive from the advantages in fabrication and material growths that have occurred over the same period of time.

MMIs have been fabricated from most of the standard integrated optical materials. These include the materials mentioned above, i.e., InP, GaAs-based semiconductors and its alloys. Due to the advantages of low propagation as the index contrast increases, compact devices requiring small bend radii have used, see for example, the reference Levy et al. (1998). Following this same approach, more recently, there have been demonstrations of the MMI devices that have used SOI materials. These devices are described in the applications section.

The interest in constructing large photonics data systems in SOI wafers has encouraged understanding carefully any limitations on using the SOI system. Clearly its high-index contrast is an important advantage for reducing the scale size of optical circuits using such material. In addition, SOI has many of the processing and patterning advantages of commercial silicon, but use of SOI introduces disadvantages as well. One of these problems is the difficulties with tight bends and associated dispersion control. This feature will be discussed in a section below.

Specifically for silicon-on-insulator (SOI) waveguides, high refractive-index-contrast guiding layers and the sharp bend radius (Li and Henry 1996; Okamoto et al. 1992) enable the device size to be reduced by many magnitudes (Soldano et al. 1992; Okamoto et al. 1992; Dragone 1988). However, device design, e.g., MMI couplers, is more challenging. Thus, the Si guiding layers are more prone to phase errors due to high-index contrast.

These phase errors cause, for example, the smaller silicon AWGs to have higher cross talk compared to those of low-index contrast materials. Careful design and fabrication allows reduction of the phase errors and the associated cross talk to an acceptable level (Dragone 1988).

A recent device which shows the importance of solving this problem has been described by Pathak et al. (2013). This devices is a compact ($560 \times 350\mu$m) 12-channel 400 GHz arrayed waveguide-grating wavelength demultiplexers (AWG) in silicon. It has a flattened spectral response due to the use of an MMI. The most important feature is the flattened spectral response, which is a result of an optimized mode shaper, utilizing the multimode interference (MMI) coupler as the input of the AWG. This use of the MMI approach in SOI has led to not only small size but in addition high performance as well. In particular, a critical feature of an AWG is the spectral response of its channel waveguide. In a standard AWG, this response is Gaussian-like (see next chapter). But, in fact, for many applications, a flat spectral response is needed. In response to this need, (Deri et al. 1992) proposed using a MMI (in fact, in this case, an InP AWG) with the AWG to fulfill this requirement.

9.2.6   Applications

In this section, various applications of MMI couplers are summarized. A brief discussion on the principle, performance, and development for each class of application is also presented. Application of MMI-type devices to wavelength routing and splitting is discussed in Chap. 11 and the corresponding references, respectively.

MMI 3 dB Splitters

The most basic class of applications of MMI devices are splitters or couplers. This application is important and makes use of the fact that a well-designed MMI device is efficient and relatively simple to fabricate. The "apex," which is a problematic area in conventional Y-branches, is not used in MMI splitters. Planar optical couplers were the earliest applications for multimode imaging devices (Pennings et al. 1991; Soldano et al. 1992; Jenkins et al. 1992; Heaton et al. 1992).

In addition, 3 dB MMI couplers can be designed to be extremely small. For example, Spiekman et al. (1994) demonstrated a miniaturized $\sim 107\,\mu$m, MMI 3 dB coupler in InP using deep etching. In this 3 dB coupler, restricted interference and tapered input waveguides were used to achieve the short length and low loss. Further length shortening reported by Levy et al. (1998) was realized by using a tapered MMI

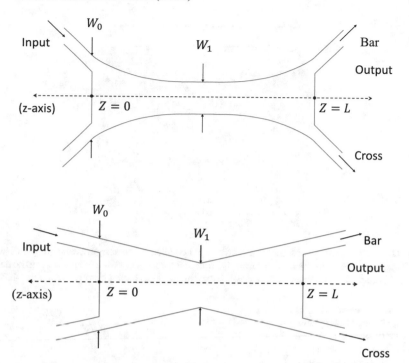

Fig. 9.7 Two comparable 3 dB tapered MMI structures:(top) The parabolically tapered device. (bottom) The linearly tapered device [Adapted from (Levy et al. 1998)]

design (see Fig. 9.7). This tapered 3 dB MMI coupler can be designed to be as short as $\sim 30\,\mu$m. A tapered geometry can also be applied to 4×4 MMI couplers.

More Complex MMI Power Splitters/Couplers

Large N (input/output waveguide number), $N \times N$, or center-fed $1 \times N$ MMI splitters/couplers have been designed and demonstrated. For example, Heaton et al. (1992) demonstrated a 1×20 power splitter ($W = 120\mu$m, $L = 2374\mu$m) with a splitting uniformity of $\pm 4\%$ on GaAs/ AlGaAs, while Rasmussen et al. (1995) fabricated and measured the performance of a 1×64 MMI power splitter, using a planar silica waveguide platform. Smaller values of N have been reported for InP-based systems; thus, a 4×4 device has been reported by Pennings et al. (1993). The performance degradation of large N devices can ultimately be shown to be due to intrinsic phase error in all MMI devices, including conventional MMI devices with deep-etched structures. This important limitation is more severe for $N \times N$ devices than for center-fed $1 \times N$ devices. A detailed discussion of this point is presented in Huang et al. (1998a), Huang et al. (1998b).

MMI devices can also be designed so as to achieve an arbitrary output splitting ratio; this is possible by making simple changes in the devices geometry from those used in conventional structures. One method uses a nonuniform butterfly MMI structure in conjunction with overlap imaging (Besse et al. 1996). As shown in Fig. 9.8, by

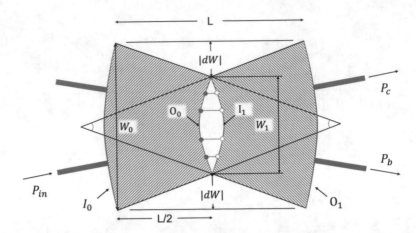

Fig. 9.8 The geometrical design of a butterfly MMI coupler. In the design, as shown by Besse, the self-imaging properties remain with carefully adjusting the length of each section during the design process [Adapted from (Besse et al. 1996)] *Source* https://ieeexplore.ieee.org/document/541220

Fig. 9.9 The geometrical design of a bent MMI coupler. $0° < \theta < 1°$ (figure not to scale). By changing the bending angle between two adjacent MMI 3 dB couplers and, hence, the modal path length at each of the two couplers, a predetermined splitting ratio at the output can be achieved

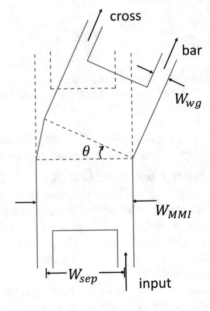

changing the taper of the multimode section, the relative phase differences for imaging points between the two sections can be altered, thus changing the splitting ratio. Also, an arbitrary splitting ratio can be realized by using both the butterfly structure and an overlapping imaging mechanism. A second approach uses a bent MMI structure (Levy et al. 1997). By changing the bending angle between two adjacent MMI 3 dB couplers and, hence, the modal path length at each of the two couplers, a predetermined splitting ratio at the output can be achieved (see Fig. 9.9).

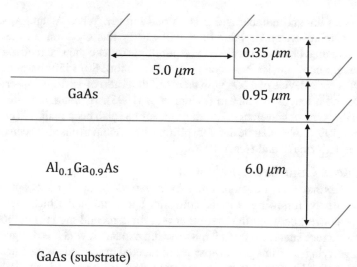

GaAs

5.0 μm

0.35 μm

0.95 μm

Al$_{0.1}$Ga$_{0.9}$As

6.0 μm

GaAs (substrate)

Fig. 9.10 Cross section of the fabricated waveguide geometry

Generalized Mach–Zehnder Interferometer

MMI couplers have been used for the splitting of phases in the propagating waves of Mach–Zehnder interferometers due to their modal uniformity and phase stability. For example, a passive polarization splitter made of two MMIs has been demonstrated by Soldano et al. (1994), yielding strong extinction over 60 nm (Fig. 9.10). Similarly, Bachmann et al. (1993) have shown that electro-optical Mach–Zehnder switches can be made to have high extinction ratios in III–V materials. In all of these designs, the choice of MMI couplers played a crucial role in attaining the wide bandwidth and polarization-independent operation of the integrated interferometer.

As will be explained in more detail in Chap. 13, these Mach–Zehnder devices, with thermo-optical or electro-optical control of the phases in one, have been used as switching elements. For example, ban a silicon MMI switch based on thermo-optic control of modal interference. The device is both compact and a high-performance silicon MMI switch, which uses thermo-optic control of symmetric-interference modes by heating of the mode-peak regions. The direct heater is formed with n-i-n-i-n resistors whose regions are placed at the peak regions of the first two-folded image.

Two Wavelength-Channel (980/1550 nm, 1310/1550 nm) Multi/ Demultiplexer

Since the characteristic beat length L_π in (9.6) for an MMI coupler is related to wavelength, an MMI coupler can separate two very different wavelengths, λ_1 and λ_2, if it is a bar coupler for one wavelength and a cross coupler for the other wavelength. The length L_c of such a demultiplexing coupler has to satisfy the following relation:

$$L_c = p(3L_\pi) = (p+q)(3L_\pi) \tag{9.38}$$

where p is an integral number and q is an odd integer. When λ_1 and λ_2 are well separated, a relatively small value of p will satisfy this equation. In this case, a very compact dual channel wavelength (de)multiplexer can then be realized using a MMI device. Such a device has been demonstrated for 980/1550 nm, pump/signal wavelengths for a device of $Si\,O_2$ with an extinction ratio of 18 dB, an insertion loss of 0.5 dB, and a length of 458 μm (Paiam et al. 1995). A polarization-insensitive design on this type of wavelength (de)multiplexer has also been realized (Paiam and MacDonald 1998). A similar design for splitting 1.3/1.55 μm has also been designed on a silica platform (Li and Henry 1996).

Other More Complex Applications

MMI couplers have been demonstrated for a variety of other novel applications. These include the following: a mode converter and combiner, which uses overlap imaging to convert between the first-order and fundamental mode (Leuthold et al. 1996); an efficient integrated optical mode-width expander, which is designed based on the magnifying, self-imaging properties of tapered multimode waveguides. MMI mirrors have also been used to provide a low loss, uniform-splitting-ratio stability, and compact size for a square-ring-laser diode (Kim et al. 1997). MMIs have also been used to broaden the spectral range of the front end of a coherent receiver (Deri et al. 1992). In this case, the wavelength-insensitive behavior of the 3 dB coupler in combination with the compact design of the photodetectors resulted in a broad spectral operating range for the receiver.

Specifically for silicon-on-insulator (SOI) waveguides, high refractive-index-contrast guiding layers and the sharp bend radius (Li and Henry 1996; Okamoto et al. 1992) enable the device size to be reduced by many magnitude (Soldano et al. 1992; Okamoto et al. 1992; Dragone 1988). However, device design, itself, e.g., in the MMI couplers, is more challenging. Thus, the Si guiding layers are more prone to phase errors due to their very high-index contrast. These phase errors cause, for example, the smaller silicon AWGs to have higher cross talk compared to those of low-index contrast materials. Careful design and fabrication allow reduction of the phase errors and the associated cross talk to an acceptable level (Dragone 1988).

A recent device, which shows the importance of solving this problem, has been described by Pathak et al. (2013). This device is a compact (560×350 μm) 12-channel 400 GHz arrayed waveguide-grating (AWG) wavelength demultiplexer in silicon. This most important feature is a result of an optimized mode shaper, utilizing the multimode interference (MMI) coupler as the input of the AWG. This use of the MMI approach in SOI has led to not only small size but in other forms of high performances as well, such as low insertion loss, etc. in comparison to that seen in other standard AWGs. In particular, a critical feature of an AWG is the spectral response of its channel waveguide. In a standard AWG, this response is Gaussian-like (see next chapter). But, in fact, for many applications a flat spectral response is needed. In response to this need, (Deri et al. 1992) proposed using a MMI (in fact, in this case, an InP AWG) with the AWG to fulfill this requirement.

Other approaches to deal with high lateral refractive index contrast have also been presented. For example, nanofabrication in the lateral cladding region, e.g., fabrication of a subwavelength grating (SWG), of a MMI has been shown to lower index contrast. This approach reduces the mode phase error and is also a single (etch) process. Using a periodic lateral SWG, a 2×4 MMI one group (Ortega-Monux et al. 2011) designed and fabricated this and used it in a coherent optical receiver. Compared to MMI with a homogenous lateral cladding, this approach increases the receiver bandwidth from 36 to 60 nm.

9.3 Star Couplers

Star couplers evenly distribute light from any one input guide to all output guides; thus, in an $N \times N$ device, any of N-independent input ports may couple light to any of N output ports. The device uses guided-wave optics for the I/O ports and free-space optics in the power-distribution region. The central problem in designing such a device is to have the power distribution be truly independent of which of these ports are chosen for the input/output waveguide. Notice one important point regarding star couplers: since the device works by power division, one encounters an automatic, intrinsic "loss" of $1/N$ when using the device. Any signal in an input point is divided by $1/N$. *Note that this loss is not encountered in MMI devices.* Nonetheless, because of their simplicity, star couplers are extremely important in integrated optics. Recently, high-quality integrated star couplers have been developed for communication applications (Cao et al. 2005). These devices use coupled single-mode waveguides and careful free-space-optics design.

A drawing of a complete star coupler is shown in Fig. 9.11. The figure shows the wider spacing of the input ports at the edge of the coupler layout; this feature is needed for fiber pigtailing. To reduce the waveguide spacing requires bends in the coupler waveguides. Also note the dummy waveguides on the coupler; their function will be discussed below. A sketch of the central region of the optical layout used for the design of the star couplers is given in Figs. 9.13 and 9.14, along with a graphical definition of the symbols used (Dragone 1988). Note that the input and output waveguides are spaced by a distance, a, from each other and the input and output arrays have angular width of a and θ, respectively. The imaging of an input waveguide is carried out by the beam undergoing Fraunhofer diffraction, which is equivalent to a spatial Fourier transform of the dominant mode, $\Psi(u)$, emitted from each input port,

$$\Phi(w) = \frac{1}{2} \int_{-\infty}^{\infty} \Psi(u) e^{jwu} du \tag{9.39}$$

where the wavevector, w, normalized by the interwaveguide spacing, a, required for diffraction of the waveguide into an angle, θ', is

$$w = \frac{\pi a \sin \theta'}{\lambda} \tag{9.40}$$

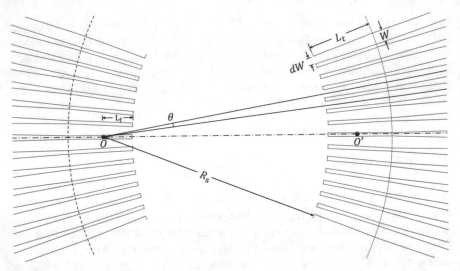

Fig. 9.11 A drawing of a complete star coupler. The figure shows the wider spacing of the input ports at the edge of the coupler layout; this feature is needed for fiber pigtailing [Adapted from (Okamoto et al. 1992)]

where θ' is defined as the angle from the axis of the device. The normalized distance u is given by

$$u \approx \frac{x}{2a} \tag{9.41}$$

where x is the distance from the center of the waveguide receiver or radiating array.

Now assuming that the mode is approximately Gaussian, then the Fourier transform of such a Gaussian-distribution feature is also a Gaussian. Thus, without further modification, such a Gaussian input source will not produce the desired uniform "rectangular" illumination of the output waveguides, that is, one seeks a distribution such that $\Phi(w) \sim rect(w2w_\alpha)$, where $w_\alpha = (\pi a \sin \alpha)/\lambda$.

To obtain this rectangular illumination distribution, it is necessary to have a near-field distribution radiating from each of the input array wavegudies with sidelobes on the large central lobe. The Fourier transform of this sinc-like function is a rectangular function. The sidelobes are generated by an input-waveguide coupling light into its neighboring guides, which then radiates into free space along with the light in the original input guide. The waveguide coupling is achieved by fanning out the waveguides so as to achieve a closer spacing over the length of the waveguide leading into the radiating end. Because of the importance of this coupling for obtaining a uniform output aperture, it is essential to adjust carefully the interwaveguide coupling. For the edge input guides, dummy waveguides are inserted to make this coupling close to that in the central part of the array (Okamoto et al. 1992).

As an example of this approach, consider a beam-propagation-method calculation of a well-designed 8×8 star coupler, which in this case uses an input from

Fig. 9.12 A sketch of the central region of the star couplers. Two linear arrays are separated by free-space region. Array elements are located on two circles centered at O and $0'$ with coordinates θ, θ' specified by angular displacements from axis. Marginal elements are displaced from axis by α [Adapted from (Dragone 1988)]

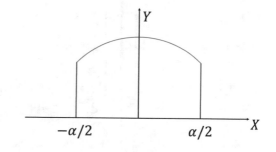

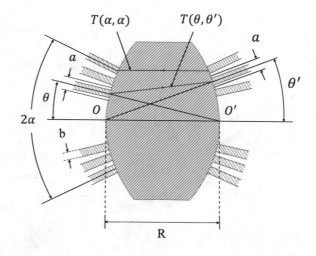

only the center waveguide of the radiating array; see Fig. 9.13. Observe that the sine radiation pattern evolves into a rectangular distribution as the light propagates toward the output guides (Okamoto et al. 1992). As an example of one well-defined device, Fig. 9.14 shows schematic of 144 × 144 wavelength-insensitive star coupler fabricated in SiO_2/Si. It had an excess loss of 2 dB, with a splitting uniformity of $\sigma = 1.47$ dB. The device had 64 input guides, a taper length of $L_T = 2mm$, and a radius of $R_s = 2.1mm$, $L_f = 0.1mm$, $\theta = 0.2°$ (Okamoto et al. 1992). The waveguide coupler is shown in Fig. 9.12. Its waveguides are arranged on circular arcs to enable good matching of the waveguide supermode, which has a curved wavefront with the output guide. Thus, each guide is approximately one focal distance from the other side.

Understanding of these devices thus requires examination of the optical physics of coupled waveguides. This discussion is provided in the preceding chapter on couplers. Recall that a waveguide array has a grating-like angular distribution, such as shown in Fig. 9.15, for infinitely narrow sources. This grating distribution can be broken into zones, called Brillouin zones in analogy with the Bloch function

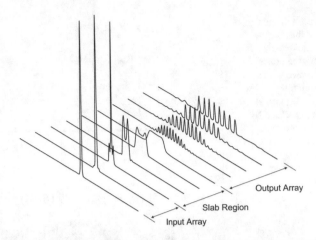

Fig. 9.13 Waveform transients of optical power obtained by Beam Propagation Method simulation

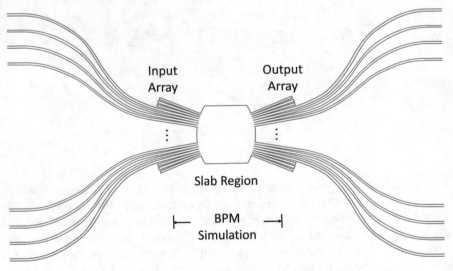

Fig. 9.14 Schematic configuration of 144 × 144 star coupler [Adapted from (Okamoto et al. 1992)]

encountered in solid-state physics. For each supermode excited, the maximum has a different angular position within each of the Brillouin zones. In general,

$$|a \sin \theta_m - m\lambda| < \frac{\lambda}{2} \tag{9.42}$$

where θ_m is the angle of the zone edge. Thus, for the $m = 0$ zone,

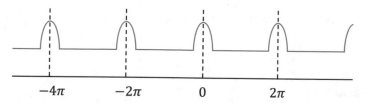

Fig. 9.15 A waveguide array has a grating-like angular distribution

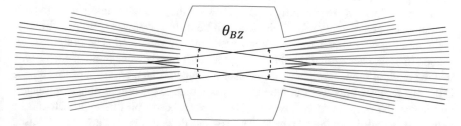

Fig. 9.16 Schematic configuration of a star coupler has a curved radius of R such that $\frac{Na}{R} \approx \theta_{BZ}$

$$ka \sin \theta_0 < \pi \tag{9.43}$$

The angle subtended by this is $2\theta_0 \equiv \theta_{BZ} \approx \lambda/aN_s$, where we have now explicitly specified the dependence on substrate effective index, N_s.

The star coupler shown in Fig. 9.16 has a curved radius of R such that

$$\frac{Na}{R} \approx \theta_{BZ} \tag{9.44}$$

where N is the number of waveguides. Typically, the angular aperture of the set of output guides is somewhat smaller than θ_{BZ}.

The curved arrangement of the output guides matches the curved diffracted wavefront from any input guide in the far field, thus ensuring that the $1/N$-divided guided waves have a uniform phase at the entrance to the output guides. Note that typically, there is a small, $1 \sim 5\%$, displacement of the phase center from the edge of the input guides.

One of the most important considerations in designing such a device is, in fact, the taper of the waveguides as they approach the free-space region. One of the reasons that this region is important is that, first, the guides must couple together to form the ideal radiation supermode pattern. This process makes the illumination uniform across the receiving array. Second, the waveguides must taper properly in the process of merging. If the tapering process is not adiabatic, higher order local modes will be excited. These higher order modes cause radiation into the second and third Brillouin zones, i.e., beyond the angular acceptance of the receiving array. Finally, the taper design must be adjusted so that it is as insensitive to wavelength as possible. In one case, the star coupler was designed to work equally well at 1.3 and 1.5μm (Okamoto et al. 1992).

9.4 Summary

This chapter has presented two important integrated devices, which operate by modal excitation and which are generally oriented toward the goal of imaging through periodic interference of modes. Both devices can be used for power splitting, although many other applications have been conceived and demonstrated. The first, the multimode imager, uses excitation of a finite width multimode waveguide in such a way that the excited modes form a series of useful images. The advantage of this device is that the transfer of power from the input waveguide to the output image or images, and hence the output can be lossless. The devices based on this approach can be extremely small, i.e., tens of $(\mu m)^2$. The second device, the star coupler, is based on excitation of an effectively infinite series of "free-space" modes. In this case, a uniform far-field pattern of the modes is employed across the output waveguides to create a uniform power splitter. This device design has intrinsic loss. However, the simplicity and robust nature of its design makes it a commonly used approach to power distribution in PICs.

Problems

1. Note: use the following parameters for this problem:
   $n_f = 3.4$;
   $\lambda_0 = 1.55$ μm.

   (a) Calculate the spatial resolution possible for 20 μm wide MMI which supports 30 modes.
   (b) Alternatively, if the input port is roughly 6 μm wide, how many modes are needed for a 30 μm guide for good resolution?
   (c) It is possible to make a 200 μm long MMI with a 20 μm width. What is the needed δL for a 0.1 μm lithography patterning system?

2. What is the first focal point for a 1×2 MMI splitter? Use the following parameters:
   $n_f = 3.4$
   $\lambda_0 = 1.55$ μm;
   $w_{eff} = 15$ μm.

3. If 100 mW power enters a high-quality star coupler. How much power leaves on each port?

4. We make an MMI using a buried-waveguide geometry. In this case, $n_c \approx n_s$. We know free-space wavelength $\lambda = 1.55\,\mu m$, $n_s = 3.35$ and $n_f = 3.40$.

   (a) Design a single TE-mode buried-waveguide structure, in which d and w are such that they are 20% below the $m = 1$ cutoff, i.e., $w = 0.8 w_{cutoff}^{m=1}$, etc.

   (b) Design the shortest 1×2 MMI using general imaging with $m_{max} = 25$, with d the same as in (a) above. Specifically find

   (1) The parameter L_π and width W_{fs}, assume that $W_e \approx W_{fs}$.

   (2) The minimum length of the imaging region needed for 1×2 imaging.

References

Bachmann, M., Besse, P., & Melchior, H. (1995). Overlapping-image multimode interference couplers with a reduced number of self-images for uniform and nonuniform power splitting. *Applied Optics*, *34*(30), 6898–6910.

Bachmann, M., Besse, P. A., & Melchior, H. (1994). General self-imaging properties in n× n multimode interference couplers including phase relations. *Applied Optics*, *33*(18), 3905–3911.

Bachmann, M., Smit, M., Besse, P., Melchior, L., et al. (1993). Polarization-insensitive low-voltage optical waveguide switch using InGaAsP/InP four-port mach-zehnder interferometer. In *Optical Fiber Communication Conference*, page TuH3. Washington, D.C.: Optical Society of America.

Besse, P. A., Bachmann, M., Melchior, H., Soldano, L. B., & Smit, M. K. (1994). Optical bandwidth and fabrication tolerances of multimode interference couplers. *Journal of Lightwave Technology*, *12*(6), 1004–1009.

Besse, P. A., Gini, E., Bachmann, M., & Melchior, H. (1996). New 2/spl times/2 and 1/spl times/3 multimode interference couplers with free selection of power splitting ratios. *Journal of Lightwave Technology*, *14*(10), 2286–2293.

Bryngdahl, O. (1973). Image formation using self-imaging techniques. *JOSA*, *63*(4), 416–419.

Cao, G., Dai, L., Wang, Y., Jiang, J., Yang, H., & Zhang, F. (2005). Compact integrated star coupler on silicon-on-insulator. *IEEE Photonics Technology Letters*, *17*(12), 2616–2618.

Deri, R., Pennings, E., Scherer, A., Gozdz, A., Caneau, C., Andreadakis, N., et al. (1992). Ultra-compact monolithic integration of balanced, polarization diversity photodetectors for coherent lightwave receivers. *IEEE Photonics Technology Letters*, *4*(11), 1238–1240.

Dragone, C. (1988). Efficient n× n star coupler based on fourier optics. *Electronics Letters*, *24*(15), 942–944.

Dragone, C. (1990). Optimum design of a planar array of tapered waveguides. *JOSA A*, *7*(11), 2081–2093.

Heaton, J., Jenkins, R., Wight, D., Parker, J., Birbeck, J., & Hilton, K. (1992). Novel 1-to-n way integrated optical beam splitters using symmetric mode mixing in gaas/algaas multimode waveguides. *Applied Physics Letters*, *61*(15), 1754–1756.

Huang, J., Hu, M., Fujita, J., Scarmozzino, R., & Osgood, R. (1998a). High-performance metal-clad multimode interference devices for low-index-contrast material systems. *IEEE Photonics Technology Letters*, *10*(4), 561–563.

Huang, J., Scarmozzino, R., & Osgood, R. (1998b). A new design approach to large input/output number multimode interference couplers and its application to low-crosstalk wdm routers. *IEEE Photonics Technology Letters*, *10*(9), 1292–1294.

Jenkins, R., Devereux, R., & Heaton, J. (1992). Waveguide beam splitters and recombiners based on multimode propagation phenomena. *Optics Letters*, *17*(14), 991–993.

Kim, H., Kwon, Y., & Hong, S. (1997). Square ring laser diode with mmi coupler cavity. *IEEE Photonics Technology Letters*, *9*(5), 584–586.

Leuthold, J., Hess, R., Eckner, J., Besse, P., & Melchior, H. (1996). Spatial mode filters realized with multimode interference couplers. *Optics Letters, 21*(11), 836–838.

Levy, D. S., Li, Y., Scarmozzino, R., & Osgood, R. (1997). A multimode interference-based variable power splitter in gaas-algaas. *IEEE Photonics Technology Letters, 9*(10), 1373–1375.

Levy, D. S., Scarmozzino, R., Li, Y. M., & Osgood, R. M. (1998). A new design for ultracompact multimode interference-based 2 × 2 couplers. *IEEE Photonics Technology Letters, 10*(1), 96–98.

Li, Y. P., & Henry, C. (1996). Silica-based optical integrated circuits. *IEE Proceedings-Optoelectronics, 143*(5), 263–280.

Okamoto, K., Okazaki, H., Ohmori, Y., & Kato, K. (1992). Fabrication of large scale integrated-optic n* n star couplers. *IEEE Photonics Technology Letters, 4*(9), 1032–1035.

Ortega-Monux, A., Zavargo-Peche, L., Maese-Novo, A., Molina-Fernández, I., Halir, R., Wanguemert-Perez, J., et al. (2011). High-performance multimode interference coupler in silicon waveguides with subwavelength structures. *IEEE Photonics Technology Letters, 23*(19), 1406–1408.

Paiam, M., Janz, C., MacDonald, R., & Broughton, J. (1995). Compact planar 980/1550-nm wavelength multi/demultiplexer based on multimode interference. *IEEE Photonics Technology Letters, 7*(10), 1180–1182.

Paiam, M., & MacDonald, R. (1997). Polarisation-insensitive 980/1550 nm wavelength (de) multiplexer using mmi couplers. *Electronics Letters, 33*(14), 1219–1220.

Paiam, M., & MacDonald, R. (1998). A 12-channel phased-array wavelength multiplexer with multimode interference couplers. *IEEE Photonics Technology Letters, 10*(2), 241–243.

Pathak, S., Vanslembrouck, M., Dumon, P., Van Thourhout, D., & Bogaerts, W. (2013). Optimized silicon awg with flattened spectral response using an mmi aperture. *Journal of Lightwave Technology, 31*(1), 87–93.

Pennings, E., Deri, R., Bhat, R., Hayes, T., & Andreadakis, N. (1993). Ultracompact, all-passive optical 90 degrees-hybrid on inp using self-imaging. *IEEE Photonics Technology Letters, 5*(6), 701–703.

Pennings, E., Deri, R., Scherer, A., Bhat, R., Hayes, T., Andreadakis, N., et al. (1991). Ultracompact, low-loss directional couplers on inp based on self-imaging by multimode interference. *Applied Physics Letters, 59*(16), 1926–1928.

Rasmussen, T., Rasmussen, J. K., & Povlsen, J. H. (1995). Design and performance evaluation of 1-by-64 multimode interference power splitter for optical communications. *Journal of Lightwave Technology, 13*(10), 2069–2074.

Saleh, B. E., & Teich, M. C. (1991). *Fundamentals of photonics*. Hoboken: Wiley

Smit, M. K., & Van Dam, C. (1996). Phasar-based wdm-devices: Principles, design and applications. *IEEE Journal of Selected Topics in Quantum Electronics, 2*(2), 236–250.

Soldano, L., De Vreede, A., Smit, M., Verbeek, B., Metaal, E., & Green, F. (1994). Mach-zehnder interferometer polarization splitter in ingaasp/inp. *IEEE Photonics Technology Letters, 6*(3), 402–405.

Soldano, L. B., & Pennings, E. C. (1995). Optical multi-mode interference devices based on self-imaging: Principles and applications. *Journal of Lightwave Technology, 13*(4), 615–627.

Soldano, L. B., Veerman, F. B., Smit, M. K., Verbeek, B. H., Dubost, A. H., & Pennings, E. C. (1992). Planar monomode optical couplers based on multimode interference effects. *Journal of Lightwave Technology, 10*(12), 1843–1850.

Spiekman, L., Oei, Y., Metaal, E., Green, F., Moerman, I., & Smit, M. (1994). Extremely small multimode interference couplers and ultrashort bends on inp by deep etching. *IEEE Photonics Technology Letters, 6*(8), 1008–1010.

Ulrich, R. (1975). Image formation by phase coincidences in optical waveguides. *Optics Communications, 13*(3), 259–264.

Ulrich, R., & Kamiya, T. (1978). Resolution of self-images in planar optical waveguides. *JOSA, 68*(5), 583–592.

Weinert, C., & Agrawal, N. (1995). Three-dimensional finite difference simulation of coupling behavior and loss in multimode interference devices. *IEEE Photonics Technology Letters, 7*(5), 529–531.

Zucker, J., Jones, K., Chiu, T., Tell, B., & Brown-Goebeler, K. (1992). Strained quantum wells for polarization-independent electrooptic waveguide switches. *Journal of Lightwave Technology, 10*(12), 1926–1930.

Chapter 10
Diffraction Gratings

Abstract A diffraction grating is an increasingly important component in integrated optics. They are used in integrated optics for such applications as in and out coupling for integrated photonics chips, for on-chip multiplexing and demultiplexing, on-chip reflecting elements, and wavelength filters. Analysis of the propagating wave is described and carried out efficiently by use of the coupled wave equations.

10.1 Introduction

Diffraction gratings are important components in several types of devices used in PICs (Yariv 1997) including those shown in Fig. 10.1: distributed feedback (DFB) lasers, Bragg filters (demultiplexing), channel-dropping filters, DFB reflectors, and reflection (spectrometer) grating demultiplexers. We will discuss these applications in later chapters. Here, we will concentrate on presenting the fundamentals of gratings used as general optical elements.

This chapter will first use coupled-mode theory to analyze several typical configurations for gratings in integrated optical devices. We then consider computation of the coupling coefficients for gratings having different surface-relief structures. Finally, we will discuss the use of gratings as reflectors or feedback elements.

10.2 Collinear Coupling

When a grating is patterned on waveguides or waveguiding lasers, the dielectric perturbation of the grating causes coupling of the waveguide modes. The "generic" configurations for such applications—discussed in Chap. 5—are shown in Fig. 10.2. In the upper panel, the waveguide modes of the grating coupler travel in opposite directions within the waveguide. Typically, the two modes which interact most

© Springer Nature Switzerland AG 2021
R. Osgood jr. and X. Meng, *Principles of Photonic Integrated Circuits*,
Graduate Texts in Physics,
https://doi.org/10.1007/978-3-030-65193-0_10

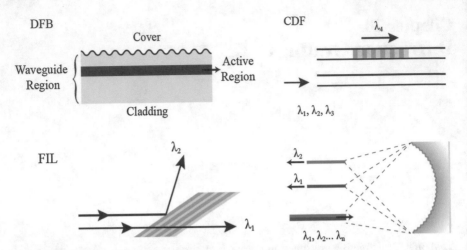

Fig. 10.1 The sketches of distributed feedback (DFB) lasers (top left), Bragg filters (demulti-plexing) (bottom left), channel-dropping filters (top right), and reflection (spectrometer) grating demultiplexers (bottom right)

strongly have the same transverse profile. Only their propagation constants are oppo-site in sign. In the second panel, the guided waves travel in the same direction. Here, the propagation constants of the two modes are generally different, and the grating enables coupling of dissimilar nodes.

As discussed in Chap. 5, the grating allows efficient phase matching between the guided modes for both codirectional and contradirectional coupling. Thus, if the dominant Fourier component of the grating is characterized by a spatial wavenumber of $K = 2\pi / \Gamma$, then the phase-matching conditions for each of the two cases shown in Fig. 10.2 are

$$\vec{\beta}_1 = \vec{\beta}_2 + \vec{K} \tag{10.1}$$

where $\vec{\beta}_{1,2}$ is the propagation constant of the 1, 2 mode, where only first-order coupling is assumed, i.e., if the order is q then $\vec{K}$ is replaced by $q\vec{K}$.

This phase-matching condition causes two modes to dominate the equations. In addition, it also allows us to neglect any coupling to the radiative modes of the guide. Recall that the degree of resonance of a grating is conveniently quantified by the detuning constant, given by

$$\delta = \frac{\beta_1 - (\beta_2 + K)}{2} \tag{10.2}$$

where β_1, β_2 are positive if they are oriented in the same direction.

Consider now the solution to the coupled-mode equations for the two cases of contradirectional and codirectional traveling waves.

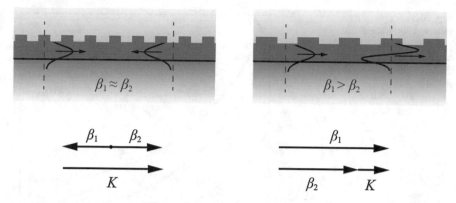

Fig. 10.2 The generic configurations for a patterned grating causing coupling of the waveguide modes. The left panel shows the situation when the waveguide modes of the grating coupler travel in opposite directions within the waveguide. The right panel shows the situation when the guided waves travel in the same direction

10.2.1 Contradirectional Waves of a Surface Grating

Contradirectional coupling allows a surface grating to be used as a wavelength-selective reflecting element for integrated optics. As mentioned above, typically in this case $\beta_1 = -\beta_2$. We seek the amplitudes for a_1' and a_2'; recall their definition as given in (5.26) relative a_1 and a_2. The boundary conditions are that for a grating of length L, $a_1(0) = 1$ at $z = 0$, and $a_2(L) = 0$ at $z = L$, and thus only one wave is incident on the grating. In this case, the solutions to the coupled-mode equations are

$$a_1'(z) = e^{-j\delta z} \left(\frac{\beta_d \cosh[\beta_d(z - L)] + j\delta \sinh[\beta_d(z - L)]}{\beta_d \cosh \beta_d L - j\delta \sinh \beta_d L} \right) \tag{10.3}$$

$$a_2'(z) = e^{j\delta z} \left(\frac{j\kappa \sinh[\beta_d(z - L)]}{\beta_d \cosh \beta_d L - j\delta \sinh \beta_d L} \right) \tag{10.4}$$

where $\beta_d = \sqrt{\kappa^2 - \delta^2}$. These amplitudes can be manipulated to give the z-dependent transmission and reflection coefficients of the grating

$$T = \left(\frac{a_1(z)}{a_1(0)} \right)^2 = \frac{1 + (\kappa/\beta_d)^2 \sinh^2[\beta_d(z - L)]}{1 + (\kappa/\beta_d)^2 \sinh^2(\beta_d L)} \tag{10.5}$$

$$R = \left(\frac{a_2(z)}{a_2(0)} \right)^2 = \frac{(\kappa/\beta_d)^2 \sinh^2[\beta_d(z - L)]}{1 + (\kappa/\beta_d)^2 \sinh^2(\beta_d L)} \tag{10.6}$$

Figure 10.3 shows a plot of the reflected and transmitted power versus αL. Note that the reflectivity at $z = 0$ is given by

Fig. 10.3 A plot of the reflected and transmitted power versus αL. The solid and dashed lines show two different situations as shown in the legends

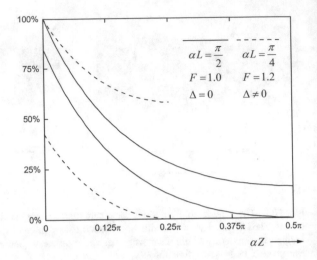

$$R = \frac{(\kappa/\beta_d)^2 \sinh^2 \beta_d L}{1 + (\kappa/\beta_d)^2 \sinh^2 \beta_d L} \tag{10.7}$$

Similarly, the transmission of the grating is given by

$$T = 1 - R \tag{10.8}$$

$$= \frac{1 - \tanh^2 \beta_d L}{1 + (\delta^2/\beta_d^2) \tanh^2 \beta_d L} \tag{10.9}$$

$$\approx 4 \frac{\beta_d^2}{|\kappa|^2} \exp(-2\beta_d L) \tag{10.10}$$

Returning to reflectivity, it is useful to plot this quantity versus the detuning, δ, assuming κ is approximately constant over this range, i.e., $d\kappa/d\delta \approx 0$. This assumption is reasonable for frequencies near the relatively narrow stop band. This plot is shown in Fig. 10.4. Notice that the distribution has a flat maximum for $\delta < \kappa$, and that there are specific zero points for the distribution. These zero points are a result of the fact that the backward wave has a set of nodal points within the grating, including the two at either end. In effect, the grating has become a resonant cavity very much like a lumped Fabry–Perot cavity, except in the case of the grating. The reflection is distributed "within" the cavity. Note that as the detuning is increased, the width of the grating resonances in Fig. 10.4 increases due to the fact that the distributed grating reflectivity, and hence its "Q" also decreases with detuning. Notice also that outside of the stop bands, the propagation constant is complex and not purely imaginary, as

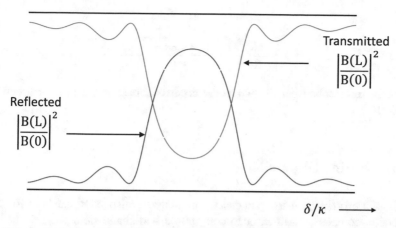

Fig. 10.4 A plot of the reflected and transmitted power versus the detuning, δ assuming κ is approximately constant over this range

it is within the stop band. A complex propagation constant allows for the existence of propagating modes.

10.2.2 Codirectional Coupling

For coupling between modes propagating in the same direction, such as needed for a mode converter, $\beta_1 > 0$ and $\beta_2 > 0$. In this case, a longer period grating than considered above is used to phase match the two modes, since the phase-matching criteria are given by the *difference* of the two propagation vectors.

The amplitudes for $a'_{1,2}$ are then

$$a'_1(z) = \exp(-j\delta z)\left[\cos\sqrt{\kappa^2 + \delta^2}\,z + \frac{j\delta}{\sqrt{\kappa^2 + \delta^2}}\sin(\sqrt{\kappa^2 + \delta^2}\,z)\right] \quad (10.11)$$

$$a'_2(z) = \exp(-j\delta z) - \frac{j\delta}{\sqrt{\kappa^2 + \delta^2}}\sin(\sqrt{\kappa^2 + \delta^2}\,z) \quad (10.12)$$

These solutions show that the equation is identical to that of a directional coupler except that the definition of δ now includes the grating period K as shown in (10.2). As a result, the maximum power transfer to mode 2, P_2^{max}, is given by

$$P_2^{max} = \frac{1}{1 + (\delta^2/\kappa^2)} \quad (10.13)$$

and

$$\frac{|a_2(L)|^2}{|a_1(0)|^2} = \frac{\sin^2(\sqrt{\kappa^2 + \delta^2}L)}{1 + (\delta^2/\kappa^2)} \tag{10.14}$$

that is, the transfer between the two modes is periodic in length just as in a directional coupler!

10.3   Grating Dispersion

Gratings are often used to compensate for material dispersion, such as for non-dispersion-compensated fiber, or to compress a wide bandwidth pulse. Thus, it is important to consider explicitly the conditions for grating dispersion.

For a Bragg grating at center frequency, ω_0,

$$\beta(\omega_0) = \frac{\pi}{\Gamma} \ (or \ K) \tag{10.15}$$

Thus, if we expand the propagation constant around this frequency, we obtain

$$\beta = \beta(\omega_0) + \frac{d\beta}{d\omega}(\omega - \omega_0) \tag{10.16}$$

but $d\beta/d\omega = \nu_g$, and thus if $\beta(\omega) - \beta(\omega_0) = \delta$

$$\delta = \frac{\omega - \omega_0}{\nu_g} \tag{10.17}$$

A plot of a typical grating dispersion is shown in Fig. 10.5.

The presence of a grating also affects the propagation of light in a more profound manner. This phenomenon has already been shown in the case of coupled waveguides, where it was shown that in the case of near-synchronous waveguides, an optical or photonic "bandgap" was excited. In the region of optical frequencies, light does not propagate. Similarly, for the case of grating near the zero-detuning point, the frequencies of the two normal modes of the grating are different. This behavior is illustrated in Fig. 10.6, which shows a plot of β versus ω/c for the two normal modes, and which are approximately the forward and backward propagating waves. The two propagation constants are

$$\beta_{1,2} = \pm\sqrt{\delta^2 - |\kappa|^2} \tag{10.18}$$

or at $\delta = 0$,

$$\beta_1(\approx \beta_2) = \pm j\kappa \tag{10.19}$$

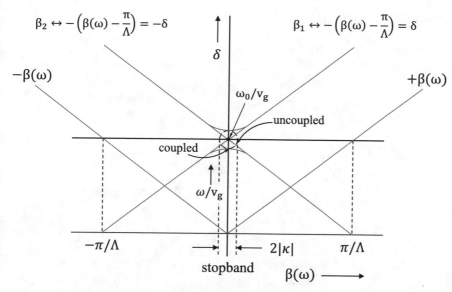

$$\beta_2 \leftrightarrow -\left(\beta(\omega) - \frac{\pi}{\Lambda}\right) = -\delta \qquad \qquad \beta_1 \leftrightarrow -\left(\beta(\omega) - \frac{\pi}{\Lambda}\right) = \delta$$

Fig. 10.5 A plot of grating dispersion from (10.15) and (10.16) with unperturbed propagation constant $\beta(\omega)$ proportional to ω

Fig. 10.6 A plot of β versus ω/c for the two normal modes when grating is near the zero-detuning point. The frequencies of the two normal modes of the grating are different

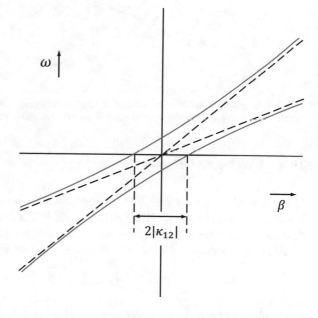

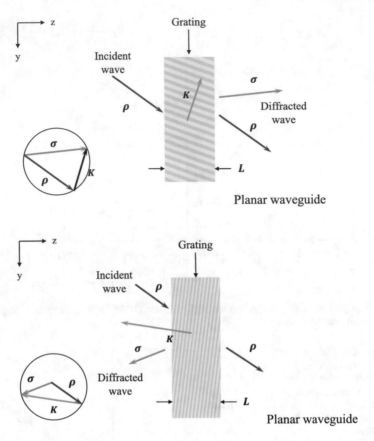

Fig. 10.7 A physical picture of coplanar coupling for the Bragg condition. The upper panel shows the transmission type while the bottom panel shows the reflection type

10.4   Coplanar Coupling

In many applications, the direction of propagation of the grating is not normal to the grating grooves. However, if the grating is within the plane, it is termed "coplanar coupling"; a physical picture of this case for the Bragg condition is shown in Fig. 10.7.

The Bragg condition for the lightwave now must be stated in vector form, i.e.,

$$\vec{\beta}_d = \vec{\beta}_i + q\vec{K} \tag{10.20}$$

where q, an integer, is the "order" for the diffracted ray, $q = 0, \pm 1, \pm 2$; $\vec{K}$ is the grating period $|\vec{K}| = 2\pi/\Gamma$; and $\vec{\beta}_d$, $\vec{\beta}_i$ is the diffracted, incident wavevector with a magnitude

$$|\vec{\beta}_d| = |\vec{\beta}_i| = \frac{2N_{\text{eff}}\pi}{\lambda}, \tag{10.21}$$

see Fig. 10.7 for an example. In this case, if the projection of a unit vector of the grating vector, $|\vec{K}|$, with (10.20) is computed, i.e., the dot product, then

$$\frac{\vec{K}}{|\vec{K}|} \cdot \left(\vec{\beta}_b = \vec{\beta}_a + q\vec{K}\right) \tag{10.22}$$

$$\frac{2\pi N_{\text{eff}}}{\lambda} \cos(90 - \theta) = -\frac{2\pi}{\lambda} N_{\text{eff}} \cos(90 - \theta) + a\frac{2\pi}{\Gamma} \tag{10.23}$$

which can be written as

$$2\Gamma \sin\theta = \frac{\lambda}{N_{\text{eff}}} q \tag{10.24}$$

More generally, it can be shown that for the case of coplanar coupling, it is again possible to write a set of coupled-mode equations. However, in this case, the equations are written in terms of the amplitudes of the incident and diffracted modes $I(z)$, $D(z)$. The equations are

$$\cos\theta_d \frac{dI(z)}{dz} = -j\kappa^* D(z) \exp(-jz\delta_z) \tag{10.25}$$

$$\cos\theta_i \frac{dD(z)}{dz} = -j\kappa I(z) \exp(+jz\delta_z) \tag{10.26}$$

and

$$2\delta_z = \beta_d \cos\theta_d - (\beta_i \cos\theta_i + qK\cos\theta) \tag{10.27}$$

and κ is the coupling coefficient between the incident and diffracted waves. The angles $\theta_{i,d}$, 0, and θ are the angles of incident, the diffracted, and the grating grooves, shown in Fig. 10.7.

These equations can be used to obtain useful working equations for both transmission and reflection gratings. For example, in the case of a diffraction grating of length L, the efficiency of diffraction, η, is given by

$$\eta = \frac{\cos\theta_i}{\cos\theta_d} \frac{|D(L)|^2}{|I(0)|^2} \tag{10.28}$$

$$= \frac{\sin^2\left(\sqrt{\delta_z^2 |\kappa|^2 / \cos\theta_d \cos\theta_i}\, L\right)}{1 + \cos\theta_d \cos\theta_i \delta_z^2 / |\kappa|^2} \tag{10.29}$$

or, exactly at the Bragg condition, $\delta_z = 0$,

$$\eta|_{\delta z} = \sin^2 \left(\frac{|\kappa|L}{\sqrt{\cos \theta_d \cos \theta_i}} \right) \tag{10.30}$$

Clearly, the efficiency increases with both κ and L for the grating. Notice also that the grating efficiency varies as a $\sin x / x$ behavior and with detuning δ_z, just as seen for collinear coupling. Although we will discuss coplanar reflection gratings, similar expressions can be derived in that case as well (Nishihara et al. 1989).

10.4.1   Coupling Coefficients for Gratings

In Chap. 5, we examined the coupling coefficient for TE–TE coupling with a shallow grating structure. In this section, we provide simplified approximate formulae for other formula of grating profiles or guided-wave polarization.

Previously, it was shown that the coupling coefficient for shallow sinusoidal gratings on a single slab waveguide is operating in the fundamental waveguide modes,

$$\kappa_{TE-TE} = \frac{\pi a}{2\lambda} \frac{(n_f^2 - N_{\text{eff}}^2)}{N_{\text{eff}}} \frac{1}{d + (1/\gamma) + (1/\delta)} \tag{10.31}$$

where d is the waveguide film thickness, γ is the substrate decay constant, δ is the cover decay constant, a is the groove depth, and λ is the free-space wavelength. In addition, it can be shown that this expression applies equally to the case of TM waves on a single-mode slab waveguide. This equation may be rewritten using an even simpler equation:

$$L_c = \frac{\lambda}{2\Delta N_{\text{eff}}} \tag{10.32}$$

$$\kappa \simeq \frac{\pi \Delta N_{\text{eff}}}{\lambda} \tag{10.33}$$

where

$$\Delta N_{\text{eff}} = \frac{1}{2} \left(N_{\text{eff}}(max) - N_{\text{eff}}(min) \right) \tag{10.34}$$

For the case of a shallow rectangular grating in which coupling for the $TE_0 - TE_0$ and the $TM_0 - TM_0$ modes occurs

$$\kappa_{TE-TE} = \frac{2\pi}{\lambda} \frac{a}{d + (1/\gamma) + (1/\delta)} \frac{\sin(qc\pi)}{q\pi} \frac{(n_f^2 - N_{\text{eff}}^2)}{N_{\text{eff}}} \tag{10.35}$$

and

$$\kappa_{TM-TM} = \frac{2\pi}{\lambda} \frac{a}{d + (1/\gamma) + (1/\delta)} \frac{\sin(qc\pi)}{q\pi} (n_f^2 - N_{\text{eff}}^2) \frac{N_{\text{eff}}}{q_s} \tag{10.36}$$

Fig. 10.8 The sketch of a shallow rectangular grating

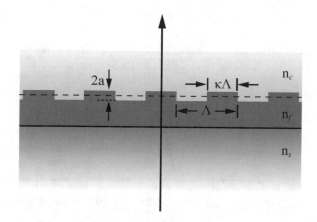

where the grating is as shown in Fig. 10.8, q is the diffraction order, c is the fraction of surface in one period which is not covered by a groove, a is the depth of the groove, and

$$q_s = \left[\left(\frac{N_{\text{eff}}}{n_f} \right)^2 + \left(\frac{N_{\text{eff}}}{n_s} \right)^2 - 1 \right] / \left[\frac{n_c^2}{2} \left(\frac{1}{n_c^4} + \frac{1}{n_f^4} \right) \right]$$

As an example of coupling coefficients in actual devices, consider a typical value of κ' for distributed feedback in single- and double-heterostructure diode lasers. In the case of GaAs, for example, for $\lambda = 1.53\,\mu\text{m}$, $A = 500\,\text{nm}$, $T = 1.3\,\mu\text{m}$, and $N_{\text{eff}}(\lambda = 1.3\,\mu\text{m}) = 3.348$, we obtain $\kappa = 24.96\,\text{cm}^{-1}$ from (10.31), and $\kappa = 25.11\,\text{cm}^{-1}$ from (10.32). Alternatively, in the case of a grating at an air interface with $GaAs$, κ typically is between 50 and 100 cm^{-1}.

10.5 High-Q Bragg Grating Structures

A simple Bragg grating has a "stop band" in the vicinity of its Bragg wavelength. That is, as shown in Fig. 10.9, on a linear plot of reflectivity versus wavelength or frequency detuning from its Bragg value, a Bragg grating acts like a high reflectivity mirror. In fact, Bragg mirrors are the most satisfactory method of making high reflectivity structures on a planar surface. They can be made by lithographic means and can incorporate a variety of useful optical features including wavelength chirping or apodization and wave front curvature as was just mentioned above. In addition, it is also possible to introduce a small change in this grating structure to make a high-Q narrow passband transmission filter using Bragg gratings. This latter structure has a very high quality factor, Q, and thus is an extremely narrow bandpass filter with low loss.

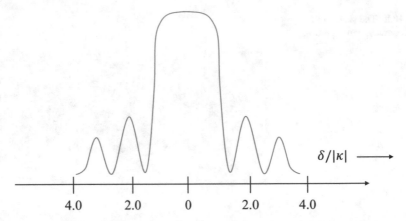

Fig. 10.9 A linear plot of reflectivity versus wavelength or frequency detuning from its Bragg value. Notice that the Bragg grating acts like a high reflectivity mirror

Ultimately, this filter will be found to be equivalent to the Fabry–Perot filter encountered in bulk optics, except that in this case, the two mirrors are replaced by distributed Bragg reflectors. Thus, by analogy with the Fabry–Perot device, it would be expected that at certain resonances, the cavity formed by the two mirrors would have a transmission maximum. In fact, as was pointed out above, such maxima exist even in a simple Bragg resonator; note, however, that for a uniform diffraction grating, the transmission resonances are not at $\delta = 0$, i.e., the Bragg frequency of the grating.

In order to understand why a resonance does not lie at the center of the stop band, consider making such a two-grating mirror filter as short as possible by placing the two mirrors adjacent to each other with only an infinitesimal spacing. In addition, for simplicity, allow each grating to be of length $L \gg \kappa^{-1}$, or effectively infinite in length. The center of the separated grating is at $z = 0$.

Now consider excitation of the infinite grating lying in the negative region at a frequency such that $\delta \ll \kappa$ and $z = 0$. Since the grating is infinite, the solutions for excitation at center are given by a special case of the earlier more general grating problem, which is obtained from the coupled-mode equations. For the grating envelope amplitudes $A_1(z)$, where as defined earlier in (5.43), $a_1(z) = A_1(z) \exp(-jKz)$. Then,

$$A_1 = A_1^-(0)e^{+|\kappa|z} \tag{10.37}$$

$$A_2 = \frac{|\kappa|}{\kappa} A_1^-(0)e^{|\kappa|z} \tag{10.38}$$

Similarly, execution of the positive half plane $z > 0$, again at $z = 0$, yields

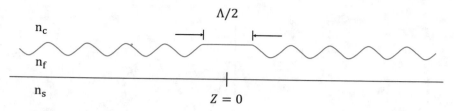

Fig. 10.10 The sketch of two gratings spaced by a quarter-wave section, $\Lambda/2$

$$A_1 = A_1^+(0)e^{-|\kappa|z} \tag{10.39}$$

$$A_2 = -\frac{|\kappa|}{\kappa}A_1^+(0)e^{-|\kappa|z} \tag{10.40}$$

By continuity, we expect the two solutions to match at $z = 0$. However, if the reflectivity or A_2/A_1 is examined for the two regions, it is clear that the reflectivity polarities differ: one is negative, one is positive. Clearly this is not the case for the two half infinite gratings; in effect, the fields at the boundary are such that their coupled reflectivity, r, is 180° out of phase. Thus, without a change in the structure, a standing wave solution is not possible.

Consider now separating the two gratings by a small spacing. In particular, if a quarter-wave section, i.e., $\Gamma/2$ in length, is inserted into the structure (see Fig. 10.10), the two reflected waves will then be matched in phase. The origin of this phase shift can be seen by now including the rapidly varying part of the solution, $\exp(-jKz)$, which is not included in the envelope terms, A_1 and A_2, of the solution, see Chap. 5. If an extra grating length of a $\Delta z = \Gamma/2$ is included in the interface, the relative phase of the two outward moving waves will change and cause the two reflection coefficients to be in phase. Note that this term is not important if there is no "phase-slip" at $z = 0$ because it is identically null under that condition.

A plot of the envelope of the forward-moving wave is plotted in Fig. 10.11, again for the case of $\delta \ll \kappa$. The envelope function given in Fig. 10.11 for $\delta \ll \kappa$ shows that most of the modal energy is stored in the field near the center of the grating. In fact, at the edges of a grating with $L_\kappa \gg 1$, the field has decreased exponentially, $\sim e^{-|\kappa|L}$. As a result of this decrease, very little power "leaks" out of the grating as it circulates back and forth near the grating center, that is, the grating structure has a very high Q.

In fact, a coupled-mode analysis of the problem (see Haus et al. 1989) shows that the modal amplitude at the grating edge, $z = L$, can be written as

$$A_1(L) = 2A_1^+(0)e^{-|\kappa|L} \tag{10.41}$$

where again the solution is for optical frequencies at the center of the stop band. Hence, the power lost out of the grating end is

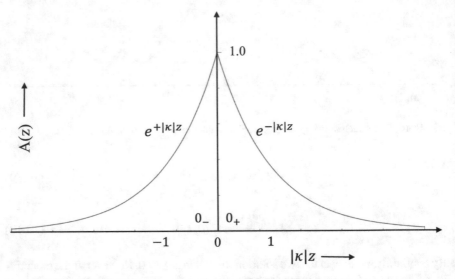

Fig. 10.11 The envelop of the forward-moving wave for the case of $\delta \ll \kappa$

$$P_2 = |A_1^+(0)|^2 e^{-2|\kappa|L} \tag{10.42}$$

The energy stored within the grating, u, is

$$u \approx \int_{\infty}^{\infty} \frac{|a_1|^2 + |a_2|^2}{\nu_g} dz = \frac{2|A^+|^2}{\nu_g|\kappa|} \tag{10.43}$$

where the solution is for optical frequencies at the center of the stop band. Thus, the number of periods required for the stored energy to be lost from the cavity, or the cavity Q_{esc} due to this loss, is

$$Q_{esc} = \frac{\pi c}{\nu_g \lambda |\kappa|} e^{+2|\kappa|L} \tag{10.44}$$

Clearly, also, by symmetry, identical power loss occurs through the other end of the grating. This is equivalent to a decay time

$$(\tau)^{-1} = \frac{\nu_g|\kappa|}{2} e^{+|\kappa|L} \tag{10.45}$$

for the cavity due to its finite length, where the fact that there are two grating ends is included in τ. Notice that a longer L and larger κ decrease the decay time or increase the cavity Q, where internal losses are assumed to be negligible.

The reflectivity of such a quarter-wave grating of length $2L$ for a wave moving from the left can be calculated by employing a more powerful matrix approach

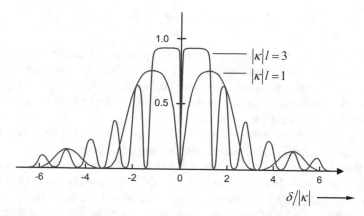

Fig. 10.12 A plot of reflectivity versus wavelength or frequency detuning from its Bragg value of a 2L-long quarter-wave grating

(McCall and Platzman 1985). The result of the calculation gives the reflectivity of a wave entering the grating from $z = -L$ to $z = L$ as

$$R = \frac{|\frac{4\kappa\delta}{\beta_d}|^2 \sinh^4 \beta_d L}{[(\frac{\kappa^2}{\beta_d})^4 - \frac{\delta^2}{\beta_d} 2 \cosh 2\beta_d L]^2 + \delta^2 \sinh^2 2\beta_d L} \tag{10.46}$$

This reflectivity is plotted in Fig. 10.12. Notice that this grating provides a very narrow passband at $\delta = 0$. The narrowness of the passband is attributable to the high reflectivity of the grating at $\delta = 0$ and is thus related to the cavity lifetime τ.

The width of this passband can be found using the expression for reflectivity or simply using the expression for Q derived above.

$$\Delta\omega_{1/2} = 4\nu_g |\kappa| e^{-2|\kappa|L} \tag{10.47}$$

or, using the normalized units discussed earlier,

$$\Delta\delta_{1/2} = 4|\kappa| e^{-2|\kappa|L} \tag{10.48}$$

This passband is also shown in the reflectivity curve of Fig. 10.12. The $1/4\lambda$-step Bragg grating described here is at the heart of many Bragg laser optical cavities. Its narrow passband allows light to leak out at specific wavelengths. Tuning such a grating either by fabrication or with temperature allows on to make multi-wavelength sources on the laser chip. In addition, Chap. **??** describes the use of these structures in a very narrowband channel-dropping filter.

10.6   Applications of Bragg Gratings

A Bragg grating is a very basic functional component for integrated optical circuits. Bragg gratings have found wide applicability in filters, as wavelength selective reflectors, dispersive elements, and in more complex devices. In this section, we discuss a few representative devices made with Bragg reflectors; other examples are distributed throughout the text.

Bragg gratings can be used to achieve simple optical functionality for manipulating two-dimensional guided waves, such as lensing. These devices are desirable because their fabrication, while exacting, is not as complex as that required to fabricate a geodesic lens. In one example, focusing Bragg gratings were made using electron-beam holography to have an efficiency of 40% (Nishihara et al. 1989). In another example, a series of two Bragg gratings was used to make a two wavelength demultiplexer in a As_2S_3, SiO_2 PIC. In this device, 70% grating efficiencies were obtained with $-15\,dB$ suppression of cross talk. The device used two tilted gratings to make wavelength-selective beam splitters which focused light into two detectors (Suhara et al. 1982).

More recently, Bragg grating structures have been used for polymer integrated optical devices (Eldada and Shacklette 2000). In this case, the gratings are made by printing through a lithographic mask to cross-link the polymer. The simplest device used a planar Bragg grating, which was tuned using the thermo-optic effect to make a wavelength-variable optical filter, 2 cm in length. By using care in designing the structure, the wavelength tunability around 1.5 μm was ∼0.04 nm/°C. The device had a bandwidth utilization of 0.92 for 75 GHz channel spacing. A second device printed such a grating across the arms of a Mach–Zehnder containing either two 3 dB couplers or two MMI couplers to form a tunable add/drop wavelength demultiplexer. The device was 4 cm-long, with a 6 μm-core and 10 μm-thick cladding layer, with a $\Delta n = 0.5\%$ in $n = 1.5$ polymer. The device output/input parts were spaced 250 μm apart. The functionality of the devices is shown in Fig. 10.13. The insertion loss for each device was 2.5 dB. These devices have been integrated to form a large-scale switch array, see Chap. **??**.

A simple wavelength-dependent reflecting filter can be made using the Bragg reflectors, such as that described in this chapter; however, an improved version can be made in conjunction with the use of a Mach–Zehnder interferometer. A device based on this principle is shown in Fig. 10.14. The device uses two 3 dB couplers in conjunction with a Mach–Zehnder interferometer. As a result, two passes through the coupler causes the power to exit fully through the opposite output port. When a DFB grating is inserted in the device, the reflected light is coupled to port 2, while transmitted (out of band) light is switched to port 4. Since a wavelength can be added on port 3, the device functions as an add/drop device. An integrated form of this device has been made, but its performance was limited by polarization splitting of the peak wavelength.

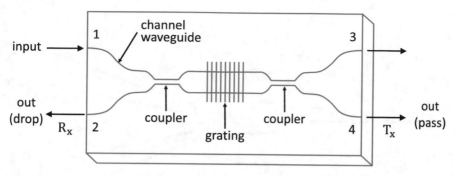

Fig. 10.13 A sketch of the device printing a grating across the arms of a Mach–Zehnder containing two 3 dB couplers or two MMI couplers to form a tunable add/drop wavelength demultiplexer

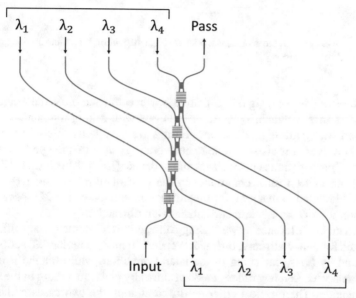

Fig. 10.14 A wavelength-dependent reflecting filter based on the Bragg reflectors [Adapted from (Eldada and Shacklette 2000)]

10.7 Bragg Diffraction and Raman–Nath Diffraction

Thus far, in this chapter, we have only considered gratings which have a relatively well-defined spatial periodicity in the z-direction. In this case, the grating wave number is established over a long length. In general, coupling from gratings can be characterized by a Q-parameter, defined as follows:

$$Q = \frac{K^2 L}{\beta} \tag{10.49}$$

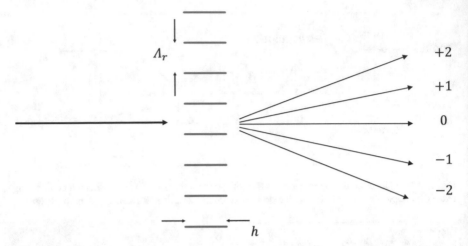

Fig. 10.15 An example of the geometry of Raman–Nath diffraction for a transmission grating near a normal angle of incidence

where $K = |\vec{K}|$ is the magnitude grating wavevector and L is the length of the grating. Gratings in which $Q \gg 1$, such as one with a long interaction length L, are termed Bragg gratings; these are gratings we have considered thus far in this chapter. However, it is also possible to have gratings such that the periodicity in the direction of propagation is not well established, and $Q \sim 1$. This form of diffraction phenomenon is seen most commonly in the case of ultrasonic devices. But it is also closely related to that seen for the diffraction phenomena in 2D spectrometers. Diffraction with $Q \ll 1$ is termed Raman–Nath diffraction.

Thus, a major difference between Bragg diffraction and Raman–Nath diffraction is that for the Raman–Nath case, diffraction occurs in many angular diffraction orders, determined by phase matching in the lateral direction, while for the Bragg case, diffraction occurs in a few orders, as determined by phase matching in the direction of propagation. The physical criterion differentiating the two cases is the relative length of the grating region.

An example of the geometry of Raman–Nath diffraction for a transmission grating is shown in Fig. 10.15 for near-normal angle incidence. For Raman–Nath diffraction, the phase-matching condition is required

$$\beta \sin \theta_i + qK = \beta \sin \theta_q \tag{10.50}$$

where is the angle of incidence relative to the z-direction, and q is the diffraction order, i.e., $q = 0, \pm 1, \pm 2$, etc.

In the case of Raman–Nath diffraction, it is also possible to analyze the grating fields using the coupled-mode equations. In this case, however, the coupling is between the same mode in different angular diffraction orders, say $q \to q + 1$. In the case of $Q \sim 1$, diffraction occurs in many orders. The solution at the exit of the

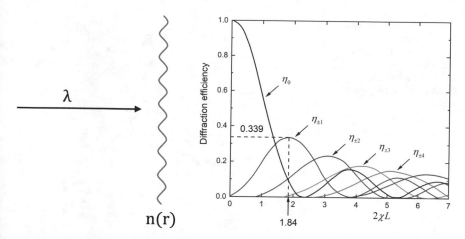

Fig. 10.16 Diffraction efficiency of Raman–Nath diffraction

grating region, i.e., $z = L$, is then found to be

$$I_l = |\phi|^2 = J_l^2 \left(2\kappa L \frac{\sin(Q\alpha/2)}{Q\alpha/2} \right) \tag{10.51}$$

where J_l is the ordinary Bessel function of order l. At normal incidence, $\theta_i = 0(\alpha = 0)$, then $\eta_q = J_q^2(2\kappa L)$, see Fig. 10.16. Because Raman–Nath theory is not typically used in integrated optics, except in the subfield of integrated acousto-optics, we shall not pursue a longer discussion of it in this text. However, a more extensive discussion can be found in Chap. 4 in Nishihara et al. (1989).

10.8 Wavelength Selecting via Reection from a Grating and Grating Fabrication

Grating "spectrometers," which use Raman–Nath diffraction, are often used in PICs for wavelength selection. They operate in a manner which is very similar to that of a large-scale commercial grating spectrometer, with the central differences being the two-dimensional geometry and small size. A top view of a slab-waveguide spectrometer was shown in the fourth panel of Fig. 10.1. In this case, the spectrometer is fed by a single-mode waveguide, which is allowed to propagate in free space before impinging on the grating. Definition of the grating grooves is usually done via etching and is a major fabrication challenge.

Although the grating can be analyzed formally by Raman–Nath coupled-mode theory, it is readily analyzed here based on a simple diffraction method. That is, it can be easily shown via a simple phase-front argument that the diffraction angle of

Fig. 10.17 A sketch of a
slab-waveguide
spectrometer. The incident
wave is at angle θ_i and the
diffracted wave is at angle
θ_D. λ is the grating spatial
wavelength

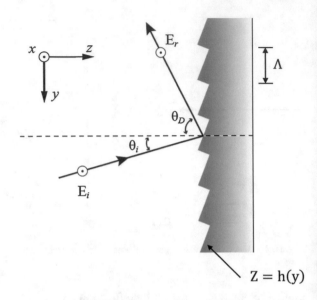

the qth order. If an incident wave is at angle θ_i and the diffracted wave is at angle θ_d,
then

$$\sin \theta_d^q = \sin \theta_i + \frac{a\lambda}{\Gamma N_{\text{eff}}} \tag{10.52}$$

where Γ is the grating spatial wavelength, see Fig. 10.17. The wave diffracted from
a reflection grating then has an amplitude pattern

$$I(\theta) = \frac{\sin^2[N_{\text{eff}} \frac{M\pi}{\lambda} \Gamma (\sin \theta_d - \sin \theta_i)]}{\sin^2[N_{\text{eff}} \frac{\pi}{\lambda} \Gamma (\sin \theta_d - \sin \theta_i)]} \tag{10.53}$$

where M is the total number of grooves which are illuminated, and is thus dependent
on the cross section of the feeder waveguide end facet and its spacing from the
grating. Notice that this angular distribution is, thus, that of a $\sin Nx / \sin x$ function,
which peaks at $x = 0$ and oscillates in the "wings." The principal maximum occurs
at $\theta = \theta_d$. The intensity of the diffracted light falls off to zero when θ is increased
from θ_d by a value $\delta\theta$ such that

$$N_{\text{eff}} \frac{M|pi}{\lambda} \Gamma (\sin \theta_d - \sin(\theta_d + \delta\theta)) = \pi \tag{10.54}$$

This spectrometer has an angular resolution of

$$\delta\theta = \frac{\lambda}{N_{\text{eff}} M \Gamma \cos \theta_d} \tag{10.55}$$

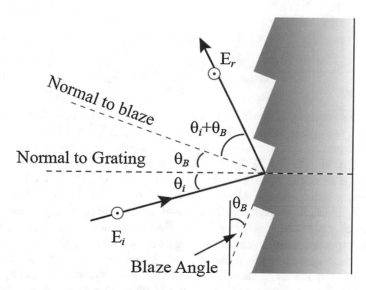

Fig. 10.18 A sketch of a blazed grating. The incident wave is at angle θ_i and the diffracted wave is at angle $(\theta_B + \theta_i)$ where θ_B is the blaze angle

Two spectral lines begin to be resolved when the maximum of one wavelength coincides with the first zero of the other:

$$\theta_d(\lambda_1 + \delta\lambda) - \theta_d(\lambda_1) = \delta\theta \tag{10.56}$$

which yields

$$\frac{m\delta\lambda}{N_{\text{eff}}\Gamma\cos\theta_d^q} = \frac{\lambda}{N_{\text{eff}}M\Gamma\cos\theta_d^q} \tag{10.57}$$

or

$$\frac{\delta\lambda}{\lambda} = \frac{1}{mM} \tag{10.58}$$

The ratio $\delta\lambda/\lambda$ is termed the chromatic resolving power of the grating. To give a concrete example, consider illuminating a $100\,\mu\text{m}$ region of a grating with $\Gamma = 1\,\mu\text{m}$ and $M = 100$. For $\lambda = 1.5\,\mu\text{m}$, and the $m = 1$ diffraction order, $\delta\lambda = 1500\,\text{nm}/100 = 15\,\text{nm}$.

Typically, a high-efficiency grating may have its grooves contoured such that the shape enhances a particular order while suppressing other diffraction orders. Such a grating is called a "blazed" grating, and can be accomplished by a variety of fabrication methods, see Sect. 10.8. To find the condition for the best blaze, for a particular order, at a particular wavelength, the diffraction angle should be equal to that corresponding to specular reflection from blaze surface (see Fig. 10.18):

$$\sin \theta_d^q = \sin(\theta_i + 2\theta_b) \tag{10.59}$$

$$\sin(\theta_i + 2\theta_b) = \sin \theta_i + \frac{a\lambda}{N_{\text{eff}}\Gamma} \tag{10.60}$$

Thus, for example, if $\theta_i = 0$, $\lambda = 1.55\,\mu\text{m}$ and for $\Gamma = 1\,\mu\text{m}$, then for a thick multimode waveguide in $GaAs$, in which $n_f \approx n_{GaAs}$, $\theta_b^{q=1} = 13.6°$.

Because grating spectrometers entail a considerable amount of free-space imaging, it is crucial to design the optical paths such that diffraction is minimized and aberration is avoided. In addition, the use of curved surfaces allows for focusing of free-space beams. An extensive discussion of these points is described in März (1995). In general, however, typically the familiar Rowland mounting scheme is used since it is a simple design and since it minimizes aberrations.

Grating spectrometers have been used for a number of PIC applications, including optical chemical sensors and wavelength demultiplexing. The latter application was of particular importance in early WDM systems. Multi-wavelength grating WDM demultiplexers were demonstrated in InP and SiO_2. These devices are illustrated here by an SiO_2 device (Tong 1998). This device was designed for two-dimensional "free-space" manipulation of the light. The spectrometer used a Rowland layout consisting of a large curved grating focused on a smaller curved region with a detector spacing of $140\,\mu\text{m}$ for having detection at each wavelength channel. The receiver detected 32 separate channels. A primary concern in fabricating the grating was the versatility and precision of the etched grating groove, $16\,\mu\text{m}$ wide by $20\,\mu\text{m}$ deep. The difficulty in achieving the desired profile caused, in part, the 6–7 dB loss in the device. In addition, residual stress in the device caused 0.5×10^{-4} birefringence in the TE and TM modes. Despite these difficulties, the device was useful in a commercial network system.

Fabrication is a major consideration in using gratings as optical elements in PICs. Because of their size and dimensional tolerances, as well as the "analog" nature of grating spatial profiles, fabrication can be exacting, and hence costly. Two different methods of patterning are used in practice: lithographic printing of an electron-beam-written master mask and direct laser holography. These patterns must then be transferred by an etching step.

In-plane gratings are made using either of the above two methods (Nishihara et al. 1989). For example, for Bragg gratings, the period structure is often fabricated by interfering two coherent laser beams, generated with a beam splitter, at the surface of a photoresist-covered wafer. The field pattern of interfering waves, at wavelength λ on a sample surface shown in Fig. 10.19, requires that the wavevector along the surface, k_x, is given by the angle of the two beams, θ,

$$k_x = \frac{2\pi}{\lambda} \sin \theta \tag{10.61}$$

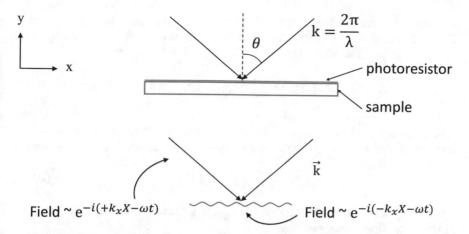

Fig. 10.19 A sketch showing fabrication of an in-plane grating. Two coherent laser beams interfere at the surface of a photoresist-covered wafer. The field pattern of interfering waves can thus be printed on the sample surface

Since the total field is proportional to the standing wave created, i.e., $2\cos k_x x$ then $I \sim \cos k_x x$ where Γ is the distance between two maxima, and

$$\Gamma = \frac{\lambda}{2\sin\theta} \tag{10.62}$$

WDM components have often been made from a grating fabricated on the side of an etched feature (see the fourth panel in Fig. 10.1). In this case, the grating is typically made in photoresisting using an e-beam-lithographic master, followed by pattern transfer via dry etching. The controlling issue in making such a grating is the quality of the etched features (see the discussions in the previous subsection).

10.9   Gratings for Output or Input Coupling with Guided-Wave Structures

A recurring problem in integrated optics is the coupling of light into a waveguide. One class of solutions simply couples in by "butting" a fiber guide against a waveguide facet. This approach, which is discussed in Chap. 7, is important and successful in many applications. However, in some cases such as that of very high-index waveguides or rapid testing, the butt coupling method is not satisfactory. Instead, gratings are more readily used to couple light into and out of the waveguide. Because of their importance, grating couplers have received a considerable amount of emphasis (Tamir and Peng 1977; Tamir 1975).

Consider a grating having a regular surface structure such as that depicted in Fig. 10.20 or earlier in Fig. 10.8. This structure gives rise to a spatially varying index or dielectric function, which can be expanded in a Fourier series:

$$\Delta\epsilon(x, z) = \sum_q \Delta\epsilon_q(x) \exp(-jqKz) \tag{10.63}$$

where q is the order of a harmonic. Each of these harmonic $\Delta\epsilon_q$ components modulates the guided wave and gives rise to a z-dependent optical mode with a propagation constant given by

$$\beta_q = N_{\text{eff}}k + qK \tag{10.64}$$

where N_{eff} is the effective index of waveguide without the grating and where we assume a shallow surface grating.

These harmonic optical modes may be guided or radiative modes depending on their propagation constants. Thus, the qth optical mode will radiate when its propagation constant, β_q, is such that $\beta_q < n_c k < n_s k$. Significant amounts of power will be lost radiating to the gth mode if phase matching occurs in the z-direction. To have phase matching for radiation in the cover or substrate requires that

$$n_{c,s}k \sin \theta_q^{c,s} = N_{\text{eff}}k + qK \tag{10.65}$$

Obviously, depending on the relative values of, $n_{c,s}$, and N_{eff}, several values of the radiation angles $\theta_q^{c,s}$ may be allowed. Notice also that in our previous discussion of coupled-mode behavior in grating, we assumed that such radiative coupling is non-existent or small. Notice, further, that the magnitude of this coupling depends on the presence of special spatial harmonics in the surface dielectric of the Bragg grating. The values of these harmonics vary with the profile and the depth of the grating and are not important in, for example, a sinusoidal grating.

For a given grating period, several different radiative beam orders can exist simultaneously. For example, Fig. 10.20 depicts a grating for which two beams in the $q = -1$ order couple with the propagating mode through one grating vector. In this case, both a radiative (i.e., in air) and a substrate beam are present; in addition, more complex multibeam coupling can be present.

The coupling of light into different radiative modes is important for two reasons. First, radiative modes can obviously be a significant loss mechanism in a grating designed for another purpose, say, as a reflector. In addition, radiation from gratings is directly useful as a way of coupling light out of a photonic integrated circuit into a free-space interconnect beam. Out coupling is best understood by using coupled-mode theory, that is, using the same approach as discussed earlier in the chapter for coupling between propagating modes.

The full coupled-mode derivation will not be discussed in this text. Basically, the approach we do use is, however, very similar to that discussed earlier and includes a phase term, which requires phase matching, and a coupling coefficient, κ^q, indexed to a particular order, q, and to the modes having phase matching. These equations lead

Fig. 10.20 A sketch of a grating for which two beams in the $q = 1$ order couple with the propagating mode through one grating vector

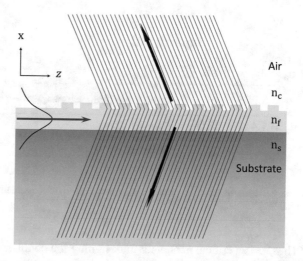

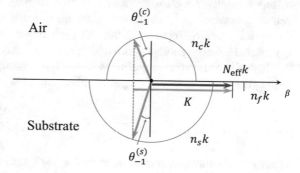

to the following equation for the amplitude of the propagating mode in the grating region, i.e., $0 < z < L$:

$$\frac{da(z)}{dz} = -\alpha a(z) \tag{10.66}$$

where α is a loss constant given by $\alpha = \pi |\kappa^\tau|^2$, and where the coupling coefficient is summed over the order, q, and the two possible beam directions: cover or substrate. Thus, as the propagating mode passes through the grating, it decays exponentially along distance z:

$$a(z) = a(0) \exp(-\alpha z) \tag{10.67}$$

Similarly, the total radiative power decays at the same rate. The power in the propagating mode decays at twice this rate:

$$P(z) = P(0) \exp(-2\alpha z) \tag{10.68}$$

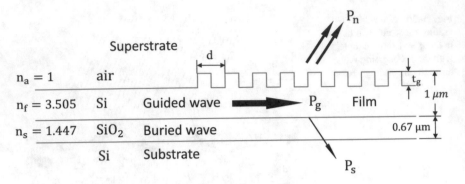

Fig. 10.21 A sketch of an example using gratings to out-couple the beam from thin silicon waveguide

Thus, α is a critical parameter in designing the coupler.

Input couplers are also extremely important for a surface grating, since they enable coupling between an external beam, which is particularly useful for probing and the optical circuity. Input coupling can be handled using a similar coupled-mode approach. However, another approach has also been used, which is based on the applicability of time reversal to problems in electromagnetics and optics. This approach shows that the efficiency of output coupling can be written as

$$\eta = \frac{P_{\text{guided}}}{P_{\text{input}}} = P_q^i \frac{|\int \tau(z) i(z) \mathrm{d}z|^2}{\int \tau^2(z) \mathrm{d}z \int i^2(z) \mathrm{d}z} \tag{10.69}$$

where $\tau(z)$ is the spatial distribution of a beam radiating out of the grating from guided mode incident on the grating and $i(z)$ is the actual profile of the beam being coupled with the grating coupler. Note that this equation assumes that the coupler has a power distribution of P_q^i for each of its output beams, and summation over q and i is understood.

The efficiency of the coupler clearly depends on the nature of the beam overlap. Clearly the best overlap is achieved when the excitation beams have spatial distributions, which are identical to the corresponding output beam for that mode in the coupler. For a one beam coupler, all input power is in one beam and order and thus $P_q^i = 1$. If this input beam has a Gaussian distribution and the grating is long enough to complete the power transfer $\eta_{max} \approx 80\%$. In practice, the grating length, L, is determined by the requirements that $L \gg 1/\alpha_\tau$. A thorough discussion of the practical aspects of input coupling, including beam shaping, experimental setup, etc. is given by Nishihara et al. (1989).

Grating coupling is an important technology in many different PIC materials and device designs. An excellent example of grating coupling is that used for silicon-on-insulator devices by Ang et al. (2000). In this example, the high-index and thin vertical dimension of the silicon waveguide made grating coupling the best choice for out coupling of the beam. The scheme used the $q = -1$ order with a substrate

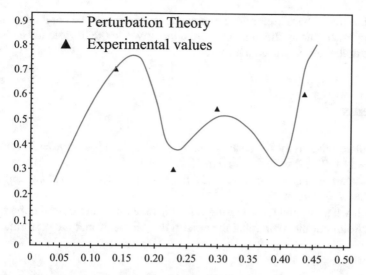

Fig. 10.22 A plot of theoretical perturbation output efficiency curve versus measured output efficiency data points with various grating heights at Si film thickness of 1 μm

and cover beam; the cover beam was the desired beam. A picture of the device (Ang et al. 2000) is shown in Fig. 10.21. The grating spacing, d, for phase matching is given by

$$d = \frac{\lambda}{N_{\text{eff}} - \sin \phi_{-1}} \qquad (10.70)$$

In the paper, $d = 0.4\,\mu m$ for a Si thickness of $\sim 1.1\,\mu m$. In this coupler, $\eta \equiv P_c/(P_c + P_g)$, and by careful adjustment of the height of the rectangular grating, i.e., $0.15\,\mu m$, a maximum efficiency of 70% was obtained; see also an experimental measurement of efficiency versus groove depth in Fig. 10.22. This adjustment was necessary since a change of 0.15–0.22 μm caused η to drop to 30%.

10.10 Summary

Diffraction gratings are widely used passive components in the design of integrated optical circuits. While they are "costly" to make, their wavelength selectivity makes them attractive in many multi-wavelength devices. Further, because gratings can manipulate the phase front of guided light wavelengths, they provide a convenient approach to make two-dimensional optical components such as lenses or reflectors. Without gratings, these devices would have to rely on components with very challenging fabrication procedures. A very important and a closely related use is that of coupling light beams from one medium to another, from free spaces to guiding medium, and from one guided medium to another. The wavelength selectivity of grating has made them important components in separating and combining multiple wavelengths. In this application, they are alternatives to the devices to be discussed in the next chapter on wavelength filters.

Problems

1. Calculate reflectivity R versus detuning δ for a grating with $\kappa = 200\,\text{cm}^{-1}$ and $\kappa L = 1/3$. Plot R with δ from 0 to $30\,\kappa$.
2. Calculate the angle ϕ_c in Fig. 10.21 from the text, for $\lambda = 1.3\mu\text{m}$ and grating spacing $d = 0.4\mu\text{m}$.
3. Sketch what an integrated grating reflector looks like and describe the function of it. Point out the main relation between the geometry and the performance.

References

Ang, T., Reed, G., Vonsovici, A., Evans, A., Routley, P., & Josey, M. (2000). Effects of grating heights on highly efficient unibond soi waveguide grating couplers. *IEEE Photonics Technology Letters, 12*(1), 59–61.

Eldada, L., & Shacklette, L. W. (2000). Advances in polymer integrated optics. *IEEE Journal of Selected Topics in Quantum Electronics, 6*(1), 54–68.

Haus, H. A., Huang, W.-P., & Snyder, A. W. (1989). Coupled-mode formulations. *Optics Letters, 14*(21), 1222–1224.

März, R. (1995). *Integrated optics: design and modeling.* Artech House on Demand.

McCall, S., & Platzman, P. (1985). An optimized π/2 distributed feedback laser. *IEEE Journal of Quantum Electronics, 21*(12), 1899–1904.

Nishihara, H., Haruna, M., & Suhara, T. (1989). Optical integrated circuits (pp. 7–8). New York.

Suhara, T., Handa, Y., Nishihara, H., & Koyama, J. (1982). Monolithic integrated micrograting and photodiodes for wavelength demultiplexing. *Applied Physics Letters, 40*(2), 120–122.

TTamir, T. (1975). Beam and waveguide couplers. *Integrated Optics* (pp. 83–137). Berlin: Springer.

Tamir, T., & Peng, S.-T. (1977). Analysis and design of grating couplers. *Applied Physics, 14*(3), 235–254.

Tong, F. (1998). Multiwavelength receivers for wdm systems. *IEEE Communications Magazine, 36*(12), 42–49.

Yariv, A. (1997). Optical electronics in modern communications, oxford series in electrical and computer engineering.

Chapter 11
Wavelength Filtering and Manipulation

Abstract The use of multiple wavelengths has many applications in integrated optics. For example, these beams can be used for high data rate optical systems or for contoured wavelength beams in a nonuniform amplifier. In this chapter, we consider the functionality of wavelength filtering and manipulation. These are illustrated with several approaches for filtering, separation, or manipulation of wavelengths using different device types. More generally there are many wavelength filters, which are widely used, such as phased-array waveguide routers, Mach–Zehnder interferometric devices, various asymmetric coupler-based devices, and free-space grating demultiplexers. This chapter examines in detail three device types. One of these, phased-array routers, are currently the most important commercial devices for wavelength selection. However, the other two device types (coupler filters and 1/4 Bragg channel-dropping filters) have also each been the subject of extensive investigation of their properties. Each of these three device types illustrates one of the major device techniques, which we will be presenting later in this chapter.

11.1 Introduction

Use of multiple-wavelength light beams allows an important extended capability for designing integrated optical systems. This capability allows spatial or angular separation of the light beam into its well-defined spectral components. For instance, very coarse spectral division, such as used in a $1.3\,\mu\text{m}/1.5\,\mu\text{m}$ transceiver (mentioned earlier in Chap. 9), can be used to separate transmission from detection in transceivers. This separation prevents signal "swamping" in an application where large variation in signal intensities is present. In other instances, e.g., WDM communication, multiple-wavelength signals allow expansion of the effective system bandwidth or permit sophisticated signal routing.

In this chapter, we will discuss the important closely related functionality: wavelength filtering and manipulation. Our short introductory section will be followed

R. Osgood jr. and X. Meng, *Principles of Photonic Integrated Circuits*,
Graduate Texts in Physics,
https://doi.org/10.1007/978-3-030-65193-0_11

233

by extensive illustrations of three different approaches for filtering, separating, or manipulating wavelengths using specific device types.

More generally, wavelength filters, which are widely used, include phased-array waveguide routers, Mach–Zehnder interferometric devices, various asymmetric coupler-based devices, and free-space grating demultiplexers. The latter category is discussed in a previous chapter on diffraction gratings and will not be discussed further here. Of the remaining three device types, phased-array routers are currently the most important commercial devices for wavelength selection. However, the other two device types (coupler filters and $1/4\lambda$ Bragg channel-dropping filters) have also each been the subject of extensive investigation of their properties. Each of these three device types illustrates one of the major device techniques, which we will present later in this chapter.

11.2   General Classes of Wavelength Filtering and Manipulation

Selecting, separating, or interchanging wavelengths can be accomplished by a wide variety of devices. However, it is possible to separate these operations into several general classes (see Fig. 11.1). **For example, in the case of wavelength filtering, there are two very broad classes: in-line and channel-dropping filters.** In-line filters simply pass a single wavelength or wavelength band by means of reflection or transmission from the device. Channel-dropping filters pick one wavelength, or channel, out of a multi-wavelength stream and drop it into a different transmission line; channel addition is the reverse process.

More complex wavelength manipulation is also often needed. For example, multiplexing or demultiplexing of multiple wavelengths to or from a bus line is common in WDM applications. These operators can be distinguished from the functionality used in a star coupler which simply separates a signal into many bus lines irrespective of wavelength. Alternately, wavelength reshuffling or routing is required, which resembles in part mux/demux.

Fig. 11.1 General classes of wavelength filtering and manipulation

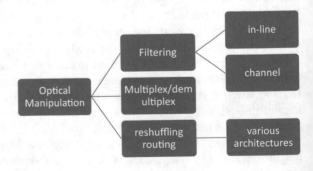

11.3 Asymmetric Coupler Filters

Many integrated coupler-based filters use asymmetric couplers as the basic functional element (Venghaus et al. 1992). These devices have been realized with a wide variety of designs. These include devices based on two nonsynchronous waveguides: on symmetric waveguides with dispersive delay lines, usually Mach–Zehnder-like in physical appearance and on grating-enabled copropagating couplers, examples of each of these devices will be discussed here.

The essential idea of these devices is best described with a simple grating-mediated coupler. In this case, two asymmetric waveguides with different propagation constants β_1 and β_2 are coupled to permit power transfer in the copropagating direction. The coupling is accomplished by using a grating on or near one of the waveguides so as to "phase match" the power transfer. The spatial periodicity of the grating, denoted by K, must be such that

$$\beta_1 - \beta_2 = K \tag{11.1}$$

and the grating length must be such that it is one coupling length L_c,

$$L_c = \frac{\pi}{2\kappa} \tag{11.2}$$

where κ is the overall coupling constant between the two waveguides. This device acts as a wavelength filter because both waveguides are dispersive: $\beta_1(\lambda)$, $\beta_2(\lambda)$. Thus, the condition in (11.1) is exactly satisfied only at one resonant wavelength; in actuality, of course, the relation holds over a narrow frequency band because the finite length of the grating broadens the spectral response. This basic device can be enhanced by fabricating it of electro-optical or semiconducting material (Alferness et al. 1992) so that it may be tuned, and by symmetrically tapering the grating depth or spacing so as to suppress sidelobes in its wavelength response (Sakata 1992).

The basic physical principal of the grating-assisted coupler is that the grating modulates the coupling between the two asymmetric (or nonsynchronous) waveguides at a spatial frequency which exactly compensates for the dephasing of the propagating waves in the two different waveguides. This same modulated-coupling technique is realized through other means in the other coupler-based devices. For example, the meander coupler shown in Fig. 11.2 uses a more macroscopic variation of the coupling region, and was one early realizations of such a spatially varying coupler filter (Bornholdt et al. 1990). This device uses two asymmetric waveguides with different effective indices, realized through two different rib heights. As described above, the power transfer at a wavelength in an asymmetric coupler dephases over a spatial period length of Γ,

$$\Gamma(\lambda) = \pi \left(\kappa^2(\lambda) + \frac{[\Delta\beta(\lambda)]^2}{4} \right)^{-1/2} \tag{11.3}$$

Fig. 11.2 The schematic top view (**a**) and cross section (**b**) of a meander coupler including definition of geometrical device parameters. Notice that the vertical scale is expanded

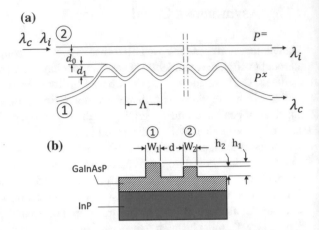

To achieve phase matching of the two waveguides in the coupler then requires that coupling be modulated with a spatial periodicity of $\Gamma(\lambda_c)$. In the meander coupler, phase matching is achieved by modulating (see below) the coupling constant between the two waveguides. This modulation was first achieved by "meandering" one of the waveguides toward and away from the second waveguide with a $\cos^2(\pi z / \Gamma(\lambda_c))$ spatial dependence. This device has been realized in the GaInAsP/InP materials system; the device had a $10 - 15nm$ wavelength bandwidth at $1.3\,\mu$m (Bornholdt et al. 1990). Sidelobe suppression was later demonstrated by smoothly and symmetrically varying the separation of the two coupled waveguides along the meander length, that is, the $\cos^2(\pi z / \Gamma)$ variation in κ had an overall symmetric taper imposed on it (Venghaus et al. 1992).

Finally, another version of the coupler filter uses periodic coupling of symmetric wavegudies, i.e., $\beta_1 = \beta_2$, but achieves asymmetry in the coupler structure by placing delay lines in one arm of the coupler. The wavelength dispersion in this device is again a result of the dispersion properties of the waveguide itself. From a second perspective, this device resembles a series of delay sections joined together, with 3 dB couplers (see Fig. 11.3) to form a series of Mach–Zehnder-like interferometers. This structure is easily analyzed mathematically and it is a valuable device structure to understand how filters can be built up. We will therefore discuss it here in more detail (Kuznetsov 1994).

In essence, this latter device functions as follows: the phase difference between the two arms of the device depends apparently on input wavelength and device geometry. One can design wavelength-selective device to separate different wavelengths to different output ports by a geometry design. For example, as shown in Fig. 11.3b, output ports P_1 and P_2 have different wavelengths.

For an easier understanding, the basic operation of a single element of this device as a wavelength filter is shown by the simplified interferometer-like device depicted in Fig. 11.3a. It has two separate paths, with each having a different path length, L_a

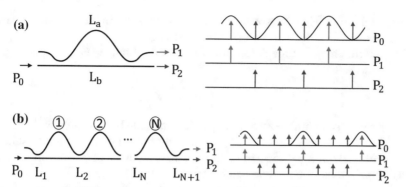

Fig. 11.3 **a** A sketch of a single delay section with a 3 dB coupler; **b** A sketch of a cascaded Mach–Zehnder-like interferometers consists of a series of delay sections joined together, with 3 dB couplers

and L_b, and two $3dB$ couplers, which are assumed to be ideal and λ independent; notice that the path length difference serves to make a delay line.

The transmission, t_a, into one arm, a, of the single-element filter, is given by

$$t_a = \cos^2(\Delta\phi/2) \tag{11.4}$$

where

$$\Delta\phi = \beta(L_a - L_b) \tag{11.5}$$

and β is the common propagation constant of the two waveguides. Since the propagation constant for waveguides is typically dispersive, the phase factor varies with λ. The operation of this device can be seen by considering the two coupled modes, or supermodes, which are excited in the two coupler regions. The introduction of the delay in one waveguide changes the phase of each of the supermodes, thus allowing light to exit on either arm a or b, depending on the phase. The power into arm b is given by $t_b = (1 - t_a)$.

However, this single device element displays only a $\cos^2$ response, and thus does not have a sharp enough spectral response to act as a useful filter. Higher discrimination or resolution can be achieved, however, by using a series of couplers, each with its own delay line. This multiple-section device is designed such that at one wavelength, the length difference, δL, in each section of the device is identical, i.e.,

$$\delta L = \frac{m\lambda_r}{N_{\text{eff}}(\lambda_r)} \tag{11.6}$$

where m is the integral number of the order of interference of the device, $N_{\text{eff}}(\lambda_r)$ is the effective index of the waveguide, and λ_r denotes the resonant wavelength for which all the path lengths are identical. At the resonant wavelength, the light is completely coupled into the cross state of each element as well as that of the full device. Light, which is off-resonance, remains dominantly in the bar state, the exact percentage of which depends on the number of stages, etc.

The interferometer of the device has a free spectral range, F_{FSR}, which is given by

$$F_{FSR} = \frac{c}{N_{eff}(L_a - L_b)} \tag{11.7}$$

Two wavelengths within the same free spectral range can be measured without ambiguity. Notice that these devices can be relatively long. For example, to achieve a $F_{FSR} = 400\,\mathrm{GHz}$ at $\lambda = 1.55\,\mu\mathrm{m}$ in InP requires a device of $\sim 4.5\,\mathrm{mm}$ in length, with a 2 mm radius-of-curvature bend (Kuznetsov 1994).

This behavior can be treated analytically. Consider the output of a sequence of delay-line coupler units also shown in Fig. 11.3b, which act together as a cascaded filter. To determine the output of the filter, it is necessary to write the transmission matrix of the two elements—the coupler, $[t_C(L_i)]$, and the delay-line section (behaving like an asymmetric Mach–Zehnder section), $[t_D]$. The total length of the coupling region, $\sum L_i$, must be specified at the resonant wavelength, λ_r, $\kappa \sum L_i = \pi 2$, that is, to have the bar state for λ_r requires that $\sum L_i$ should be equal to one coupling length. Then, if the input (i) and output (o) fields in the top waveguide, a, and that in the bottom waveguide, b, are written as a vector using transfer matrices (see Chaps. 6 and 5),

$$\begin{bmatrix} a^o \\ b^o \end{bmatrix} = [t_C(L_{N+1})]\cdots[t_D][t_C(L_C)][t_D][t_C(L_1)]\begin{bmatrix} a^i \\ b^i \end{bmatrix} \tag{11.8}$$

where

$$[t_C(L_i)] = \begin{bmatrix} \cos(\kappa L_i) & -j\sin(\kappa L_i) \\ -j\sin(\kappa L_i) & \cos(\kappa L_i) \end{bmatrix} \tag{11.9}$$

and

$$[t_D] = \begin{bmatrix} \exp(j\Delta\phi/2) & 0 \\ 0 & \exp(-j\Delta\phi/2) \end{bmatrix} \tag{11.10}$$

where we have used the fact that the transfer matrix of the waveguide delay section depends on the difference in path length or phase.

A multiple-section filter is shown in the solid-line curve in Fig. 11.4 for the case of five sections. Note that the filter is peaked but with strong multiple sidelobes. This distribution can be further sharpened by employing additional elements. The sidelobes in the response curve result from small but finite coupling at other wavelengths. The sidelobes can be supressed by "weighting" of each section. This adjustment of the weighting may be done, for example, via the coupler length, L_i, such that

$$\omega_i = \frac{L_i}{L_{total}} \sum \omega_i \tag{11.11}$$

where $L_{total} = \sum L_i$. There are many possible weighting schemes to use for a filter, just as is common practice, for example, in the design of electronic filters. Three

Fig. 11.4 A plot of $N = 5$ stage coupled Mach–Zehnder (CMZ) filter transmission curve for the uniform (solid line), cosine (a $= 0.8$) (dashed line), and binomial (dotted line) coupler weight distributions

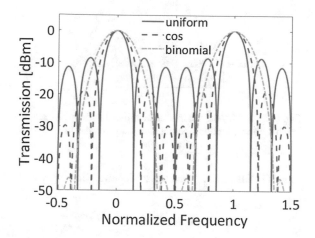

possible schemes are as follows:

$$Uniform \ \omega_i = 1 \quad (i = 1, 2, \ldots, m + 1),$$
$$Cosine \ \omega_i = \cos(\pi a((i - (m + 2)/2/m))),$$
$$Binomial \ \omega_i = \frac{m!}{(i - 1)!(m - i + 1)}$$

Note the trade-off between the width of the central maximum in transmission and the side lobe suppression among the various distributions. Thus, in going from the uniform to the binomial distribution, the width of the central peak widens; however, the sidelobe levels decrease in intensity, being -8 dB for the uniform distribution and -47 dB for the binomial distribution.

In other optical filters, this cascaded coupler filter can be characterized by its finesse, F, where

$$F = \frac{F_{FSR}}{\Delta\nu} \tag{11.12}$$

where $\Delta\nu$ is the width of the central transmission maximum, measured at a characteristic level, denoted by T_f. The finesse of a filter increases with the number of stages. Thus, for a binomial distribution,

$$\frac{1}{F} = \sqrt{\frac{4}{\pi^2 m} \ln\left(\frac{1}{T_f}\right)} \tag{11.13}$$

where, again, T_f is the transmission level used to define the width of the central lobe and m is the number of stages. In fact, the transmission function, T_{bin}, for a binomially weighted series is $T_{Bin} \approx \cos^{2m}(\Delta\phi/2)$, i.e., the response of the single device raised to the mth power. Interestingly, use of a truncated binomial distribution, i.e., dropping the first and last element, makes it easier to achieve higher finesse, since

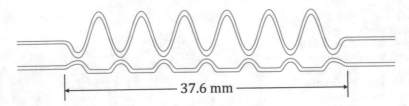

|← ———————————— 37.6 mm ————————————→|

Fig. 11.5 The layout of a resonant coupler device having seven directional coupler and six cascaded Mach–Zehnder sections. Notice that the vertical scale is expanded by 50X

then it is found that the finesse depends linearly on the number of stages,

$$F \sim \frac{1}{m} \tag{11.14}$$

Generally, device loss and fabrication variability make it difficult to use truly large numbers of stages, i.e., $m > 8$. An extensive theoretical discussion of apodized delay-line coupler filters is presented in (Kuznetsov 1994).

Despite these difficulties, there are several examples of multiple-element delay-line couplers which have been fabricated and tested. An excellent example is shown in Fig. 11.5, which shows a six-element cascaded Mach–Zehnder filter (Yaffe et al. 1994). The large length of these filters ($\sim$4 cm) makes it necessary to use only very low loss optical materials, viz., SiO_2 on Si or polymeric layers; the device in Fig. 11.5 uses SiO_2 on Si.

The device in Fig. 11.5 also incorporates two important design elements. First, the waveguides feeding the couplers are both curved away with equal radius to cancel out any spurious coupling in this region, and, second, the waveguides in the actual coupler region are tapered to enhance out-of-band rejection. The latter capability is similar to the use of adiabatic tapering discussed in Chap. 5. The response of this filter, using an inverse-cosine weighting and seven sections, is shown in Fig. 11.6, along with the output calculated from the product of the cascaded transfer functions. The measured total insertion loss at $\sim$1.65 µm was -0.25 dB.

The same technology as used for these multi-element filters has been applied to other WDM PICs. For example, a 6 cm SiO_2 on Si device (0.6mm width) has been used for mux/demux of 1.30 and 1.55 µm signals (Li et al. 1995a). This device used the wavelength layout geometry shown in Fig. 11.7. In addition, an application designed to enable an erbium-doped fiber amplifier (EDFA) gain equalization has also been described, which uses a similar filter design (Li et al. 1995b). This filter was fabricated using SiO_2/Si technology and found to have an insertion loss of 1 dB and an out-of-band wavelength rejection of 50 dB.

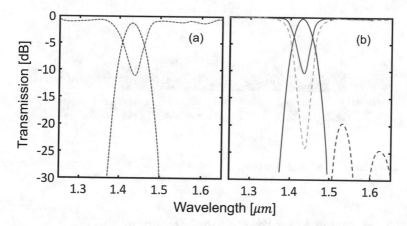

Fig. 11.6 A plot of the transmission curve of **a** a resonant coupler filter(RCF) with seven tapered couplers: solid curve bar state, dashed curve cross state. **b** simulated RCF: dashed curve—perfect couplers

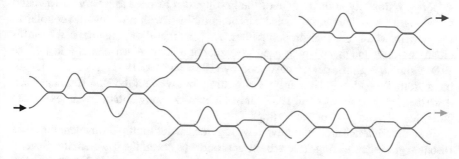

Fig. 11.7 Layout of a 1.3/1.55 μm EDFA gain equalizer. Arrows represent input and output ports. The total length of the device is 60 mm and the width is 0.6 mm. Notice that the vertical scale is stretched 20 times the length

11.4 Devices Based on Diffraction Gratings

11.4.1 In-Line Filters Based on Bragg Gratings

Recall that in Chap. 10 several applications of Bragg gratings in wavelength filtering were discussed. Two of these applications considered in-line filters. In the first, a simple Bragg filter can be used to block a range of wavelengths. For a grating with coupling coefficient κ, the bandwidth is proportionate to κ. On the other hand, if a $1/4\,\lambda$ step is located in the midst of the grating, the grating will act as a narrowband filter, e.g., only a simple telecommunications λ will be passed. More details on these filters are given in Chap. 10.

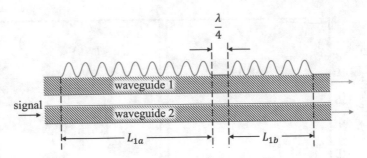

Fig. 11.8 A sketch of a Bragg filter consists of two waveguides with a coupling coefficient κ. The drop waveguide (upper) has a $1/4\ \lambda$ step Bragg diffraction grating

11.4.2   A Bragg Filter for Channel Dropping

As explained in Chap. 10 on diffraction gratings, Bragg gratings containing a $1/4\ \lambda$ step, approximately in the middle of the grating, can be used as a very narrowband wavelength filter. This same device can be combined with a waveguide coupler to form a narrowband channel-dropping filter. This channel-dropping filter can be sufficiently narrowband that it can be used to select out individual wavelengths in a dense WDM signal, e.g., the device described by (Haus and Lai 1991) has a wavelength bandwidth of ~ 0.5Å. A similar type of grating, but without the $1/4\ \lambda$ step, has been described by (Kazarinov et al. 1987). It has a narrow bandwidth but requires a more complicated geometry for the gratings.

In such a channel-dropping filter, a waveguide carries multiple wavelengths and a resonant grating reflecting structure extracts or drops one of the wavelengths. Several different and very advanced forms of this device have been described. However, here, the device proposed by Haus and Lai (1991), shown in Fig. 11.8, will be discussed. In the device, many wavelengths propagate along the bus waveguide but only one wavelength couples resonantly to the drop port for detection or other use. The device uses two waveguides, with a coupling coefficient κ, in conjunction with a $1/4\ \lambda$ step Bragg diffraction grating on one of the waveguides. The coupled structure has been the subject of many theoretical studies and several experimental realizations (Kazarinov et al. 1987; Levy et al. 1992). Recall that this basic grating structure was discussed in detail in Chap. 10.

In order to illustrate the basic principles, we will examine only the simplest version, namely, which is shown in Fig. 11.8. In this structure, the "bus" waveguide, i.e., waveguide 1, will only couple to the drop waveguide, waveguide 2, if one of its wavelengths lies within the passband of the $1/4\ \lambda$ step grating arm, that is, into the central resonance of the Bragg resonator. Wavelengths outside of this band, but still in the stop band of the Bragg resonator, will be reflected and will not couple. The expected narrow width of the central resonance of $1/4\ \lambda$ step grating suggests that a narrow filter can be made.

This structure may be analyzed numerically by solving the following coupled-mode equations, which are written for the case of identical waveguides:

$$\frac{dA_1}{dz} = -j\delta A_1 + \kappa B_1 - j\mu A_2 \tag{11.15}$$

$$\frac{dB_1}{dz} = j\delta B_1 + \kappa A_1 - j\mu B_2 \tag{11.16}$$

$$\frac{dA_2}{dz} = -j\delta A_2 - j\mu A_1 \tag{11.17}$$

$$\frac{dB_2}{dz} = j\delta B_2 + j\mu B_1 \tag{11.18}$$

where $A_{1,2}$ and $B_{1,2}$ denote the forward and backward amplitudes of the optical wave in the two waveguides shown in the figure, $\delta = (\omega - \omega_0)/\nu_g$ is the detuning parameter, ω_0 is the frequency at band center, ν_g is the group velocity, μ is the waveguide coupling coefficient, and κ is the grating coupling coefficient. Notice that the detuning parameter, δ, is in the same form as in Sect. 5.3.2, and field amplitudes are those without the rapid grating-induced variation (Haus and Lai 1991).

However, a simpler approach has been adopted by (Haus and Lai 1991) to determine the behavior of these filters, which uses only power-loss arguments. This analysis shows that filter bandwidth, $\Delta\omega_f$, and the stop band, $\Delta\omega_s$, are given as follows:

$$\Delta\omega_f = 8\nu_g\kappa|\frac{\mu}{\kappa}|^2 \tag{11.19}$$

and

$$\Delta\omega_s = 2\nu_g\kappa \tag{11.20}$$

Clearly the two quantities are related by the ratio $|\mu/\kappa|^2$; in fact, these quantities give the ratio of the power stored in the DFB resonator versus the power in the bus waveguide. As an example of the results obtained from such an analysis, typical plots of the output from such a filter are shown in Fig. 11.9 for the case of $\kappa = 200\,\text{cm}^{-1}$, $\mu = 5\,\text{cm}^{-1}$, $L_{1a} = 250\,\mu\text{m}$, $L_{1b} = 150\,\mu\text{m}$, (Levy et al. 1992).

An analytic solution to the equations has been proposed by (Haus and Lai 1992). Among other results, this solution showed that in a single-stage device, such as the type shown in Fig. 11.10, a maximum of 30% power output coupling could be achieved. As a result, more complex multistage devices have been proposed to drop a higher fraction of the input power.

Several experimental versions of these channel-dropping filters have been made including those in $GaAs/AlGaAs$ (Levy et al. 1992) and in SOI (Yu et al. 2009). The results are very close to those predicted from the analysis given above. The chief impediment to the practical implementation of this device is fabrication of reproducible directional couplers.

Fig. 11.9 Theoretical
response of the
channel-dropping filter with
$^1/_4\,\lambda$ step grating. The inset is
the zoomed figure [Adapted
from (Levy et al. 1992)]

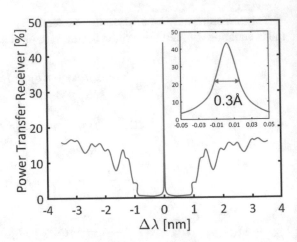

Fig. 11.10 The 3D design of
the Bragg grating for channel
dropping [Adapted from
(Levy et al. 1992)]

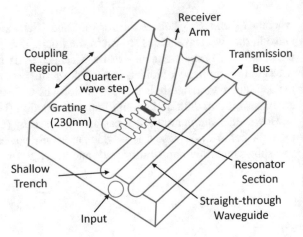

11.5 Waveguide-Grating Router

11.5.1 Introduction

One very important and widely used waveguide routing device (Smit 1988), is a
PHASAR, or waveguide-grating router. Such a device makes use of calibrated delay
lines, which are formed via a curved, parallel array of waveguides. These delay lines
are "fed" via uniform illuminated light from the input waveguide, which has first been
expanded in angle via diffracting in a free-space region. After passage through the
delay lines, the outputs from the delay waveguides are focused into a set of specific
wavelength channels formed by the PHASAR output guides. A "waveguide-grating"
router can serve as a simple $1 \times N$ demultiplexer (Smit 1988) or a $N \times N$ wavelength
router (Dragone 1991).

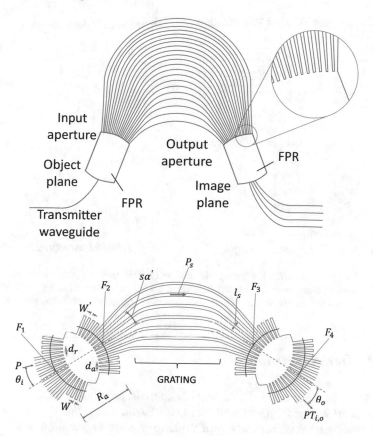

Fig. 11.11 A sketch of a $1 \times N$ and an $N \times N$ arrayed waveguide-grating (AWG) router

The basic device is shown in Fig. 11.11. It consists of two symmetrically placed star couplers separated by an uncoupled waveguide array with a graduated path difference between each waveguide. Because this graduated path difference leads to a precise phase relationship between neighboring guides at the array output, it is sometimes called a "waveguide grating" since a diffraction grating also has such a wavelength-sensitive delay. If the device receives a specific input wavelength, then, as a result of the optical imaging and phase delays, it steers this wavelength to a specific output guide. The graduated path difference in the waveguide array causes a tilting of the wavefront emerging from the grating array at an angle, which is distinctive for each input wavelength.

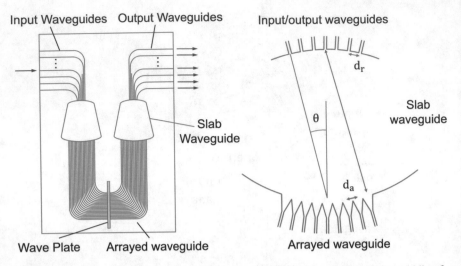

Fig. 11.12   a The schematic of a waveguide with polyimide $1/2$-wave plate in the middle of a waveguide-grating array to equalize the path lengths for the TE-polarized and the TM-polarized guided waves. **b** A detailed schematic of the slab waveguide [Adapted from (Takahashi et al. 1995)]

11.5.2   Device Structure

PHASARs have been made of a variety of materials, principally SiO2 on Si (Taka-hashi et al. 1993) or InP (Smit and Van Dam 1996). The actual device, on which Fig. 11.11 is based, is made of SiO_2 on a Si platform. For comparison, a second design for such a waveguide router is shown in Fig. 11.12. As discussed below, polarization insensitivity is essential for realizing a useful router. Thus, in the router in Fig. 11.12, a polyimide $1/2$-wave plate can be included in the middle of a waveguide-grating array to equalize the path lengths for the TE-polarized and the TM-polarized guided waves (Takahashi et al. 1993).

11.5.3   Basic Functions of the Waveguide Array

A waveguide array provides two important functions for a PHASAR: focusing of the light onto one of the receiver waveguides and providing angular dispersion for wavelength routing. Consider first the focusing function.

Focusing is accomplished by reversing the optical transformation achieved when the mode from the input waveguide diffracts into the free-space region and is received at the waveguide array. Thus, a uniform phase distribution fed by the curved end face of the waveguide array is the very reverse of the input section, and will thus yield a focused beam. This phase front will have a uniform phase when the length difference,

Fig. 11.13 A measured transmission spectrum of a PHASAR (de)multiplexer from central input port to central output port (Takahashi et al. 1995) *Source* https://ieeexplore. ieee.org/document/372441

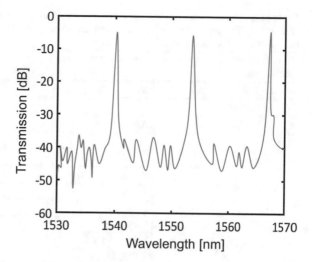

ΔL, between adjacent waveguides, is equal to an integral number, m, of wavelengths in the waveguide; thus

$$\Delta L = \frac{m\lambda_c}{N_{eff}^f} \qquad (11.21)$$

where λ_c is the central wavelength of the array and N_{eff}^f is the effective index of the waveguides. Further, in order to reduce overlap of multiple orders, it is necessary that the waveguides be as closely spaced as possible, i.e., typically within one waveguide width of each other. This condition is similar to the design principle that a diffraction grating uses its lowest diffraction order when its groove spacing is small. The input and output apertures are spaced at a distance R_a from the receiver waveguides. In this case, using an analogy to a Rowland mounting for diffraction gratings, the output/input waveguides must be on a circle of $R_a/2$. One of the most important advantages of Rowland mounting is that its image does not have coma distortion. It also can be designed to eliminate spherical aberration for a linear phase distribution (März 1995). Figure 11.13 shows a measured transmission spectrum from central input to central output.

Second, the waveguide array provides angular or spatial dispersion at each wavelength. In order to discuss this dispersion, we use the simplified geometry shown in Fig. 11.12. This chapter focuses on the essential physics of the device dispersion and thus ignores some design elements of the device, such as the waveguide mutual coupling near the input/output aperture. As described in Chap. 9 in conjunction with star couplers, this coupling is important to provide uniform illumination of the receiver waveguides and, thus, to obtain the best performance in each of these imaging devices. From Fig. 11.12, the equivalent grating equation, for order m, of the system is given by

Table 11.1 Typical quantities for a reported near-commercial waveguide-grating array (Takahashi et al. 1993). The parameters apply to the device shown in Fig. 11.12. Typical parameters used for PHASARs are InP, SiO2, and Si

Basic materials and optics parameters	Device parameter	
$\lambda = 1.5538$	N	16
$N_{eff}^s = 1.4529$	R_a	1 cm
$N_{eff}^f = 1.4513$	ΔL	130 μm
$n_g = 1.4752$	d_a	25 μm
$\omega_o = 4.5$ μm	m	118
	d_r	25 μm
	$\Delta\nu$	100 GHz (0.8*nm* at 1.55 μm)
	FSR	1600 GHz
	Cross talk	−30 dB
	Insertion loss	5 dB
	$FWHM$	30 GHz (∼4 μm spot size)

$$N_{\text{eff}}^s d_a \sin\theta_i + N_{\text{eff}}^f \Delta L + N_{\text{eff}}^s d_a \sin\theta_o = m\lambda \tag{11.22}$$

where N_{eff}^s and N_{eff}^f are the effective indices of the free propagation region (FPR) and the waveguide, respectively; θ_i and θ_o are angles of the diffracted wavefront at the input and output slabs from the emitting and the receiving (output) waveguides, respectively; d_a is the spacing of the array guides; and ΔL is the length difference between successive waveguides in the grating array. The angles $\theta_{i,o}$ are given as follows:

$$\theta_i = \frac{i \cdot d_r}{R_a} \tag{11.23}$$

$$\theta_o = \frac{o \cdot d_r}{R_a} \tag{11.24}$$

where i, o is the number index of the input/output waveguides, measured from the center of the receiver waveguide array for wavelength λ, and d_a is the waveguide spacing as shown in the figure. Note that (11.21) assumes that these angles were zero. Typical numbers for the parameters used in (11.22) are given in Table 11.1. Observe that for the center wavelength of the router, λ_c, (11.22) reduces to $N_{eff}^f \Delta L = m\lambda_c$, i.e., (11.21) holds.

Differentiating (11.22) gives the angular dispersion of the grating in the vicinity of the center wavelength, where $\theta_{i,o} \approx 0$; thus

$$\frac{d\theta}{d\nu} = -\frac{m\lambda^2 n_g}{N_{\text{eff}}^s d_a c N_{\text{eff}}^f} \tag{11.25}$$

and where the group index, n_g, of the waveguide is given by

$$n_g = N_{\text{eff}}^f - \frac{dN_{\text{eff}}^f}{d\lambda} \cdot \lambda \tag{11.26}$$

and c is the speed of light in vacuum. The quantity in (11.25) is easily converted into the spatial dispersion of the device, $dx/d\nu$, near the center frequency by multiplying that equation by R_a to give

$$\frac{dx}{d\nu} = -\frac{R_a m\lambda^2 n_g}{N_{\text{eff}}^s d_a c N_{\text{eff}}^f} \tag{11.27}$$

The channel spacing is then given by multiplying $d\nu/dx$ by d_r, the spacing or pitch of the receiver arrays, to yield

$$\Delta\nu = \frac{d_r}{R_a} \left(\frac{m\lambda^2 n_g}{N_{\text{eff}}^s d_a c N_{\text{eff}}^f} \right)^{-1} \tag{11.28}$$

Figure 11.14 demonstrated an 18-channel AWG (de)multiplexer with channel spacing of 200 GHz. The cross talk is -18 dB and the insertion loss is -6 dB.

Fig. 11.14 An 18-channel AWG (de)multiplexer with channel spacing of 200GHz. The cross talk is -18dB and the insertion loss is -6dB.

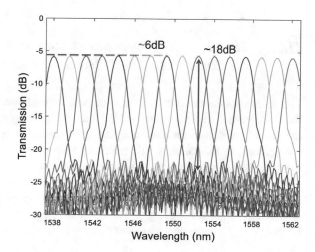

11.5.4   Free Spectral Range

In order for the device to operate as a useful filter, it is necessary for the router to operate over the entire spectrum of interest within a single "diffraction" order, m. Alternatively stated, the frequency or wavelength range of the device must be such that the spectral response does not overlap on itself; this requirement is guided by the free spectral range (FSR) of the device. If the channel spacing is $\Delta \nu$ and the total channel number is N, the minimum useful free spectral range (FSR) is $N \Delta \nu$. Thus, when $\nu \to \nu + FSR, m \to m + 1$. At this frequency, $N_{\text{eff}}^s \to N_{\text{eff}}^s + \Delta N_{\text{eff}}^s$ and $N_{\text{eff}}^f \to N_{\text{eff}}^f + \Delta N_{\text{eff}}^f$, where the differences $\Delta N_{\text{eff}}^{s,f} = -\frac{FSR}{\nu}(\lambda(dN_{s,f}/d\lambda))$. Substitution of these quantities into (11.22) gives a difference equation for FSR as

$$(FSR)^{-1} = \frac{1}{c}\left[\left(N_{\text{eff}}^f - \lambda\frac{dN_{\text{eff}}^f}{d\lambda}\right)\Delta L + \left(N_{\text{eff}}^s - \lambda\frac{dN_{\text{eff}}^s}{d\lambda}\right)(d_a \sin\theta_i + d_a \sin\theta_o)\right] \tag{11.29}$$

$$\approx \frac{1}{c}\left[n_g(\Delta L + d_a \sin\theta_i + d_a \sin\theta_o)\right] \tag{11.30}$$

where c is the speed of light in vacuum. For the diagonal ports in the device $(0, 0)$, $(1, -1)$, $(2, -2)$, etc., the inverse of the free spectral range is given by $(FSR)^{-1} = n_g \Delta L/c$.

11.5.5   The Frequency-Dependent Transmission Function of the Router, i.e., The PHASAR Frequency Response

The focal spot centered at each output port depends on the focusing properties of the waveguide array. As described in Chap. 9, the imaging quality and transmission can be enhanced by introducing a degree of mutual coupling between neighboring waveguides. In this section, this coupling will be neglected and the waveguides in the array and the two apertures will be assumed to be the same! In addition, use of a Rowland optical layout geometry is assumed since its usage leads to high-quality images. The focal spot size of the grating on the output port can be fit with a Gaussian distribution, and the coupling of the input and output waveguides is then that of two coupled Gaussian distributions of spot size ω_0, laterally separated by a displacement x, i.e.,

$$T(\delta\nu) \sim \exp\left(-\left(\frac{x}{\omega_0}\right)^2\right) \tag{11.31}$$

Fig. 11.15 A plot of measured performance of the 16-port PHASAR wavelength demultiplexer near a resonance peak (Takahashi et al. 1995) *Source* https://ieeexplore. ieee.org/document/372441

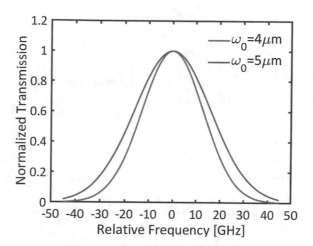

The frequency-dependent transmission function of the focal spot thus depends on the frequency difference, $\delta\nu$, from the center frequency.

$$T(\delta\nu) \approx \exp\left(-\frac{\delta\nu}{\omega_0}\frac{dx}{d\nu}\right)^2 \qquad (11.32)$$

where $d\nu/dx$ is the spatial linear dispersion of the device and ω_0 is the focal spot size. As pointed out in Takahashi et al. (1995), this response function is the same as a Gaussian bandpass filter, which has a FWHM or $\Delta\nu = 2\sqrt{\ln 2}\,\omega_0(d\nu/dx)$. The spectral response or resolution of the device is thus set by the imaging quality. Typical numbers for device resolution are ~10 s of GHzs. A plot of the calculated and measured value for the 16-part device of Takahashi et al. (1995) is given in Fig. 11.15. Since the frequency response is directly related to the focal spot size, ω_0, factors which give well-defined focal spot, such as using the Rowland mount, etc., are important for good frequency filtering as well. A typical first approximation, described by Smit and Van Dam (1996), is that the focal spot size is equal to the effective width of the modal field times a constant, ω_e, and thus $\omega_0 = \omega_e\sqrt{2/\pi}$.

11.5.6 PHASAR Uniformity and Insertion Loss

An issue related to, but separate from, the resolution of each wavelength channel is the overall wavelength response envelope of the device (Smit and Van Dam 1996). A diagram showing an example of this response is given in Fig. 11.16. The diagram shows both the overall insertion loss at the central wavelengths plus the envelope function that causes nonuniformity of the wavelength channels of the device. Notice that in this example a portion of the light passing through the device appears in orders

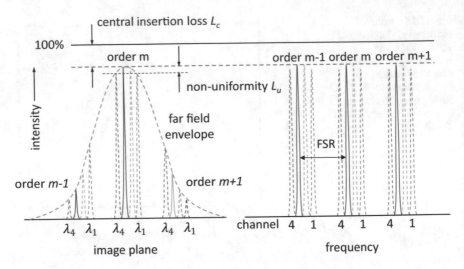

Fig. 11.16 A diagram showing an example of the overall wavelength response envelope of the device. The diagram shows both the overall insertion loss at the central wavelength plus the envelope function that causes nonuniformity of the wavelength channels of the device

different from m; the existence of these other orders results in the insertion loss for the desired light in order m.

This envelope function of the wavelength channels can be found by considering a Gaussian transformation to be

$$I(\theta) = I_o \exp(-2\theta^2/\theta_0^2) \tag{11.33}$$

where $\theta_0 = \pi\lambda/N_{\text{eff}}^s w_0$. Note that w_0 is related to the effective modal half width in the waveguide, w_e, by $w_0 = w_e\sqrt{2/\pi}$. The value of $2w_e$ has been discussed earlier for slab waveguides, where it was called t_{eff}. In the notation here, it is

$$w_e = \frac{\omega}{2}\left(1 + \frac{2}{\gamma}\right) \tag{11.34}$$

$$\approx \frac{\omega}{2}\left(0.5 + \frac{1}{V - 0.6}\right) \tag{11.35}$$

where ω is the geometric width of the waveguide and V is the normalized frequency of the guide. Equation 11.35 was obtained by curve fitting the variation in γ with the normalized frequency V.

This function leads to a nonuniformity L_u, i.e., $10\log(I_{\text{max}\theta}/I_{\text{central}})$, across the full spectral response, as depicted in Fig. 11.16, which is equal to

$$L_u \approx 8.7\theta_{\text{max}}^2/\theta_0^2 \tag{11.36}$$

where $\theta_{\max} \approx x_{\max}/R_a$ and $x_{\max}$ is the transverse distance in the receiver array and L_u is in dB. The uniformity is thus $\propto N d_r$. Note that this uniformity results in additional insertion loss for the outer channels compared to the central channel.

The loss of the central channel is due to diffraction into higher orders. This loss, which is due to the presence of any outer $m \pm 1$ order and has a relative magnitude to that in the center can be obtained by realizing it is displaced in frequency by $\Delta \nu \approx \nu_{FSR}$. Thus, the power in this order, say $m + 1$, will be

$$P_{m+1} \sim \exp(-2\Delta\theta_{FSR}^2/\theta_0^2) \tag{11.37}$$

where

$$\Delta\theta_{FSR} = \frac{1}{R_a}\frac{dx}{d\nu}\Delta\nu_{FSR} \tag{11.38}$$

If only the adjacent $m \pm 1$ orders are considered, the power loss in dB can be estimated to be (where we have assumed that the propagation loss is zero)

$$loss \approx 17\exp(-2\pi^2\omega_0^2/d_a^2) \tag{11.39}$$

This loss includes that occurring at both the input and output stages and the fact that light is lost into two orders. Notice that as mentioned earlier, d_a should be small to minimize coupling into higher grating orders.

11.5.7 Channel Cross Talk

Cross talk is a major consideration in designing routers. It arises from fabrication imperfections, such as from index fluctuations as well as from factors, which can be controlled by good design, including, most obviously, the overlap of the modal fields in adjacent receiver waveguides. This overlap is clearly a function of the receiver waveguide spacing, d_r, and the receiver modal width, $\sim w_e$. Second, truncation of the array aperture as a result of an abrupt end on a waveguide introduces a diffraction ripple on the image. This spatial imaging spreads light into orders other than m. Similar effects can arise from modal conversion in the array waveguides. Finally, curves in the array waveguides can also lead to conversion into higher modes. These modes radiate power at higher angles and couple into the $m + 1$ and $m - 1$ modes.

The first two sources can be placed on more quantitative footing using an analysis by Smit and Van Dam (1996). Consider first cross talk due to the overlap of fields in the receiver waveguides. This overlap is a result of the lateral "tail" of the mode in the input waveguide, which is being imaged into one receiver waveguide, overlapping onto a neighboring input waveguide. Since the "tail" portion of this mode is not well fit by a Gaussian distribution, unlike the central region of the mode, the overlap integral must consider the actual shape of the waveguide mode as a function of V

and the normalized receiver waveguide spacing, d_r/w. The results of this calculation give cross talk versus d_r/w for several values of the waveguide parameter V.

The second source of design-controllable cross talk is that due to the finite angular width of the array aperture, $2\theta_a$. At the input region of the aperture, this finite width causes loss since a finite aperture does not intercept the entire radiation lobe, characterized by θ_0, from the input waveguide. At the exit aperture of the delay array, a finite width gives rise to sidelobes on the focused image of each receiver channel. These sidelobes can then cause cross talk to their deviation from a simple single-lobe image. This source of cross talk is clearly a function of the actual relative position of the receiver waveguide. It is, however, relatively insensitive to the waveguide V.

11.5.8   PHASAR Polarization Properties

Birefringence in the waveguide router can lead to major degradation of the output. If the router responds differently to the two polarizations, then the output, as shown in Fig. 11.17, will exhibit a dual response at each of the two polarizations. This splitting, $\Delta\nu_{pol}$, can be expressed as

$$\Delta\nu_{pol} \approx \nu \left(\frac{N_{eff}^{TE} - N_{eff}^{TM}}{n_g^{TE}} \right) \qquad (11.40)$$

where it is understood that $N_{eff}^{TE,TM}$ refers to the effective index of the waveguide for TE or TM polarization. In one example, i.e., $InGaAsP - InP$ waveguides, the shift was ~5 nm. Since this shift scales as N_{eff}^f it is clearly expected to be low in low-index materials.

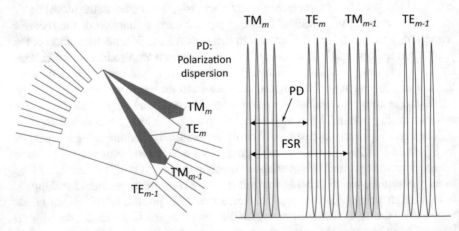

Fig. 11.17   A diagram showing the output if the router responds differently to the two polarizations. It will exhibit a dual response at each of the two polarizations [Adapted from (Smit and Van Dam 1996)]

Several approaches have been adopted to reduce birefringence, including overlapping the m and $m + 1$ orders for different polarizations, use of symmetrical waveguides, insertion of $\lambda/2$ plate in the array section, and compensation of the dispersion by surface loading of the array. Recall that a diagram showing the insertion of $\lambda/2$ plate was depicted in the router given in Fig. 11.12; in this case, the plate was a thick, free standing polyimide film.

11.5.9   Design of a Multiplexer/Demultiplexer

Practical design of a PHASAR-type device can be illustrated by one specific approach to the design of a wavelength demultiplexer/multiplexer, the "Delft" procedure (Smit and Van Dam 1996). It is assumed that the basic waveguide structure is fixed by the materials technology used: the waveguide can be described by its width, w, and lateral normalized frequency, V. Further, the device channel spacing, $\Delta\nu$, and the total number of channels are specified as a result of the system needs. Finally, the waveguide spacing d_a is chosen to be as small as possible, consistent with maintaining independent waveguides; thus again in this illustration, the coupling between waveguides is ignored. A small spacing, i.e., d_a, is important since the spacing allows a large proportion of the diffracted input beam to be captured by the array. Note that the presence of collecting horns (or reverse tapers) can be used to increase collection.

   The first design goal is to achieve an acceptable degree of receiver cross talk, which is determined by the spacing of the receiving waveguides compared to their width. This spacing is set by the channel spacing. This overlap spacing results from the finite size of the input-waveguide image between neighboring waveguides. This overlap can be adjusted via the relative receiver waveguide spacing, d_r/w. To determine d_r/w, the cross talk is set by the material and fabrication tolerance, say $-40\,\text{dB}$, and then use the geometry in figure a along with our waveguide V to obtain d_r/w.

   Next, the radius of the free-space propagating region must be selected. This dimension is found from the acceptable margin on uniformity, which in turn fixes θ_{max} as shown in (11.36). This expression gives the excess loss for the outermost waveguides, which are at angle θ_{max}, compared to that of central waveguide. Once this nonuniformity is decided, θ_{max} is specified, then using the basic geometric relation $\theta_{max} = Nd_r(2\theta_{max})$, i.e., the receiver waveguide spacing and the total number of waveguides, the value $R_a \approx Nd_r(2\theta_{max})$ is also determined.

   The array length difference, ΔL, is then obtained via the full device dispersion relation, which is equal to $d_r/\Delta\nu$, i.e., the choice of receiver waveguide spacing and the fixed channel spacing. Specifically, the required ΔL is obtained from the angular dispersion (11.25) and the fact that $N_{\text{eff}}^f \Delta L = m\lambda_c$, yielding this relation:

$$\frac{d_r}{\Delta\nu} = \frac{1}{\nu_c} \frac{n_g}{N_{eff}^s} \frac{\Delta L R_a}{d_a} \tag{11.41}$$

where use had been made of the fact that $dx/d\nu = d_r/\Delta\nu$. This expression can then be manipulated to obtain an equation for ΔL.

The angular half-width of the array aperture, θ_a, is determined by the fact that a finite angular width causes sidelobes in the focused spot at the receiving aperture. The choice of θ_a then fixes the number of array waveguides, N_a, due to simple geometric considerations,

$$N_a = 2\theta_a R_a/d_a + 1 \tag{11.42}$$

Note that two general PHASAR layouts have been used. These are shown in Figs. 11.11 and 11.12. Figure 11.11 is the most straightforward design since the bend radius is constant throughout, while Fig. 11.12 requires less waveguide transitions. These two designs have been used for SiO_2 and III-V devices, respectively.

Finally, note that this design procedure is useful only for a simple $1 \times N$ PHASAR. Wavelength routers require a somewhat different approach; this different procedure is given in the paper by Smit and Van Dam (1996).

11.5.10 More Advanced Design Features

Since the development of the routers, router design has been continually refined. As a result, many new features have been realized. For example, chirping or even double-chirping of the waveguide lengths in the router can be used to eliminate side lobes (Doerr and Joyner 1997). The reader is referred to the excellent reviews by Smit and Van Dam (1996) and by Takahashi et al. (1995) for more details and examples.

11.5.11 MMI Phased-Array Wavelength (De)multiplexer

PHASARs can also be made by replacing its star-coupler section with an MMI-based coupler. These MMI-PHASARs have an inherently periodic spectral response and are of a compact size for the typical number of channels. In addition, in comparison with conventional star-coupler-based devices, these phasers have low insertion loss, better uniformity, and simpler device structure. However, they have relatively high cross talk as a result of their narrow low crosstalk window, and their wavelength response cannot be further flattened. The crosstalk problem has prevented their application to wavelength routers, especially when N, the channel number, is large. Recently, an optimized index-contrast technique (Van Dam et al. 1995) has been proposed to improve the crosstalk performance for $N \times N$ MMI-PHASARs, especially for large N.

MMI-based PHASAR demultiplexers are particularly useful for multi-wavelength lasers since in this case cross talk is not a central issue (see below). Thus far, 1×4, 4×4 and 1×5 MMI-PHASARs have been fabricated, all using high-index contrast deep-etched structures on InP (Smit and Van Dam 1996). MMI-PHASARs

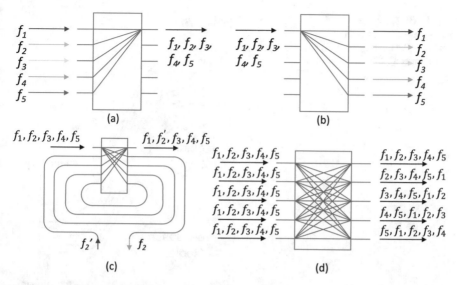

Fig. 11.18 Basic functions of arrayed-waveguide $N \times N$ multiplexer. **a** Multiplexing; **b** Demultiplexing; **c** Add-drop multiplexing operation; **d** $N \times N$ interconnect

have also been operated either passively or dynamically. Electro-optically controlled phase shifters can be added to each phase-arrayed waveguide to achieve different wavelength routing permutations. This type of MMI-PHASARs can be used in a reconfigurable add-drop (de)multiplexer (see Fig. 11.18) to greatly reduce the device complexity, compared to that in star-coupler-based PHASARs, although again, the issue of cross talk is important in this application. Other applications of arrayed-waveguide multiplexers are given in Fig. 11.1. Finally, the reader is reminded that a discussion of MMI-based PHASARs is included in Chap. 9, although the discussion in that chapter has a different motivation.

11.5.12 Applications of PHASARs

Because of their very powerful wavelength multiplexing and demultiplexing functionalities, PHASARs have become an extremely versatile component in integrated optic multi-wavelength applications. These applications are so numerous that we will only list a few examples in this section. Some of these applications are reviewed by (Smit and Van Dam 1996).

As an example, a simple but important class of waveguide-grating router applications is in integrating an semiconductor amplifier with a grating section. This combination can be used in conjunction with cleaved "chip" facets for an integrated series of laser sources. These sources have well-defined, fixed frequencies as set by the PHASER device. However, tuning of the laser source or sources can be achieved

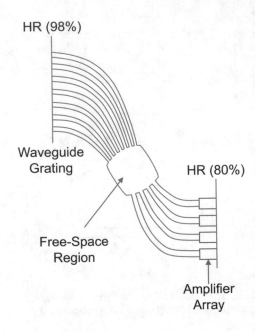

Fig. 11.19 InP-based
integrated series of laser
sources [Adapted from
(Zirngibl et al. 1994)]

by selectively turning on or off the amplifier sections. An example of such a device
is shown in Fig. 11.19 (Zirngibl et al. 1994). The device was made of InP-based
materials. It had four laser sections for CW emissions in the vicinity of $\sim$1.5 μm
into seven wavelength channels.

11.6 Summary

Wavelength manipulation, including filtering, is at the very heart of the utility of
integrated optics in WDM systems. This chapter has examined at examples of three
different technologies for accomplishing this filtering. Of these three device types,
the most important commercially is that of the wavelength router. Its ability to manipulate and route multiple wavelengths has enabled a large number of crucial system
applications to be demonstrated. Routers have been shown to be important in either
SiO_2 or III-V material platforms; the application of single-mode Si photonics to this
device is however more of a challenge due to the very strong waveguide dispersion
in Si wire devices.

In addition, each of the three device types given in this chapter illustrates a different
feature of designing PICs. These include the use of transverse matrix design for the
delay-line coupler filter, use of Bragg gratings for resonators and coupling in the $\frac{1}{4}$ λ-
shift channel-dropping filter, and phase arrays and planar optics for routers. While
each of these design areas appears in many other integrated optical devices, their use

for wavelength filtering is central for one of the most important areas of integrated
devices.

Problems

1. (a) Calculate the channel uniformity for the PHASAR having the characteristics
      of that in Table 11.1.
   (b) Calculate the spatial dispersion near $\lambda_c = 1.55$ mm.
   (c) Use your skill with waveguides to show how to obtain w_0. You will first
      need to obtain the V of the waveguide. Again use the data in Table 11.1.
   (d) Determine the loss of the device based on (11.39). How does it agree with
      the 5dB value in Table 11.1?
   (e) Derive (11.41).
2. Optical Modulator: You wish to build the following phase modulator in LiNbO3
   using proton exchange:

   (a) What is the crystal orientation for using r_{33}?
   (b) What is the width, w, for single-mode performance?
   (c) What is the estimated Γ if electrode spacing equals $1.5\,w$?
   (d) What is a typical $V_\pi L$ for phase modulation in LiNbO3?
   (e) Calculate V_π length if $V = 10$ V and $f_{RF} \sim 100$ MHz (do not use 4).
   (f) What is C/L in pF/cm using graph in notes?

3. Use what you have learnt so far about PHASOR, given the following parameters
   of a PHASAR in Silica, try to answer the questions.
   $$\lambda_s = 1.5\,\mu m$$
   $$N^s_{eff} = 1.4529$$
   $$N^f_{eff} = 1.4513$$
   $$W_0 = 4.5\,\mu m$$
   $$n_g = 1.4752$$
   $$d_r = d_a = 30\,\mu m$$
   $$R_a = 2\ cm$$

   (a) What is the required ΔL for diffraction order $m = 150$?
   (b) What is the channel spacing $\Delta\nu$?
   (c) If 16 channels are needed, what will be the nonuniformity in dB?
   (d) What is the rough power loss of this device if there is no propagation loss?

4. Derive (11.30), the expression for FSR of a PHASOR.
5. If you test a grating reflector, and find the coupling coefficient $\kappa = 200$ cm^{-1}.
   Assume the group index of this grating is in the order of unity. Answer the
   following questions:

Fig. 11.20 Cross-sectional
view of a modulator

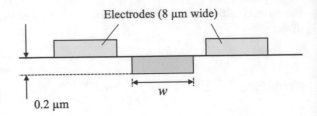

Electrodes (8 μm wide)

w

0.2 μm

    (a) How long must the grating be to get a reflection rate $R = 99\%$ at the detuning $\delta = 0$?

    (b) What is the stop band of the grating?

6. Derive the filter function as in Fig. (11.3) for a three-stage uniform filter. Assume the input fields of the input ports are a_i and b_i, respectively, and the couplers are 50:50 couplers. Hint: use matrix representation of photonic devices.

7. Consider the PHASAR shown in figure. Assume the PHASAR is designed for the following "mission": channel spacing 100 GHz, $\lambda_c = 1.55\,\mu m$ and has the following parameters: $N_{\text{eff}}^s = 1.453$, $N_{\text{eff}}^f = 1.451$, $w_0 = 4.5\,\mu m$, $N = 16$, $R_a = 1\,cm$, $d_a = d_r = 25\,\mu m$, $\Delta\nu = 100\,GHz$

    (a) Calculate the spatial dispersion in $\mu m/GHz$ used in this device.

    (b) Estimate the fractional nonuniformity of the images on the receiver array, that is, what approximately is $I_{\text{outermost}}/I_{\text{centeal}}$. Notice the answer should *NOT* be in *dB*.

    (c) Write down as an equation the condition for ΔL to have an image of order m on the receiver array (Fig. 11.20).

References

Alferness, R., Buhl, L., Koren, U., Miller, B., Young, M., Koch, T., et al. (1992). Broadly tunable ingaasp/inp buried rib waveguide vertical coupler filter. *Applied physics letters*, *60*(8), 980–982.

Bornholdt, C., Kappe, F., Müller, R., Nolting, H.-P., Reier, F., Stenzel, R., et al. (1990). Meander coupler, a novel wavelength division multiplexer/demultiplexer. *Applied physics letters*, *57*(24), 2517–2519.

Chen, S., Fu, X., Wang, J., Shi, Y., He, S., & Dai, D. (2015). Compact dense wavelength-division (de) multiplexer utilizing a bidirectional arrayed-waveguide grating integrated with a mach-zehnder interferometer. *Journal of Lightwave Technology*, *33*(11), 2279–2285.

Doerr, C., & Joyner, C. (1997). Double-chirping of the waveguide grating router. *IEEE Photonics Technology Letters*, *9*(6), 776–778.

Dragone, C. (1991). An n* n optical multiplexer using a planar arrangement of two star couplers. *IEEE Photonics Technology Letters*, *3*(9), 812–815.

Haus, H. A., & Lai, Y. (1991). Narrow-band distributed feedback reflector design. *Journal of lightwave technology*, *9*(6), 754–760.

Haus, H. A., & Lai, Y. (1992). Narrow-band optical channel-dropping filter. *Journal of lightwave technology*, *10*(1), 57–62.

Herben, C., Vreeburg, C., Leijtens, X., Blok, H., Groen, F., Moerman, I., et al. (1997). Chirping of an mmi-phasar demultiplexer for application in multiwavelength lasers. *IEEE Photonics Technology Letters, 9*(8), 1116–1118.

Kazarinov, R., Henry, C., & Olsson, N. (1987). Narrow-band resonant optical reflectors and resonant optical transformers for laser stabilization and wavelength division multiplexing. *IEEE journal of quantum electronics, 23*(9), 1419–1425.

Kuznetsov, M. (1994). Cascaded coupler mach-zehnder channel dropping filters for wavelength-division-multiplexed optical systems. *Journal of Lightwave Technology, 12*(2), 226–230.

Levy, M., Eldada, L., Scarmozzino, R., Osgood, R., Lin, P., & Tong, F. (1992). Fabrication of narrow-band channel-dropping filters. *IEEE photonics technology letters, 4*(12), 1378–1381.

Li, Y., Henry, C., Laskowski, E., Mak, C., & Yaffe, H. (1995a). Waveguide edfa gain equalisation filter. *Electronics Letters, 31*(23), 2005–2006.

Li, Y., Henry, C., Laskowski, E., Yaffe, H., and Sweatt, R. (1995b). Monolithic optical waveguide 1.31/1.55/spl mu/m wdm with-50 db crosstalk over 100 nm bandwidth. *Electronics Letters,* 31(24):2100–2101.

Lierstuen, L., & Sudbo, A. (1995). 8-channel wavelength division multiplexer based on multimode interference couplers. *IEEE Photonics Technology Letters, 7*(9), 1034–1036.

März, R. (1995). Integrated optics: design and modeling. artech house. *Inc., Norwood, MA.*

Sakata, H. (1992). Sidelobe suppression in grating-assisted wavelength-selective couplers. *Optics letters, 17*(7), 463–465.

Smit, M. K. (1988). New focusing and dispersive planar component based on an optical phased array. *Electronics letters, 24*(7), 385–386.

Smit, M. K., & Van Dam, C. (1996). Phasar-based wdm-devices: Principles, design and applications. *IEEE Journal of selected topics in quantum electronics, 2*(2), 236–250.

Takahashi, H., Hibino, Y., Ohmori, Y., & Kawachi, M. (1993). Polarization-insensitive arrayed-waveguide wavelength multiplexer with birefringence compensating film. *IEEE photonics technology letters, 5*(6), 707–709.

Takahashi, H., Oda, K., Toba, H., & Inoue, Y. (1995). Transmission characteristics of arrayed waveguide n/spl times/n wavelength multiplexer. *Journal of Lightwave Technology, 13*(3), 447–455.

Van Dam, C., Amersfoort, M., ten Kate, G., van Ham, F., Smit, M., Besse, P., Bachmann, M., and Melchior, H. (1995). Novel inp-based phased-array wavelength demultiplexer using a generalized mmi-mzi configuration. In *Proc. 7th Eur. Conf on Int. Opt.(ECIO95)*, pages 275–278.

Venghaus, H., Bornholdt, C., Kappe, F., Nolting, H.-P., & Weinert, C. (1992). Meander-type wavelength demultiplexer with weighted coupling. *Applied physics letters, 61*(17), 2018–2020.

Weber, J.-P. (1997). A new type of tunable demultiplexer using a multi-leg mach-zehnder interferometer. In *Proceedings of European Conf. Integrated Optics (ECIO 1997, Stockholm, Sweden)*, pages 260–263.

Yaffe, H. H., Henry, C. H., Serbin, M. R., & Cohen, L. G. (1994). Resonant couplers acting as add-drop filters made with silica-on-silicon waveguide technology. *Journal of lightwave technology, 12*(6), 1010–1014.

Yi-Yan, A., Deri, R., Seto, M., & Hawkins, R. (1989). Gaas/gaalas asymmetric mach-zehnder demultiplexer with reduced polarization dependence. *IEEE Photonics Technology Letters, 1*(4), 83–85.

Yu, H., Yu, J., Yu, Y., & Chen, S. (2009). Design and fabrication of a photonic crystal channel drop filter based on an asymmetric silicon-on-insulator slab. *Journal of nanoscience and nanotechnology, 9*(2), 974–977.

Zirngibl, M., & Joyner, C. (1994). 12 frequency wdm laser based on a transmissive waveguide grating router. *Electronics Letters, 30*(9), 701–702.

Zirngibl, M., Joyner, C., & Glance, B. (1994). Digitally tunable channel dropping filter/equalizer based on waveguide grating router and optical amplifier integration. *IEEE Photonics Technology Letters, 6*(4), 513–515.

Zirngibl, M., Joyner, C., & Stulz, L. (1995). Wdm receiver by monolithic integration of an optical preamplifier, waveguide grating router and photodiode array. *Electronics Letters*, *31*(7), 581–582.

Chapter 12
Electro-Optical Modulators

Abstract Electro-optic modulators are based on the control of guided waves using electro-optic variation of the phase or amplitude using an applied electric field. Different theoretical approaches can be used to describe electro-optic control including coupled-mode theory. Phase retardation of the guided light enables polarization splitting, optical switching, and wavelength-selective coupling. Two model–material systems, which are extensively used for E/O devices: GaAs (or other III–V semiconductor material) and lithium niobate. In addition, secondary material systems are mentioned throughout.

12.1 Introduction

The electro-optic effect can be used in optical waveguide devices to control the phase or amplitude of a guided wave. Typically, these devices use this effect to control polarization or interference of guided-light beams, with the most important application being in integrated optical modulators. These devices operate on the electro-optical coupling of TE and TM modes in a waveguide, or pure phase retardation of waveguide modes. This coupling can be understood and treated via coupled-mode theory. In addition, electro-optic control of phase retardation of the guided light enables polarization splitting, optical switching, and wavelength-selective coupling.

The electro-optic effect results from a redistribution of bound charges in a dielectric waveguide upon the application of a voltage across the optical guide (Saleh et al. 1991). Deformations of the lattice from the applied electric field also contribute to the effect. These perturbations cause a change in the optical impermeability tensor relating the displacement vector, $\vec{D}$, and the electric field, $\vec{E}$, of the optical wave. Equivalently, the application of an electric field causes a change in the index of refraction of the crystal.

In this chapter, we will first briefly review the fundamentals of the electro-optic effect and consider two model–material systems, which are used heavily for the E/O devices: GaAs (or other III–V semiconductor material) and LiNbO$_3$. In addi-

© Springer Nature Switzerland AG 2021

R. Osgood jr. and X. Meng, *Principles of Photonic Integrated Circuits*,
Graduate Texts in Physics,
https://doi.org/10.1007/978-3-030-65193-0_12

Fig. 12.1 The x- and
y-components of the E-field
in a crystal with different
refractive indices

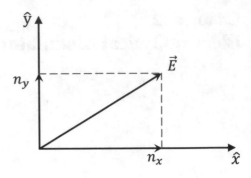

tion, electro-optical devices have been made in a variety of other material systems, including polymers, etc.; these secondary materials are mentioned throughout. The chapter will then present a detailed discussion of optical modulators.

12.2   An Overview of Electro-Optic Modulation

Consider first the application of an electric field to a bulk crystal which has an anisotropic set of refractive indices. The application of the field can change the magnitude and direction of the indices in the crystal. In subsequent sections given below, we will explain the details of this change, including its dependence on the direction of the applied external electric field, E^0, and the crystal. If we assume that the crystal has well-defined indices of refraction, say $n_{x,y}$, for an optical field polarized along with $\hat{x}$ and $\hat{y}$, respectively, where $\hat{x}$ and $\hat{y}$ are termed the principal axis, then if the optical electric field is not along $\hat{x}$, or $\hat{y}$, its components along $\hat{x}$ and $\hat{y}$ have different propagation velocities because of the different indices of refraction in the crystal; see Fig. 12.1.

This change in the principal axes can be used to make a simple modulator. If the input is linearly polarized at 45° with respect to the $\hat{x}$-axis and the input optical field amplitude is E_0, then after traveling a distance L in the crystal, the optical electric field, $\vec{E}$, will be given by

$$\vec{E} = \frac{E_0}{\sqrt{2}} \left(\hat{x} e^{-i\frac{2\pi}{\lambda} n_x L} + \hat{y} e^{-i\frac{2\pi}{\lambda} n_y L} \right) \tag{12.1}$$

The different relative phases in the $\hat{x}$- and $\hat{y}$-directions show that the wave is elliptically polarized.

A polarization analyzer at 45° to the $\hat{x}$ or $\hat{y}$ axes will "pass" optical radiation as follows:

$$|\vec{E}_{45°}| = \vec{E} \cdot \left(\frac{1}{\sqrt{2}} (\hat{x} + \hat{y}) \right) \tag{12.2}$$

Fig. 12.2 The magnitudes of components in x- and y-directions are the same. However, each of the two views of different refractive indices. Thus, the total magnitude is related to the propagation distance in material

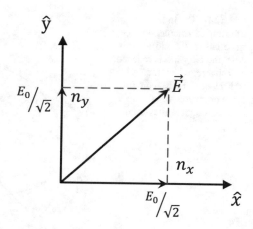

$$= E_0 e^{-i\frac{\pi}{\lambda}(n_x + n_y)L} \cos \frac{\pi}{\lambda}(n_x - n_y)L \tag{12.3}$$

where it is noted that (12.2) contains a dot product, see Fig. 12.2. The intensity (here we ignore the prefactor $1/2\sqrt{\epsilon/\mu}$) along this 45° direction is then

$$|\vec{E}_{45°}|^2 = E_0^2 \cos^2 \frac{\pi}{\lambda}(n_x - n_y)L \tag{12.4}$$

For the linear (Pockels) electro-optic effect, $n_x - n_y = \alpha \cdot E_0$, where α is a constant and E_0 is the applied electric field. Thus, by varying the applied voltage, one can modulate the transmitted guided-light intensity.

12.2.1 Basics of Propagation of Lightwaves in Anisotropic Crystals

The electro-optical effect operates by the modification of a crystal's optical properties in the presence of an electric field. Thus the starting point in describing the electro-optical effect is the propagation of light in a crystal in the absence of an external field (Saleh et al. 1991). In a material medium such as an anisotropic crystal, the derivation of the wave equation shows that light propagation is governed by the relation between the two optical fields, $\vec{D}$ and $\vec{E}$. This relation is generally given by

$$E_i = \epsilon_{ij}^{-1} D_i \tag{12.5}$$

where the sum over repeated indices, e.g., the index in (12.5), is understood, and where ϵ_{ij}^{-1} is the inverse dielectric tensor or impermeability tensor. If one chooses a set of reference coordinate axes, which lie along a special set of axes, called principal axes, the tensor is diagonal, i.e.,

Fig. 12.3 The index
ellipsoid of zincblende-type
materials (solid line). In this
case, the ellipsoid is reduced
to a sphere, which means the
indices of refraction are the
same no matter what
direction light comes from

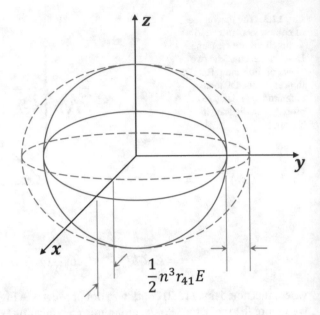

$$[\epsilon] = \begin{bmatrix} \epsilon_{11} & 0 & 0 \\ 0 & \epsilon_{22} & 0 \\ 0 & 0 & \epsilon_{33} \end{bmatrix} \tag{12.6}$$

If $\vec{E}$ is polarized along such a principal axis, $\vec{E}$ is parallel to $\vec{D}$. Light polarized along any of these principal axes will propagate with an index of refraction given by n_i, where $\epsilon_i^{-1} = 1/n_i^2$.

12.2.2   Index Ellipsoid

We can capture the directional dependence of ϵ_i^{-1} by using the index ellipsoid, which is a geometric construct that allows one to find the indices of refraction and corresponding polarization directions for an arbitrary direction of propagation (Saleh et al. 1991). This ellipse, shown in Fig. 12.4, is obtained from (12.4), Maxwell's equation, and the definition of the Poynting vector in the medium. Given a direction of propagation, one uses the index ellipsoid to find the indices of refraction by determining the ellipsoid's semi-major and -minor axes. As shown in Fig. 12.3, these quantities are obtained from the index ellipsoid and the plane normal to the direction of propagation while passing through the center of the ellipsoid. The directions of these axes define the normal modes of polarization for the directions corresponding to those indices since a wave polarized along a principal axis will remain along that axis as it propagates through the crystal. The index ellipsoid is defined as

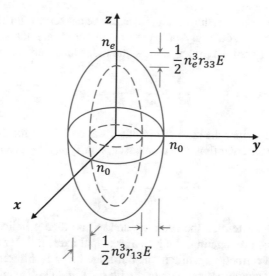

Fig. 12.4 An example of the index ellipsoid. The meaning of the index ellipsoid with dashed lines will become clear in the following section, which describes the change of refractive index due to an applied electrical field, the so-called electro-optical effect

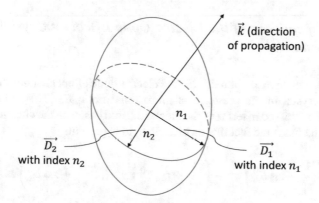

Fig. 12.5 Demonstration of how to read out the indices of refraction from an index ellipsoid

$$\epsilon_{ij} x_i x_j \equiv 1 \tag{12.7}$$

where again a sum over repeated indices is implied (Fig. 12.5).

For zincblende-type materials, such as GaAs (cubic 43 m), the crystal is isotropic. In this case, the index ellipsoid in the absence of an applied voltage is

$$\frac{x^2 + y^2 + z^2}{n_0^2} = 1 \tag{12.8}$$

Since the crystal is isotropic, the principal axes can be chosen arbitrarily; see Fig. 12.3.

However, in general, other crystal types or symmetries are not isotropic in index. Thus, if $\hat{x}$, $\hat{y}$, and $\hat{z}$ are then principal axes, the index ellipsoid is then written as

$$\frac{x^2}{n_x^2} + \frac{y^2}{n_y^2} + \frac{z^2}{n_z^2} = 1 \tag{12.9}$$

For example, consider the case of LiNbO$_3$ (trigonal 3 m); for this crystal, $n_x = n_y \neq n_z$, which is the condition for a uniaxial crystal. Typically, its ellipsoid is written as

$$\frac{x^2}{n_0^2} + \frac{y^2}{n_0^2} + \frac{z^2}{n_e^2} = 1 \tag{12.10}$$

where n_o and n_e are termed the ordinary and extraordinary indices of refraction, respectively. Thus, propagating along two principal axes, a lightwave experiences the same index of refraction, while in the third, the index is different; see Fig. 12.3. LiNbO$_3$ has only single optical axis and with $n_o > n_e$, i.e., LiNbO$_3$ is a negative uniaxial crystal.

12.2.3 Dependence of the Electro-Optic Effect on Crystal Symmetry

In the presence of an external field, $\vec{E}_0$, the electro-optic effect changes the dielectric tensor of a crystal in the presence of an external field, $\vec{E}_0$. This change can be conveniently expressed in terms of the index ellipsoid based on the changed principal axes, which includes the fact that an electric field is present,

$$\left(\frac{1}{n_x^2} + r_{1k}E_k^0\right)x^2 + \left(\frac{1}{n_y^2} + r_{2k}E_k^0\right)y^2 + \left(\frac{1}{n_z^2} + r_{3k}E_k^0\right)z^2$$
$$+2yzr_{4k}E_k^0 + 2zxr_{5k}E_k^0 + 2xyr_{6k}E_k^0 = 1 \tag{12.11}$$

where a sum over $k = x$, y, z is implied, and where $\vec{E}_k^\circ = (E_x^0, E_y^0, E_z^0)$ is the applied electric field. Note that at typical values of the electric field, the field-dependent terms are much, much smaller than the static values, thus the field-induced changes can be treated as perturbations. More generally, the electro-optic effect can be expressed as

$$\delta\epsilon_{ij}^{-1} = r_{ijk}E_k^0 \tag{12.12}$$

with a sum over the repeated index k, then

$$\delta\epsilon_{ij} = -\epsilon_i\epsilon_j r_{ijk}E_k^0 \tag{12.13}$$

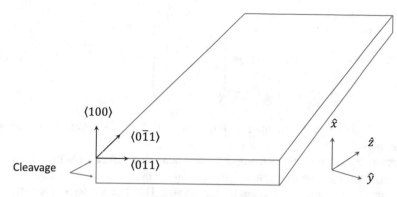

Fig. 12.6 A sketch of the GaAs crystal axes

Table 12.1 Optical electro-optical constants of LiNbO$_3$ and GaAs

Crystals	Refractive index	Electro-optical coefficient	Dielectric constant	Transparency window
LiNbO$_3$	$n = 2.286$	$r_{33} = 30.8$	$\epsilon_\perp = 43$	
(3 m)	$n = 2.269$	$r_{13} = 8.6$	$\epsilon_\parallel = 28$	
	$n = 2.237$	$r_{22} = 3.4$		0.4–5 μm
	$n = 2.157$	$r_{51} = 28$		
	($\lambda = 0.63$ μm)	($\lambda = 1$ μm)		
GaAs	$n = 3.5$	$r_{41} = 1.2$	$\epsilon = 13.2$	0.9 μm
(43 m)	($\lambda = 0.9$ μm)	($\lambda = 0.9$ μm)		

The tensor, r_{ij}, in (12.13), which gives the material response, can be redefined by the noting symmetry of the indices $ij = 11, 22, 33, 23(32), 13(31), 12(21) = 1, 2, 3, 4, 5, 6$. Then setting $i = 1, \ldots, 6$.

$$r_{ik} = \begin{pmatrix} r_{11} & r_{12} & r_{13} \\ r_{21} & r_{22} & r_{23} \\ r_{31} & r_{32} & r_{33} \\ r_{41} & r_{42} & r_{43} \\ r_{51} & r_{52} & r_{53} \\ r_{61} & r_{62} & r_{63} \end{pmatrix} \tag{12.14}$$

with $1 \equiv x$, $2 \equiv y$, and $3 \equiv z$. This tensor is called the electro-optic tensor and its form depends on the symmetry group, to which the crystal belongs. A tabulation of the electro-optical tensor elements for the most common electro-optic materials is given in Table 12.1. For example, one common class of electro-optical materials is that of the zincblende semiconductor crystals (GaAs, CdTe, etc.), which are of the cubic 43 m. In this case,

$$r_{ik} = \begin{pmatrix} 0 & 0 & 0 \\ 0 & 0 & 0 \\ 0 & 0 & 0 \\ r_{41} & 0 & 0 \\ 0 & r_{41} & 0 \\ 0 & 0 & r_{41} \end{pmatrix} \tag{12.15}$$

with $\hat{x} = [100]$, $\hat{y} = [010]$, and $\hat{z} = [001]$; see Fig. 12.6 for a sketch of the GaAs crystal axes.

Mechanical stress in electro-optical crystals, due either to lattice mismatch between cladding and waveguide epilayers or to the presence of electrodes, can affect the electro-optical properties of that crystal. This stress reduces the symmetry of the crystal. For example, for GaAs, stress along [0, 1, 1] causes the crystal to shift from cubic (43 m) to orthorhombic (2 mm), while for stress along [1, 0, 0], the symmetry shifts into yet another symmetry tensor. Since different symmetry groups have different forms for the electro-optic tensor, r_{ik}, the appropriate electro-optic tensor should be used for each orientation of the stress field. For example, a GaAs crystal which is stressed (uniaxial) along the [0, 1, 1] direction belongs to the orthorhombic (2 mm) symmetry group, with

$$r_{ik} = \begin{pmatrix} 0 & 0 & r_{13} \\ 0 & 0 & r_{23} \\ 0 & 0 & r_{33} \\ 0 & r_{42} & 0 \\ r_{51} & 0 & 0 \\ 0 & 0 & 0 \end{pmatrix} \tag{12.16}$$

with $\hat{x} = [0, 1, 1]$, $\hat{y} = [0, 1, 1]$, and $\hat{z} = [1, 0, 0]$.

A second common class of electro-optical materials is that of the ferroelectric or ferroelectric-like metal oxides, with $LiNbO_3$ being the best example. $LiNbO_3$, as mentioned above, is a negative uniaxial crystal containing ordinary n_o and extraordinary n_e indices of refraction, i.e., $n_x = n_y = n_o$, and $n_o > n_e$. In such a crystal, $\hat{z}$ is thus the optical axis, since light propagating along $\hat{z}$ experiences an index of refraction of n_o, irrespective of the transverse polarization of the lightwave. The electro-optic tensor for this crystal class, trigonal 3 m, is

$$\begin{pmatrix} 0 & -r_{22} & r_{13} \\ 0 & r_{22} & r_{13} \\ 0 & 0 & r_{33} \\ 0 & r_{51} & 0 \\ r_{51} & 0 & 0 \\ -r_{22} & 0 & 0 \end{pmatrix} \tag{12.17}$$

All three of the above examples show that typically, symmetry causes the electro-optic tensor to be "sparse," and many of the matrix elements are identical or zero.

Thus, the calculation of the electro-optical shift can be far less formidable than the full tensor in (12.14) would suggest at first glance.

With the above review in hand, this chapter will now concentrate on a discussion of electro-optic modulation of guided waves. Two materials are emphasized, GaAs and LiNbO$_3$; the former is meant to be the only representative example of the III–V crystals, which are as suggested above broadly used for different sets of electro-optical devices.

12.3 Electro-Optic Modulation Illustrated for Two Crystal Types

Consideration of (12.11) shows that the application of external electric fields can have two effects on the propagation of light in an electro-optical crystal (Saleh et al. 1991). First, the field can change the refractive indices of the crystal and, second, the field can alter the crystal's principal axes. The two points are illustrated below using our two material types.

12.3.1 III–V (Zincblende) Crystals

Consider an applied external electric field, E°, along the $[1, 0, 0]$ direction, which is vertical in the sketch in Fig. 12.7 for a $[1, 0, 0]$ crystal in unstressed zincblende crystals. In this case, the new index ellipsoid is:

$$\frac{x^2 + y^2 + z^2}{n_0^2} + 2yzr_{41}E^\circ = 1 \tag{12.18}$$

Note that the presence of a cross product term when the field is applied means that a new set of principal axes must be chosen. These transformed axes will be designated by a prime notation, e.g., y' and z'. The transformation in the alignment of the electric field involves a rotation of the principal axes about the x-axis. The new principal axes are in the $[0, 1, 1]$ and $[0, 1, 1]$ direction; see Fig. 12.7. These may be written

$$x = x' \tag{12.19}$$

$$y = \frac{1}{\sqrt{2}}(y' - z') \tag{12.20}$$

$$z = \frac{1}{\sqrt{2}}(y' + z') \tag{12.21}$$

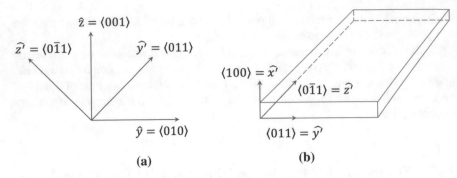

Fig. 12.7 **a** After applying an external electrical field along the [1, 0, 0] direction, the principle axes $\hat{x}$ and $\hat{y}$ transform to $\hat{x}'$ and $\hat{y}'$. **b** A sketch of the zincblende crystal

The new index ellipsoid then becomes

$$\frac{x'^2}{n_o^2} + y'^2\left(\frac{1}{n_0^2} + r_{41}E^0\right) + z'^2\left(\frac{1}{n_0^2} - r_{41}E^0\right) = 1 \qquad (12.22)$$

Since generally $1/n_o^2 \gg r_{41}E^\circ$, that is, the electro-optical index change is small, the equation may be simplified using a series approximation

$$\left(n_0 \mp \frac{1}{2}n_0^3 r_{41}E^0\right)^{-2} \simeq \frac{1}{n_0^2} \pm r_{41}E^0 \qquad (12.23)$$

and thus,

$$\frac{x'^2}{n_0^2} + \frac{y'^2}{(n_0 - \frac{1}{2}n_0^3 r_{41}E^0)^2} + \frac{z'^2}{(n_0 + \frac{1}{2}n_0^3 r_{41}E^0)^2} = 1 \qquad (12.24)$$

With the new principal axis, for light polarized at 45° to these axes and in the plane of these axes, the component along [1, 0, 0] has index n_0, while the component along [0, 1, 0] has index $n_0 - 1/2n_0^3 r_{41}E^0$. Notice that for GaAs, and other III–V materials, the change in the orientation of the principal axes occurs in the presence of any measurable field since it breaks the symmetry of an otherwise isotropic crystal.

Recall that using the intensity modulation scheme discussed earlier, the intensity, I, of an optical wave varies with the distance, L, in an electro-optical crystal as

$$I \sim \cos^2\left[\frac{\pi}{\lambda}(n_{x'} - n_{z'})L\right] \qquad (12.25)$$

Thus, for GaAs, the indices in (12.24) give

$$I = \cos^2\left[\frac{\pi}{2\lambda}n_0^3 r_{41}E^0 L\right] \qquad (12.26)$$

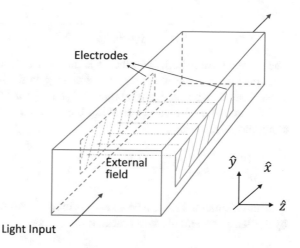

Fig. 12.8 One commonly used external-field orientations for LiNbO$_3$. z-axis is the optical axis, the light is propagating along x-direction. The external field is along z-axis

As an example, if $\lambda = 1.5\,\mu m$, tabulated data show that $n_0 \sim 3.35$, and $r_{41} \sim 1.3 \times 10^{-12}$ m/V. Thus, for a device having 5 V applied across a 3 μm gap between two electrodes, $E^0 = 1.66 \times 10^6$ V/m, and $L = 5$ mm; then the argument in (12.26), $(\pi/2\lambda)n_0^3 r_{41} E^0 L$, will be 0.43 rad $= 24.4°$, i.e., a very sizable phase shift.

In the above case, an applied E-field along $[1, 0, 0]$ gives principal axes along $[1, 0, 0]$ and $[0, 1, 1]$. By contrast, an applied E-field along $[0, 1, 1]$ will give principal axes with indices $n_0 + (1/2)n_0^3 r_{41} E^0$ and $n_0 - (1/2)n_0^3 r_{41} E^0$. For this crystal orientation, one must launch polarized light along $[0, 1, 1]$ or $[1, 0, 0]$ to get modulation.

12.3.2 LiNbO$_3$ Crystals

A similar analysis as applied above can be applied to LiNbO$_3$ and related crystals. However, here for the sake of brevity, two specific, common external electric field orientations will be examined, and only the final formulae will be stated here. In the first case, shown in Fig. 12.8, the external field is aligned along the z-axis, i.e., only $E_z^0 \neq 0$ and the z-axis is the optical axis of the crystal. Notice also that the direction of propagation of the light beam is along the x-axis. After the application of an external field, the index ellipsoid becomes

$$\left(\frac{1}{n_o^2} + r_{13}E_z^0\right)(x^2 + y^2) + \left(\frac{1}{n_e^2}r_{33}E_z^0\right)z^2 = 1 \qquad (12.27)$$

For small fields, this equation can be written in a way that shows that such an external field retains the "field-free" principal axes but changes the magnitude of the indices of refraction; that is,

$$\frac{x^2 + y^2}{n_o^2(1 - \frac{1}{2}r_{13}n_0^2 E_z^0)} + \frac{z^2}{n_e^2(1 - \frac{1}{2}r_{33}n_e^2 E_z^0)} = 1 \qquad (12.28)$$

The second term in both parentheses is the field-induced change in the indices. This electro-optic orientation causes only a change in the index, but not in the principal axes; it can thus be used to modulate the phase of the light beam.

The second case involves aligning the external field normal to the optical axis, along the y-axis. Then,

$$\left(\frac{1}{n_o^2} - r_{22}E_y^0\right)x^2 + \left(\frac{1}{n_o^2} + r_{22}E_y^0\right)y^2 + \left(\frac{1}{n_e^2}\right)z^2 + 2r_{51}E_y^0 yz = 1 \qquad (12.29)$$

This orientation causes a new set of principal axes to be formed; they are oriented by a rotation around the x-axis and have indices

$$n_{y'} = n_o - \frac{1}{2}n_o^3 r_{22}E_y^0 + n_o^3(r_{51}E_y^0)^2 \Big/ \left(\frac{1}{n_e^2} - \frac{1}{n_o^2}\right) \qquad (12.30)$$

$$n_{z'} = n_e - n_e^3(r_{51}E_y^0)^2 \Big/ \left(\frac{1}{n_e^2} - \frac{1}{n_o^2}\right) \qquad (12.31)$$

While these equations allow significant manipulation, in fact, all the terms containing the electric field are very small and to a very good approximation:

$$n_{y'} \approx n_0 \qquad (12.32)$$

$$n_{z'} \approx n_e \qquad (12.33)$$

Thus, the primary effect of the external field is simply to rotate the principle axis. This rotation introduces a small off-diagonal element in the dielectric tensor:

$$\delta\epsilon_{23} = -n_e^2 n_0^2 r_{51} E_y^0 \qquad (12.34)$$

Since it is off-diagonal, it couples the E_z to E_y fields. Hence, in this case, the application of the external field modulates the light through polarization by coupling the TE and TM modes. To obtain the modulation of the lightwave requires the use of a polarizer after the beam exits the crystal. Further, since r_{51} is as large as r_{33}, the extended field orientation can be used to make an efficient modulator.

12.4 Electro-Optic Modulators

In this section, the device aspects of electro-optic modulation in waveguide devices will be discussed. Thus, the section will introduce the most common types of modulators and discuss some of the more important design parameters of modulators, including electric and optical field overlap and the configuration of the modulator

$$
\begin{bmatrix}
0 & -r_{22} & r_{13} \\
0 & r_{22} & r_{13} \\
0 & 0 & r_{33} \\
0 & r_{51} & 0 \\
r_{51} & 0 & 0 \\
-r_{22} & 0 & 0
\end{bmatrix}
\begin{bmatrix}
0 & 0 & 0 \\
0 & 0 & 0 \\
0 & 0 & 0 \\
r_{41} & 0 & 0 \\
0 & r_{41} & 0 \\
0 & 0 & r_{41}
\end{bmatrix}
$$

$$
\begin{array}{cc}
\mathrm{LiNbO_3} & \mathrm{GaAs} \\
Trigonal(3\,\mathrm{m}) & Cubic(43\,\mathrm{m})
\end{array}
$$

electrodes. Our discussion will be presented in terms of $LiNbO_3$ modulators, chiefly because $LiNbO_3$ devices are conceptually the simplest. Lithium niobate is also the most widely engineered material for modulators because of its high performance and low optical coupling and insertion loss; although other similar materials, such as the tantalates (Kaminow 1965) also have important advantages. Finally, optical switches are closely related in function to modulators; however, this topic is sufficiently large and the device requirements are sufficiently different that it will be discussed Chap. 13.

12.4.1 Layout of a Basic Modulator

Modulators are fabricated by placing electrodes on the surface of an electro-optical waveguide structure. The design of these electrodes and the waveguide structure for modulation can be sophisticated. However, there are a set of common parameters, which apply to all types. Since the quantities, which characterize the electrode arrangements, are important for our subsequent discussion, we will briefly describe these fundamental aspects first. The layout of electrodes is critical since it determines the modulator voltage and the efficiency of interaction between the applied field and the optical beam. As we will see later in this chapter, electrode design can be different significantly for high-speed modulators; see Sect. 12.4.3. Electrode placement in $LiNbO_3$ modulators is made easy by the fact that the crystal is an insulating dielectric. Thus, any free charge present in semiconductor materials makes their design much more complicated.

The two most common electrode configurations for a $LiNbO_3$ modulator are shown in Fig. 12.9. Roughly speaking, the electric field in the waveguide region

Fig. 12.9 Upper: The configuration for an external field that is horizontal $E_\parallel$ Bottom: Two common electrode and electric field configurations for the LiNbO$_3$ modulator

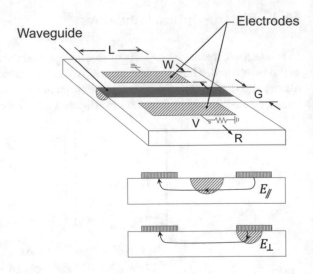

varies as $V°/d$, where d is the electrode spacing and $V°$ is the applied voltage. In fact, however, for a realistic waveguide design, the actual electric field distribution is more complex and varies in amplitude and direction between the electrodes. Further, the magnitude of the electro-optic effect depends on the overlap of the external field distribution over the cross-sections of the optical field. As a result, the idealized field, $V°/d$, is modified by an empirical factor, Γ, which accounts for the averaged overlap between the optical and external electrical fields. As an example, consider the case of an optical beam in y-cut LiNbO$_3$, which is aligned such that it propagates along the y-axis and is polarized in the z-direction. If the applied electric field is dominantly along the z-axis, then

$$\Delta n = \frac{1}{2}\left(n_e^3 r_{33}\frac{V°}{d}\right)\Gamma \tag{12.35}$$

Then the overlap factor Γ is obtained from the simple but generally nonanalytic integral:

$$\Gamma = \frac{d}{V°}\frac{\int\int E^0(x,y)E^*\mathrm{d}A}{\int\int E^2\mathrm{d}A} \tag{12.36}$$

where $E°(x, y)$ is the applied field, E is the optical field, and the integral is averaged over the cross-sectional area. The factor of $d/V°$ in (12.36) normalizes the magnitude of the applied field.

The electrode configuration for a modulator is often dependent on the practical constraints of the device, as well as the desire to make use of the largest element in the electro-optical tensor. Thus, a Ti-diffused LiNbO$_3$ waveguide modulator in an x- or y-cut crystal, such as that shown in Fig. 12.9, can make use of the large r_{33} coefficient in this material if electrodes are placed on either side of the waveguides so as to have a dominantly horizontal field. Further, the fact that the electrode is placed away from the waveguide such that optical losses in the metal electrodes are negligible. The disadvantage of this placement is that a relatively large electrode spacing is incurred thus increasing the voltage requirement. Turning now to z-cut material, a vertical field must be achieved. This can be done by placing one electrode directly on the waveguide and positioning two ground electrodes adjacent to the waveguide. This approach requires a transparent dielectric under the metal electrodes to reduce electrode loss; however, this layer also lowers the overlap integral. Despite this reduction in overlap integral, the z-cut configuration is more commonly used in practice. Finally, electrode design becomes more complex when designing for high-speed modulators. For example, the width of the electrode, w, is important in setting the impedance and, hence, the bandwidth of the modulator; see Sect. 12.4.3. Some of the issues in this case will be discussed below.

12.4.2 Modulator Types

12.4.2.1 Phase Modulation

Modulation of the phase of an optical beam is readily accomplished in waveguide modulators. A simple phase modulator is also one of the building blocks for making other forms of modulators. For example, when a phase modulator is inserted into the arm of a Mach–Zehnder (MZ) interferometer, the resultant device forms one of the most widely used intensity modulators.

Phase modulators operate by changing only the phase of the propagating guided-light wave. If the LiNbO$_3$ is y-cut and the electric field is aligned along the z-axis (or optical axis), then the induced phase is

$$\phi_z = kx \left(n_e - \frac{1}{2} n_e^3 r_{33} E_z^0 \right) \tag{12.37}$$

where the propagation direction is along x. A similar expression is obtained if the wafer is x-cut, z polarized, and propagating along y, except for the exchange of x and y in (12.37).

If a phase modulator requires that $\Delta\beta L = p\pi$, where the value of p depends on the exact type of modulator (Alferness 1982), then (12.35) shows that the voltage–length product for a modulator can be written as

$$VL - \frac{p\lambda d}{n^3 r \Gamma} \tag{12.38}$$

where the symbols are as described above. This equation suggests that to improve the voltage–length product of the device and thus minimize device power and device size (and cost), one should improve both the overlap integral, Γ, and reduce the electrode

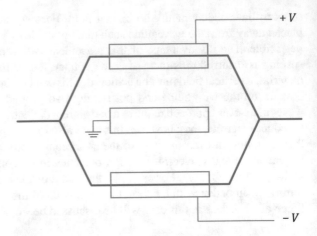

Fig. 12.10 Pairs of interferometeric modulators are operated to cancel out the harmonics of the usual nonlinearities inherent in the sinusoidal response

gap, d. Of course, these factors also control other performance considerations in the design, including parameters such as insertion loss and bandwidth (see below). Each of these parameters will be discussed below.

Finally, if the modulating electric field having an angular frequency ω_m is denoted as

$$E_z^0 \equiv E_{zm}^0 \sin \omega_m t \tag{12.39}$$

then the modulated optical field will be

$$E_z = E_z^i \cos(\omega t - \phi_0 + \delta_m \sin \omega_m t) \tag{12.40}$$

where the depth of modulation, δ_m, is given by

$$\delta_m = \left(\frac{\pi}{\lambda} n_e^3 r_{33} E_{zm}^0 \right) \tag{12.41}$$

$\phi_0 = 2\pi k n_e x$, and E_z^i is the amplitude of the z-polarized optical field on the modulator. The argument for E_z can be expanded in terms of Bessel functions; the amplitude of the first sideband is then given by

$$J_1(\delta_m)/J_0(\delta_m) \tag{12.42}$$

12.4.2.2   Interferometric Modulators

The MZ modulator, shown in Fig. 12.10 is one of the most important commercial LiNbO$_3$ waveguide modulators. In this modulator, a phase shifter is typically mounted in each arm. This layout allows a positive phase shift in one arm, which is balanced by the negative of exactly the same phase shift in the other, causing push–pull mode modulation. In this case, the total phase shift, $2\Delta\phi$, between the two arms is obtained from (12.35) to be thus

$$2\Delta\phi = 2\pi \frac{\Gamma n_e^3 r_{33} L V}{\lambda d} \tag{12.43}$$

where L is the total electrode length and the same field directions are used as for (12.35). Based on the value for V which sets the total phase shift between the arms of π, we can define the quantity, V_π, for this modulator:

$$V_\pi = \frac{\lambda d}{2\Gamma n_e^3 r_{33} L} \tag{12.44}$$

Note that V_π is an important operational parameter for modulators (see discussion in a section below) and is dependent on both material and device parameters. It scales inversely with length, but other factors such as bandwidth can also limit the full use of this quantity.

In the instances where the modulators are used for analog applications, it is often crucial to eliminate any nonlinearity in the modulator response. In one important case, pairs of interferometric modulators are often operated so as to cancel out many of the harmonics of the usual nonlinearities inherent in the sinusoidal response of a typical MZ device. In this operational mode, they are also operated around the inflection point of the sinusoid (see Fig. 12.10) and both optical and electrical parameters are adjusted to effect the cancellation (Bridges and Schaffner 1995). Unfortunately, these schemes often come with a cost of reduced bandwidth.

Interferometric modulators can also be used as switches; this type of operation will be discussed in the subsequent chapter on switches. One important fact regarding their use as switches is that in comparison to directional coupler-based switches, they have a factor of $\sqrt{3}$ lower voltage–length product.

12.4.2.3 Resonant-Cavity Modulators: Operating Principles and Voltage Dependence

Since the phase shift of an interferometric modulator scales with its path length, it is of interest to consider extending its "effective" length by adding a high-quality Q-optical cavity. The high Q in the device also makes it possible for small input fields to build up within the device. However, such a cavity also prolongs the photon interaction time by trapping the light in the structure due to the use of resonance (e.g., a cavity with two high-reflectivity mirrors or a circular ring, etc.). Further, since the V_π of a ring modulator is inversely proportional to its length, use of a resonant cavity would also reduce V_π (Gheorma et al. 2000). Recently and importantly, these same considerations have been shown to be operable and, in fact, highly developed in Si

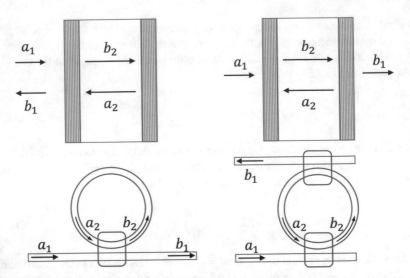

Fig. 12.11 Ring and Fabry–Perot resonators. The principle of a ring resonator (both a single-bus ring and an add-drop ring) can be illustrated by Fabry–Perot cavity

resonant rings (Xu et al. 2005). Due to the different types of resonant modulators, it is important to develop an analysis, which allows comparison of the different types. This analysis appears below.

Resonant-cavity modulators, in the form of Fabry–Perot (FP) cavities, were among the first modulators used in light modulation experiments. Resonant devices are attractive because they can have an extremely low V_π as the signal voltage is varied around a resonant voltage; however, as mentioned above, a low V_π comes with a reduced bandwidth. In addition, Si ring modulators also can have very low drive power due to their small diameter and thin Si film thickness. This attribute is crucial for data PICs in many data system applications.

The basic principle of such a resonator-based device is best illustrated by the FP modulator (see Fig. 12.11), which is identical in function to an FP filter or laser optical cavity. The modulator consists of an etalon of electro-optical material which is highly reflective coated on both sides. The relative intensity of the transmitted light through an FP cavity, fabricated from an electro-optic material, as a function of the modulating voltage, V, is

$$\frac{I(V)}{I_0} = \frac{1}{1 + (4F^2/\pi^2) \sin^2(-\pi n^2 r \Gamma L V / \lambda d)} \tag{12.45}$$

where $F = \pi\sqrt{R}/(1 - R)$ is the finesse of the cavity, L the length of the cavity, d the distance between the electrodes, Γ the RF optical field overlap coefficient, r the electro-optical coefficient, and λ the free-space wavelength. Note that the output in (12.45) is not zero due to the finite F; thus, we cannot define a V_π voltage in the same manner as is used, for example, for MZ or cross-polarizer modulators. Instead, an alternate definition can be employed as is, sometimes used for other types of modulators. This definition is based on the slope of the variation of the optical intensity with respect to voltage for the modulator at the half-transmission point, i.e., the most efficient modulation bias point. This slope-based definition gives a lower limit of the voltage either for digital or analog applications.

For the FP modulator, with a unit input intensity

$$\frac{dI_{FP}}{dV} \approx \frac{\pi n^3 r \sqrt{R} L}{\lambda (1 - R) g} \tag{12.46}$$

In comparison, the analogous quantity for an MZ modulator is $\frac{dI_{MZ}}{dV}|_{1/2} = -(\pi/2V_\pi)$. We can redefine V_π^{MZ} for an MZ using its slope

$$V_\pi^{MZ} = \frac{\pi}{2} \left(\frac{dI_{MZ}}{dV}|_{1/2} \right)^{-1} \tag{12.47}$$

This expression also allows one to define an equivalent V_π^{FP} for an FP modulator by substituting $(\frac{dI_{FP}}{dV})|_{1/2}$ to give

$$V_\pi^{FP} = \frac{\pi}{2} \frac{1}{dI_{FP}/dV|_{1/2}} = \frac{g}{L} \frac{\lambda(1 - R)}{2n^3 r \Gamma \sqrt{R}} \tag{12.48}$$

This definition makes it easy to compare the MZ modulator with an FP device.

Traveling-wave optical resonators (TWR) including ring, race-track, disk, or sphere resonators, have also been considered for integrated optic applications. In fact, an equivalence between the standing and traveling-wave resonator modulators can readily be shown using the coupled-mode approach (see Gheorma 2002). An example of this set of equations is for a reflection FP modulator, which appears in the equations below with dielectric mirrors, within the stop band of the mirrors (also, note that the indexes on both sides of the mirror are assumed equal,

$$\begin{vmatrix} b_1 \\ b_2 \end{vmatrix} = \begin{vmatrix} t_{FP} & i\kappa \\ i\kappa & t_{FP} \end{vmatrix} \begin{vmatrix} a_1 \\ a_2 \end{vmatrix} \tag{12.49}$$

where r is the mirror reflectivity and t_{FP} is the mirror transmission. In both of these equations, b_2 and a_2 can, with generality, be related by $a_2 = Ae^{+\theta}b_2$. $A = e^{-\alpha d} + \cdots$ represents the roundtrip return (including the attenuation due to transmission through the second mirror for an FP), is the roundtrip phase change, and is an equivalent exponential loss per pass.

12.4.2.4 Bandwidth

Device bandwidth is an important quality for most applications of electro-optical modulators. It is limited in practice by the magnitude of the electro-optical effect, by the lumped circuit elements of the modulator, and, at very high speeds, by phase mismatch between the electrical and the optical waves.

In order to determine the limitation on the bandwidth of the device, it is necessary to divide modulators, and their electrodes, into two classes. In the first, the electrode length is much smaller than the wavelength of the applied field. Thus the harmonic period of the field is much longer than the optical transit time, and the electrode is describable by a lumped circuit model. In the second, the conditions are the very opposite, and field propagation, or traveling-wave, effects must be considered in the modulator design. A key difference between the two electrode types is that the traveling wave type is fed by a transmission line and, hence, requires a characteristic impedance termination at the other end. Both types require consideration of the capacitance per length of the device.

For a dielectric waveguide, such as $LiNbO_3$, the key question of capacitance can be conveniently found by numerical techniques or by conformal mapping. A plot of the capacitance per length versus d/w, where w is the width of the electrode and d is again the distance between the two electrodes, obtained by conformal mapping for three common electrode types used in RF-driven systems is given in Fig. 12.12. In all cases, the capacitance per length, C/L, is found to be proportional to ϵ_{eff}, where ϵ_{eff} is an effective or average dielectric constant for the electrode structure and is independent of length. Because of the importance of ϵ_{eff}, it is clear that low dielectric materials, such as polymers, should and do yield much faster modulators (Chen et al. 1997). In addition, the figure also shows that C/L scales approximately as $(d/w)^{-1}$. Thus, as suggested in the discussion above, scaling down of the electrode area can significantly increase the device capacitance and, hence, decrease its bandwidth.

For the first type of modulator, i.e., the lumped-circuit type, the highest frequency response is, first, ensured by terminating the high-voltage electrode with 50 Ω resistor. The frequency response or bandwidth, $\Delta\nu$, is then governed by the usual RC circuit limitation, $\Delta\nu = 1/\pi\,RC$. Because of this relation and the inverse dependence of device capacitance on length, $\Delta\nu \propto 1/L$, for a fixed 50 Ω termination resistance. As a result, the product of bandwidth × length, such as that also shown in Fig. 12.12 provides one figure of merit for the modulator. For a particular modulator structure, this value is adjusted by varying the electrode geometry through the ratio d/w. Regardless of how low the capacitance is, however, ultimately the frequency cutoff for the modulator is determined by the transit time, ν_t. This transit time limitation is given by

$$\nu_t = \frac{c}{\pi\sqrt{\epsilon_{eff}}L} \tag{12.50}$$

For $LiNbO_3$, ϵ_{eff} is such that

$$\nu_t \approx 2\,\text{GHz} \cdot \text{cm} \tag{12.51}$$

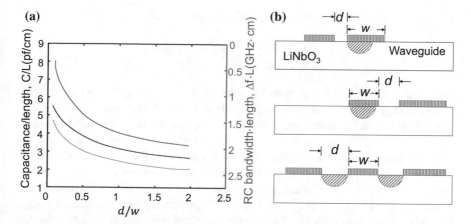

Fig. 12.12 **a** A plot of the capacitance per length and RC bandwidth times length vs. d/w. The figure shows that the capacitance per length decreases as d/w increases, proportional to $(d/w)^{-1}$. **b** A sketch of three common used electrode configuration, corresponding to the curves in (a)

At high frequencies, the electrode structure must be considered to be part of the transmission line. In this case, the bandwidth limitation of traveling modulators is typically velocity mismatch between the microwave and the optical signals. Analysis of the traveling wave problem shows that the mismatch between the optical and microwave effective index, N_{eff}^{o} and N_{eff}^{m}, respectively, controls the mismatch velocity and leads to the following limitation on the useful bandwidth, $\Delta\nu$:

$$\Delta\nu L \approx \frac{2}{\pi} \frac{c}{N_{\text{eff}}^{m}} \left(1 - \frac{N_{\text{eff}}^{o}}{N_{\text{eff}}^{m}}\right)^{-1} \tag{12.52}$$

For LiNbO$_3$, $(N_{\text{eff}}^{o})/(N_{\text{eff}}^{m}) \approx 0.5$ and, thus, $\Delta\nu L \approx 9.6$ GHz · cm. Other effects, such as voltage drop along the electrodes, due to microwave absorption, are also important. The use of thick electro chemically deposited copper helps alleviate this problem (Noguchi et al. 1995).

Consideration of (12.38) and (12.55) yields insight into a practical aspect of high-speed dielectric modulators. Specifically, these equations show that both modulation voltage and bandwidth scale as $1/L$, where again we neglect any loss in the modulator electrodes is neglected. As a result, there is an important tradeoff involving the device length and the modulator bandwidth and voltage. Specifically, low V_π can be achieved by increasing the device length. However, this comes at the cost of lower bandwidth. Similarly, high bandwidth and V_π are found for a short device.

In fact, one figure of merit for characterizing a modulator is the ratio of V_π to bandwidth $V_\pi/\Delta\nu$, since it eliminates the scaling with length. In its most general form, this ratio is a complex relationship involving the device geometric factors, the microwave and optical dielectric constants, and the modal overlap function; see the discussion by Alferness (1982) for details. As a particularly simple example, consider the case of a traveling-wave modulator. In that case,

$$\frac{V_\pi}{\Delta\nu} = \frac{p\pi\lambda}{2c}\left(\frac{d}{\Gamma}\right)\frac{N_{\text{eff}}^m - N_{\text{eff}}^o}{n^3 r} \tag{12.53}$$

where again electrode loss has been neglected. This equation gives the basic scaling laws for the figure of merit of the voltage/bandwidth ratio. A high-performance modulator will minimize this ratio, since a low V_π for high bandwidth is then desired. Thus, matching of the effective index, increasing of the overlap integral, Γ, and decreasing of the electrode gap, d, all decrease the $V/\Delta\nu$ ratio. For the case of LiNbO$_3$, values of ~0.3 V/GHz are typical in standard traveling-wave commercial modulators. This value may be compared to $V/\Delta\nu \approx 2$ for lumped circuit modulators. Unfortunately, (12.53) ignores another equally important aspect of modulator design: electrode impedance, namely to eliminate unwanted voltage reflections from the electrodes, traveling-wave modulators should have 50 Ω characteristic impedance. To achieve this electrode impedance often requires a modulator design, in which other factors, e.g., mode overlap, are compromised. As a result, achieving the best performance comes from a series of interrelated tradeoffs. Nonetheless, extremely high-performance devices have been made by paying close attention to the most crucial design limitations (Noguchi et al. 1998, 1995).

12.4.2.5   Very High-Speed Modulators

Careful materials and structural design can result in much higher speed MZ modulators, that is those with bandwidths of 50–100 GHz. These modulators have been made in LiNbO$_3$ and in polymer materials; see also Sect. 12.5 for very high-speed electro-absorption (EA) modulators in semiconductors.

A sketch of one such high-speed LiNbO$_3$ modulator is shown in Fig. 12.13. This MZ modulator has a 75 GHz 3 dB bandwidth at 1.5 μm (Noguchi et al. 1998). The LiNbO$_3$ crystal is z-cut so that the maximum electro-optical coefficient, r_{33}, is achieved with a vertical applied electric field. To prevent optical loss from the waveguide into the electrodes, a "buffer" layer of SiO$_2$ is applied between the electrodes and the LiNbO$_3$.

This device uses several design improvements to achieve its high speed. As alluded to above, the central design issue is to eliminate the index mismatch between the optical and microwave effective index. Recall that in LiNbO$_3$, wave matching is particularly serious because of the different values of the dielectric constant at these two frequencies. The problem can be reduced by lowering the microwave effective index by allowing more of the waveguide to be exposed to the air. This allows more of the microwave field to leak into the air region, giving a lower effect index. This effect is accomplished by forming an etched rib. In addition, the electrode thickness is increased and adjusted so that more of the microwave field is in the metal, thus reducing loss and inter-electrode capacitance. These two effects are shown in Fig. 12.14. However, notice that the increase in bandwidth in this device comes with a

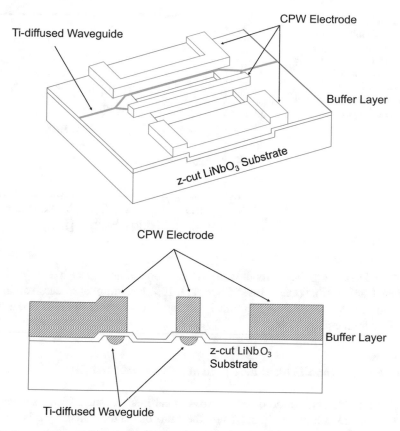

Fig. 12.13 An example of the high-speed LiNbO₃ modulator, with a bandwidth of 75 GHz at 1.5 μm [Adapted from (Noguchi et al. 1998)]

price: the overlap between the optical and the microwave field is reduced by forcing the microwave field into the air region due to the etched rib. The $V_\pi\, L$ product of this modulator ∼10.2 V · cm is relatively high (Noguchi et al. 1998). The optical bandwidth is 110 GHz; the value of $V_\pi \Delta \nu \approx 0.5$ VGHz for a 2 cm-long device. Other approaches such as the use of thin LiNbO₃ slabs can potentially give high bandwidth plus low voltage (Gheorma et al. 2000).

Of course, it is also possible to achieve closer velocity matching of the microwave and optical waves by using a material with comparable, and typically lower, dielectric constants at both optical and microwave frequencies, specifically polymers. There have been several recent demonstrations of very high modulation bandwidth in MZ modulators fabricated of low dielectric constants using polymer materials. The polymer used in these devices was designed to have a low optical loss, which can otherwise be significant at telecommunications wavelengths. The device was demonstrated at 1.3 μm and had $V_\pi \approx 2.2$ V · cm and a bandwidth of 110 GHz. The velocity matching

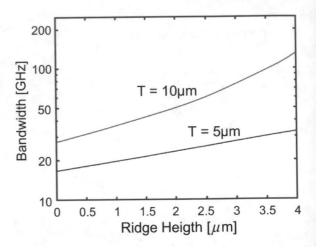

Fig. 12.14 A plot for the optical 3-dB modulation bandwidth versus ridge height (at different electrode thickness). A higher bandwidth can be achieved when the ridge is higher and the electrode is thicker

in this and in other polymer modulators is very good; numbers such as 5% velocity mismatch are feasible (Dagli 1999; Chen et al. 1997). Additional research on important engineering issues such as optical damage is now being done to make these devices commercially viable.

12.4.2.6   A Second Look at the Speed of a Resonant Modulator

For an FP modulator, based on an electro-optical index change, there is a tradeoff between bandwidth and voltage. Thus, the ratio of bandwidth/voltage is an even more useful figure of merit for an electro-optic modulator. For an electro-optical FP modulator, this ratio can be obtained from

$$\frac{f_{3\,dB}^{FP}(cl)}{V_\pi^{eq}} = 0.643 \frac{cn^2 f \Gamma}{2\pi \lambda g} \tag{12.54}$$

Note that the modulator length disappears in this ratio, and the ratio is also independent of cavity Q or finesse (Gheorma et al. 2000).

We can compare a figure of merit to that for MZ modulators, both lumped-element and traveling-wave configurations. For example, neglecting the RC constant, lumped-element MZ modulators have the following figure of merit:

$$\frac{f_{3\,dB}^{FP}(cl)}{V_\pi} = \frac{2.8cn^2 f r \Gamma}{\pi \lambda g} \tag{12.55}$$

For the same material and same electrode gap, this ratio is a factor ≥ 8 larger for the MZ modulator than for the electro-optic FP modulators.

Fig. 12.15 A sketch of the three-layer semiconductor modulator

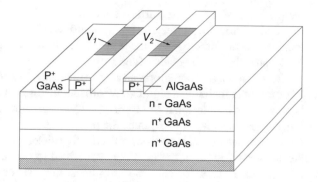

12.4.2.7 Semiconductor Electro-optical Modulators

Thus far, we have focused on the case of LiNbO$_3$ modulators. However, modulation in semiconductor materials is also important (Alferness et al. 2013). In semiconductors, thus, the existence of free carriers in semiconductor modulators can make the application of the voltage more complex to analyze, but more recently, semiconductors have been used to make a series of highly effective ultrasmall modulators, which are having a major impact on complex, dense high-data rate systems (Xu et al. 2005). Generally, a semiconductor modulator structure consists of a three-layer diode-like structure, as is shown in Fig. 12.15. This structure must be fully depleted by a strong reverse bias, or the resulting optical modulation will be due to both the electro-optical effect and free carrier absorption. The presence of free carriers does allow electrical control of modulation as will be discussed below. However, do note that in some cases free carriers can also introduce additional optical and microwave loss. Finally, note that the standard electro-optical effect for crystals and semiconductors is governed by a tensor as explained earlier in this chapter. Free carrier effects are not taken into consideration in this chapter.

Typically, GaAs or other III–V based devices use (001) wafers. For a vertical, applied field, and in the absence of carrier effects, the index change is then in the [0, 1, 1] or [0, 1, 1] direction. It can be written as

$$\Delta n = \frac{1}{2} n^3 r_{41} \frac{V^\circ}{d} \Gamma \qquad (12.56)$$

The electro-optic tensor for GaAs, for example, is small, but the large index of refraction causes the effective electro-optic response to be about 1/5 of that of lithium niobate. Adding a rib waveguide helps localize the electric field as well as providing lateral optical confinement. In this structure, the voltage is applied to reverse bias of the Schottky or p+in junction. The presence of heavily doped regions which have overlap with the optical mode introduces loss into the waveguide, e.g., 4 dB/cm for GaAs. Introduction of a local stripe of undoped dielectric under the waveguide region has been found to reduce this loss by a factor of $\sim$1/2. Simple Schottky barriers have also been used for electrodes on semiconductor modulators. However, the presence of a metallic surface contiguous to the optical mode introduces both loss and polarization effects. This difficulty can be alleviated again by placing a local dielectric directly over the waveguide but allowing the metal to contact the guide laterally on either side of the stripe.

Traveling-wave modulators can also be made in III–V materials. As shown earlier, a key factor in realizing a high-bandwidth modulator is to have $N_{\text{eff}}^o \approx N_{\text{eff}}^m$, since the bandwidth is proportional to $(1 - N_{\text{eff}}^m/N_{\text{eff}}^o)^{-1}$. At microwave frequencies, $N_{\text{eff}}^m \approx$ 3.6, compared to $N_{\text{eff}}^o = 3.4$; thus, the phase matching is relatively close. However, in the presence of real electrodes, the effective dielectric constant for the microwave signal is reduced since the field is not confined to the semiconductor and emerges into the air. In fact, for coplanar strip lines, this gives rise to the following effect index, N_{eff}^0

$$N_{\text{eff}}^0 = \left(\frac{1 + \epsilon_{sc}}{2}\right)^{1/2} \tag{12.57}$$

where ϵ_{sc} is the dielectric constant of the semiconductor; this phenomena leads to a $\sim$40% index mismatch (Dagli 1999). Despite this, velocity matching is better in III–V materials and can be achieved by using slow-wave structure, done with periodic capacitive loading, or by using "undoped" GaAs epitaxial layers in conjunction with slow-wave electrodes with local transverse phase reversal. Because of its conceptual simplicity, we will describe the latter approach in somewhat more detail. In particular, low-doped epitaxial layers behave as an insulating dielectric film and thus do not introduce microwave losses. Carefully engineered segmented microwave electrodes can allow push–pull operation, good microwave-optical wave velocity matching, and a $\sim$50 Ω characteristic impedance. In one case, a 1 cm long modulator had $V_\pi = 17$ and a $\sim$50 GHz bandwidth at 1.55 μm. While these are impressive operating parameters, they still are not as good as those in, say, LiNbO$_3$ (Spickermann et al. 1996).

12.4.3   Free-Carrier Modulations: Ultrasmall Si Semiconductor Modulators

Modulator device size is a key property of complex photonic chips for data manipulation. Silicon is preferred in these applications due to the ease of using standard Si patterning tools. If a resonant strongly light-confining structure is used, it is possible to have the ring respond to small changes in the Si refractive index and still have a high-speed operation. Under these conditions, the modulator diameter is much smaller than the previous Si designs. Earlier work had shown that it is possible to fabricate ultrasmall modulator devices in high-index materials, e.g., III–V compound semiconductors (Sadagopan 2004). Similarly, all optical silicon ring resonators were shown to work with an optical control beam Almeida et al. (2004), however, full useable modulator functionality requires having an electronically driven structure. Such

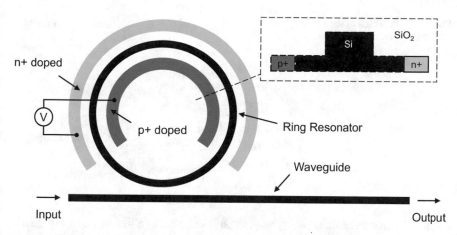

Fig. 12.16 A sketch of a micro-ring Silicon EO modulator based on free-carrier effect.

electronic driving including attainment of high-speed operation has been demonstrated by Liu et al. (2004) based on a CMOS patterned chip. However, high spatial device resolution was not reached due to the weak optical confinement (Liu et al. 2004) and in fact only relatively long device lengths of mm could be demonstrated (Xu et al. 2005).

Light-confining resonating structures and its attendant strong confinement can enhance the effect of refractive index change on the transmission response (Heebner et al. 2004; Almeida et al. 2004). In fact, using the resulting strong confinement enables a silicon electro-optical modulator of a few micrometers in the size to be fabricated. The modulator consists of a ring resonator coupled to a single waveguide, with the waveguide transmission depending sensitivity on the particular signal wavelength being used. In particular, the transmission is reduced at wavelengths, which are multiples of the ring circumference (Little et al. 1998; Almeida et al. 2004).

Figure 12.16(inset) shows a cross section of a strip waveguide. The thickness of the slab thickness (50-nm-thick Si layer) is much less than the wavelength in the 1.5 mm mode profile and thus closely approximates that of the silicon strip waveguide. The device tunes the effective index of the ring waveguide so as to strongly modulate the transmitted signal. Index modulation is via electron and hole injection.

12.5 Semiconductor Electro-Absorption Modulators

EA modulators (Dagli 1999) are a very different modulator type. They operate by using an applied voltage to modulate the absorption coefficient in a semiconductor

Fig. 12.17 A sketch of the typical EA modulator

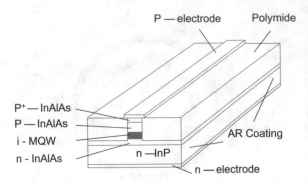

sample. EA modulators typically use the physical structure illustrated in Fig. 12.17. The optically active portion of the modulator is in a composite region containing an intrinsic semiconductor region incorporating multiple quantum wells. The contact to this region is through a p-type upper region and an n-type lower region. The complete semiconductor region is etched into a channel waveguide as shown in the figure. The refractive indices in the n and p regions are lower than in the intrinsic MQW region and thus light is confined in the optically active region.

The mechanism for absorption in the quantum wells is via excitonic bands, which have a relatively sharp narrow line. When an external voltage is applied to this structure, the hole- and an electron-wavefunction overlap is reduced and the absorption feature is reduced in strength and broadened. In effect, optical absorption in this material is controlled by the application of an external voltage. Figure 12.19 shows the absorption versus wavelength for typical EA material. Note that in this case, two excitonic features are present because of different electron–hole band combinations. Several different InP and GaAs, quarternary or ternary alloys are used for EA modulators. For example, for $\lambda = 1.55\,\mu m$ GaAsP or GaAlAs quarternary alloys have been used. In the case of ternary alloys, opposing strained-layer regions have been used to achieve thicker layers. The fact that relatively complex material structures are used for modulators means that integrated structures, which generally have simple passive waveguides, require careful etch and regrowth.

Optical absorption in these materials is shown in (Dagli 1999) for two wavelengths as a function of voltage. The voltage-dependent transmission (loss) $T(V)^0$ for an E modulator can be written as

$$T(V^0) = C \exp[-\Gamma \alpha(V)^0 L] \qquad (12.58)$$

where C is a coupling coefficient between the input and output of the device and the input and output waveguides, respectively, Γ is the overlap between the optical mode and the quantum-well cross section, L is the semiconductor pass length, and $\alpha(V^0)$ is the voltage-dependent absorption coefficient. This equation allows the on/off ratio to be written as

$$10 \log[T(V^0)/T(0)] = \frac{\Delta\alpha}{\alpha} L \tag{12.59}$$

where L is the propagation loss in dB in the modulator in the $V^0 = 0$ state and $\Delta\alpha = \alpha(V) - \alpha(0)$. The second term $\alpha(0)$ is the background loss which can be of order -3 dB. The term $\Delta\alpha/\alpha$ is typically of order -3 dB, which implies a large on/off ratio.

Electro-adsorption modulators can be extremely fast. In part, this is due to their strong on/off ratio allowing relatively short devices to be made, i.e., $50-200\,\mu\text{m}$, which have, therefore, low capacitance. Thus, "lumped circuit" devices typically can be as fast as 50 GHz with a 2 V driving voltage. Traveling-wave modulators have also been made via careful design so as to avoid microwave loss. As an example of the traveling-wave devices, a $200\,\mu\text{m}$ long 54 GHz device has been demonstrated with a 3 V drive voltage and a 20 dB on/off ratio (Kawano et al. 1997).

12.6 Summary

This chapter has described the operation of the linear electro-optic effect and showed how it may be used to make modulators in insulating and semiconducting crystals. This material is then used to focus the remainder of the chapter on how modulators can be designed to have the wide bandwidths needed for high-speed communications networks or for analog optical links. To make the discussion of modulators complete, a section on electro-adsorption modulators has been included, despite the fact that EA devices are not based on the electro-optic effect.

In the following chapter, the discussion of electro-optical devices will be extended to switches, which can also be operated via the thermo-optical effect. There are other applications of electro-optic-driven index changes, including scanners or tunable filters. And many of these convenience these devices are also mentioned in other chapters.

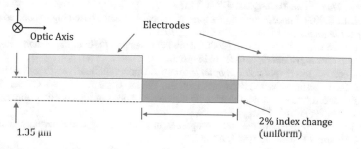

Fig. 12.18 Cross-sectional view of a modulator

Problems

1. We are to investigate a modulator based on the $LiNbO_3$ materials platform. The external electrical field (*NOT* the propagating wave) is along the optic axis. The optical field is TE polarized (along optic axis). The device is being modulated using the Pockels coefficient r_{33}. The figure shows a cross section of one arm of an MZI modulator that has a modulator:

   (a) Assume the electrode spacing is $d = 2\,\mu m$. Calculate the length of the device, assuming the applied voltage is uses 10 V to obtain a π phase shift between the two arms. Assume only one arm is being modulated. Do not calculate the modal properties; simply assume $d = 2\,\mu m$.

   (b) Given the bandwidth-length product $\Delta \nu L = 10\,GHz \cdot cm$. What is the figure of merit $V_\pi / \Delta \nu$ for this device?

   (c) List two more ways to achieve phase modulation other than using electro-optic effect, briefly talk about their advantages and disadvantages.

References

Alferness, R., Burns, W., Donelly, J., Kaminow, I., Kogelnik, H., Leonberger, F., et al. (2013). *Guided-wave optoelectronics* (Vol. 26). Berlin: Springer Science & Business Media.

Alferness, R. C. (1982). Waveguide electrooptic modulators. *IEEE Transactions on Microwave Theory Techniques, 30*, 1121–1137.

Almeida, V. R., Barrios, C. A., Panepucci, R. R., & Lipson, M. (2004). All-optical control of light on a silicon chip. *Nature, 431*(7012), 1081.

Bridges, W. B., & Schaffner, J. H. (1995). Distortion in linearized electrooptic modulators. *IEEE Transactions on Microwave Theory and Techniques, 43*(9), 2184–2197.

Burns, W., Milton, A., & Tamir, T. (1988). Guided wave optoelectronics.

Chen, D., Fetterman, H. R., Chen, A., Steier, W. H., Dalton, L. R., Wang, W., et al. (1997). Demonstration of 110 ghz electro-optic polymer modulators. *Applied Physics Letters, 70*(25), 3335–3337.

Dagli, N. (1999). Wide-bandwidth lasers and modulators for rf photonics. *IEEE Transactions on Microwave Theory and Techniques, 47*(7), 1151–1171.

Gheorma, I.-L. (2002). Fundamental limitations of optical resonator based high-speed eo modulators. *IEEE Photonics Technology Letters, 14*(6), 795–797.

Gheorma, I.-L., Savi, P., & Osgood, R. (2000). Thin layer design of x-cut linbo 3 modulators. *IEEE Photonics Technology Letters, 12*(12), 1618–1620.

Haus, H. A. (1984). *Waves and fields in optoelectronics*. Upper Saddle River: Prentice-Hall.

Heebner, J. E., Lepeshkin, N. N., Schweinsberg, A., Wicks, G., Boyd, R. W., Grover, R., et al. (2004). Enhanced linear and nonlinear optical phase response of algaas microring resonators. *Optics Letters, 29*(7), 769–771.

Kaminow, I. (1965). Microwave dielectric properties of n h 4 h 2 p o 4, and partially deuterated k h 2 p o 4. *Physical Review, 138*(5A), A1539.

Kawano, K., Kohtoku, M., Ueki, M., Ito, T., Kondoh, S., Noguchi, Y., et al. (1997). Polarisation-insensitive travelling-wave electrode electroabsorption (tw-ea) modulator with bandwidth over 50 ghz and driving voltage less than 2 v. *Electronics Letters, 33*(18), 1580–1581.

Little, B. E., Foresi, J., Steinmeyer, G., Thoen, E., Chu, S., Haus, H., et al. (1998). Ultra-compact si-sio 2 microring resonator optical channel dropping filters. *IEEE Photonics Technology Letters, 10*(4), 549–551.

Liu, A., Jones, R., Liao, L., Samara-Rubio, D., Rubin, D., Cohen, O., et al. (2004). A high-speed silicon optical modulator based on a metal-oxide-semiconductor capacitor. *Nature, 427*(6975), 615.

Noguchi, K., Mitomi, O., & Miyazawa, H. (1998). Millimeter-wave ti: Linbo/sub 3/optical modulators. *Journal of Lightwave Technology, 16*(4), 615–619.

Noguchi, K., Mitomi, O., Miyazawa, H., & Seki, S. (1995). A broadband ti: Linbo/sub 3/optical modulator with a ridge structure. *Journal of Lightwave Technology, 13*(6), 1164–1168.

Sadagopan, T., Choi, S. J., Dapkus, P. D., & Bond, A. E. (2004). *Digest of the leos summer topical meetings mc2-3.*

Saleh, B. E., Teich, M. C., & Saleh, B. E. (1991). *Fundamentals of photonics* (Vol. 22). New York: Wiley.

Spickermann, R., Sakamoto, S., Peters, M., & Dagli, N. (1996). Gaas/algaas travelling wave electro-optic modulator with an electrical bandwidth> 40 ghz. *Electronics Letters, 32*(12), 1095–1096.

Wooten, E. L., Kissa, K. M., Yi-Yan, A., Murphy, E. J., Lafaw, D. A., Hallemeier, P. F., et al. (2000). A review of lithium niobate modulators for fiber-optic communications systems. *IEEE Journal of Selected Topics in Quantum Electronics, 6*(1), 69–82.

Xu, Q., Schmidt, B., Pradhan, S., & Lipson, M. (2005). Micrometre-scale silicon electro-optic modulator. *Nature, 435*(7040), 325.

Yariv, A. (1973). Coupled-mode theory for guided-wave optics. *IEEE Journal of Quantum Electronics, 9*(9), 919–933.

Chapter 13
Integrated Optical Switches

Abstract This chapter presents one related modulator-like functionality; i.e., switching of a propagating light wave. This functionality is important for many applications, including signal multiplexing in time and wavelength, signal routing, and signal encoding. An effective photonic switch requires a high extinction ratio, low loss, and a short response time. Switches are shown to be based on mechanical movement, thermo-optic, electro-optic, or electron-plasma optical to change the refractive index. Our discussion focuses on electro-optical switches since they were the most important type for integrated switches particularly those for ultrafast lithium niobate crystals. In addition, the chapter includes a short section on thermo-optical-based switches as well. Finally, methods of improving switch performance are also presented in the chapter.

13.1 Introduction

In the previous chapter on the fundamentals of electro-optical devices, our discussion focused on optical modulators operating via the electro-optic effect (with a short discussion on electro-absorption). In this chapter, we will focus specifically on one important modulator function: switching. Switching is used for a number of important applications, including signal routing, signal multiplexing in time and wavelength domain, and signal encoding. Basically, in these applications, the photonic switch turns a particular optical channel on or off. Thus, an effective device requires a high extinction ratio, low loss, and a short response time that, is dependent on the interval between bitstreams. Switches can use simple mechanical movement, thermo-optic, electro-optic, or electron-plasma optical to change the refractive index. Our initial discussion will focus on electro-optical switches since they played an essential role in the evolution of integrated switching technology particularly those using $LiNbO_3$; however, we also include a brief section on thermo-optical switches.

© Springer Nature Switzerland AG 2021
R. Osgood jr. and X. Meng, *Principles of Photonic Integrated Circuits*,
Graduate Texts in Physics,
https://doi.org/10.1007/978-3-030-65193-0_13

13.1.1   Electro-Optical Switches

As suggested in the previous chapter, most integrated electro-optical switches rely on phase modulation in their waveguiding structure. These switches use an external voltage to cause a shift in the material refractive index, thus the propagation constant of the waveguide. Electro-optical switches are available in several well-studied forms: directional couplers, intersecting waveguides, and interferometers (Alferness 1982; Campbell et al. 1975; Soldano et al. 1994).

Despite the fact that these switches can be quite different in structure, their electro-optical function can be expressed in equations that show their close similarity. The common equations stem from the fact that most, but not all, switches rely on some form of phases shifting in a waveguide. Thus, as shown in Chap. 12, the phase shifter, $\Delta\phi$, in one arm of the waveguide containing an electro-optical element of length L and electro-optic coefficient, r, can be written quite generally as

$$\Delta\phi = \Delta\beta L = -\pi n^3 r \Gamma \frac{VL}{d\lambda} \tag{13.1}$$

where the symbols are defined, for example, as in the case of (12.35). Since the phase shift is usually related to some multiple of π, empirically, one can then also write

$$|\Delta\beta L| = p\pi \tag{13.2}$$

where p is a real number with a characteristic value, which varies from one modulator type to another. For example, for a directional coupler switch $p = \sqrt{3}$, for a Mach–Zehnder (MZ) $p = 1$, and for a polarization switch $p = 1/2$. Thus, (13.1) and (13.2) gives the voltage–length product, as determined by electro-optics, as

$$V_\pi L = \frac{p\lambda d}{n^3 r \Gamma} \tag{13.3}$$

This more general formulation of the phase shift (Haus 1984) allows one to compare the performance of switches, which are based on different modulation types.

13.1.2   The $\Delta\beta$ Directional Coupler Switch

One of the most basic electro-optical switches, shown in Fig. 13.1, uses a typically symmetric, directional coupler formed in electro-optical material. This device was one of the earliest demonstrated electro-optical switches; both $GaAs$-based Campbell et al. (1975) and $LiNbO_3$ Papuchon et al. (1975) versions were fabricated. In the device, the coupling length was adjusted so that in the absence of an applied voltage, light was coupled, with coupling coefficient κ, into the adjacent waveguide, a condition termed the cross ($\times$) state in switches. Thus, the device length L is such

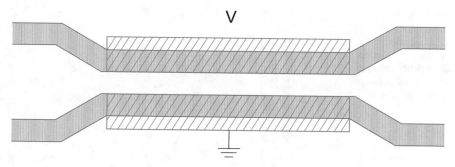

Fig. 13.1 A sketch of the electro-optical switch that uses a symmetric, directional coupler. Applying voltage on the electrodes changes the refractive index of the coupling region, thus the coupling ratio sketch of the electro-optical switch that uses a symmetric, directional coupler.

that $L = L_c$, or

$$\kappa L = \frac{\pi}{2} \tag{13.4}$$

When a voltage is applied to the structure, a shift in the waveguide refractive index is induced. This shift, reflected in the waveguide propagation constant, β, causes the waveguides to be no longer phase-matched and light remains in the original waveguide; hence, the switch is called a $\Delta\beta$ switch. In principle, the application of a voltage can also change the value of the coupling coefficient, κ; however, typically this effect is minimal and not important for this type of switch.

Recall from (5.37) and (5.38) that for a symmetric directional coupler, the power transfer efficiency, η_x, from an input waveguide to its cross waveguide is given by

$$\eta_x \simeq \left(\frac{\kappa}{\beta_c}\right)^2 \sin^2 \beta_c L \tag{13.5}$$

and the efficiency of remaining in the input (or bar) waveguide is

$$\eta_{\parallel} = 1 - \eta_x = 1 - \left(\frac{\kappa}{\beta_c}\right)^2 \sin^2 \beta_c L \tag{13.6}$$

where $\beta_c = \sqrt{\kappa^2 + \Delta^2}$ and where $\Delta = \Delta\beta/2$. Figure 13.2 gives a plot of the normalized output of the switch, that is, its cross waveguide power versus $\Delta\beta$, and hence applied voltage. From relations (13.5) and (13.6), it is apparent that complete power transfer to the cross state, $\eta_x = 1$, can only occur if $\beta_c = \kappa$; that is, when $\Delta = 0$ or when the voltage is off (see (13.1)). In addition, the cross state requires that $\sin^2 \beta_c L$ must be unity, or $\beta_c L = (2n + 1)\pi/2$, where n is an integer. However, as is also shown in Fig. 13.2, when $\Delta \neq 0$, the power transferred to the opposite waveguide is reduced, and in fact for certain specific values of the voltage-controlled

Fig. 13.2 A plot of normalized output of the switch versus the propagation-constant discrepancy in the two waveguides due to the applied voltage

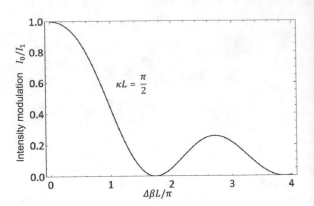

propagation-constant mismatch of the two waveguides, $\Delta\beta$, all entering power exits through bar state, that is, on the input waveguide. Thus, as is shown in the figure, a typical $\Delta\beta$ switch is designed so that in its bar state the voltage is on, and in its cross state the voltage is off. In electro-optical switches, the value of $\Delta\beta$ is controlled with an external voltage via the electro-optical effect applied to one arm of the directional coupler; if the device is operating in push–pull, voltage is applied to each arm. As suggested by the first two maxima in Fig. 13.2, the light exiting the device on the cross-state waveguide has sidebands of decreasing magnitude at higher values of $\Delta\beta$, i.e., external voltage. These sidebands, which lead to undesired leakage through an imperfectly fabricated device in its nominally off state, may be decreased by using electrode tapering Schmidt and Kogelnik (1976).

Switches can also be made for devices, which are multiples of the coupling length. Accessing the first maximum, that is, switching to the bar state, requires $\Delta\beta L = \sqrt{3}\pi$, or p in (13.2) is $\sqrt{3}$. However, as a consequence of the voltage dependence of (13.5), the switching voltage from the cross to the bar state increases with the number m of coupling lengths, $L = mL_c$, in the device. Specifically, if the device length, L, is $(2m + 1)\pi/2\kappa$, i.e. such that with no voltage applied $\eta = 0$, then the voltage for switching increases with m. This can be shown by noting that in each case after the voltage is applied, switching from the cross to the bar state occurs when

$$\beta_c L = ((2m + 1) + 1)\pi/2. \tag{13.7}$$

Since $\beta_c = \sqrt{\kappa^2 + \Delta^2}$ and $\Delta = \Delta\beta/2$, the (13.7) can be written as

$$\Delta\beta L = 2\pi \left((m + 1)^2 - (m + 1/2)\right)^{1/2}. \tag{13.8}$$

Thus, the longer the length, the higher the voltage, and in the limit of large m, $\Delta\beta L \approx 2\sqrt{m}\pi$. A plot of intensity in the cross channel is shown in Fig. 13.3 for two values of length $L = \pi/2\kappa$ and $3\pi/2\kappa$. Note that the voltage required for the cross

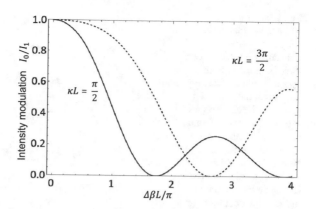

Fig. 13.3 A plot of normalized output in the cross channel for two different lengths, $L = \pi/2\kappa$ and $3\pi/2\kappa$

state, i.e., the first null point in the plot, increases with the length of the device, L, as indicated by (13.8).

In designing $\Delta\beta$ switches, including those having multiple electrodes, which are discussed below, it is necessary to have a more methodical approach for the dependence of the switch geometry on variations in L and voltage (Kogelnik and Schmidt (1976)). A convenient approach to this problem, and an approach, which is transferable to more complex $\Delta\beta$ switches, is that provided by the use of the transfer matrix for a directional coupler (here a switch) that was introduced in Chap. 6. Recall that if such a switch is characterized by inputs in the two waveguides of a_1^i and a_2^i, respectively, then the output a_1^o, a_2^o, can be obtained via the transfer matrix U for a coupler of length L as follows:

$$\begin{bmatrix} a_1^o \\ a_2^o \end{bmatrix} = \begin{bmatrix} u_{11} & -ju_{12} \\ -ju_{12}^* & u_{11}^* \end{bmatrix} \begin{bmatrix} a_1^i \\ a_2^i \end{bmatrix} \tag{13.9}$$

where the matrix elements have been given earlier in (6.16) and (6.17). This matrix is written for a positive voltage, u^+, for the device shown in Fig. 13.1. The voltage for the opposite polarity, u^-, is identical to u^+ except that the diagonal elements are reversed, i.e., $u_{11} \rightarrow u_{11}^*$ and $u_{11}^* \rightarrow u_{11}$.

For a simple directional coupler switch of length L, one input, say, a_2^i, is always zero. In addition, the device is designed such that in its "off" condition, i.e., $\Delta\beta = 0$ or $\Delta = 0$, the device is in its cross state, that is $a_1^o = 0$ at $z = L$. Consideration of (13.9) shows that these conditions require that $u_{11} = 0$. Using the explicit form of the matrix element given in (6.16) and (6.17) shows that $u_{11} = 0$ only when

$$\kappa L = (2m + 1)\pi/2 \tag{13.10}$$

or

$$\frac{L}{L_c} = 2m + 1 \tag{13.11}$$

Fig. 13.4 A plot of operation points of the simple $\Delta\beta$ switch, illustrating one of the two limitations of discrete operation points. This figure shows that the operation points on the $\Delta\beta$ axis are actually only the bar state, cannot be compensated for the length discrepancy. Notice that here the electro-optical effect is used to change the refractive index of the coupling region, thus the wavevector. Symbol HV means high voltage

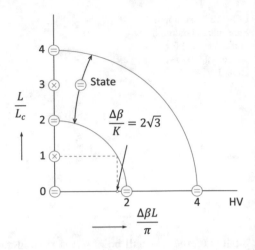

i.e., the exact conditions specified in the earlier discussion in this section and where m is again an integer, starting at $m = 0$. In a similar manner, it can be shown that after application of an external voltage, the bar state, which requires $a_2^o = 0$, also implies that $u_{12} = 0$. This condition holds when

$$(\kappa L)^2 + [(\Delta)(L)]^2 = (m\pi)^2 \tag{13.12}$$

or

$$\left(\frac{L}{L_c}\right)^2 + \left(\frac{\Delta\beta L}{\pi}\right)^2 = 4m^2 \tag{13.13}$$

This equation has multiple solutions, in contrast to that for the cross state. The two conditions are displayed in Fig. 13.4 in a plot of L/L_c versus $\Delta\beta L/\pi$ for both bar and cross states. In this diagram, switching occurs by moving horizontally, at a fixed value of L/L_c, on the diagram. Notice, in addition, the condition (13.11) corresponds to a series of fixed points along the ordinate, regardless of the voltage. As a result, the device can attain the cross state only if L is precisely L_c; adjustment in L/L_c via voltage trimming (through $\Delta\beta$) is impossible in the cross state. However, the condition for the bar state, given in (13.13), may be satisfied for a set of lengths L and voltage; thus, if L is not precisely L_c, the bar state can be reached by voltage trimming. The physical basis for this behavior is discussed earlier in this section in conjunction with (13.5) and (13.6). As examples, the operating points for two devices, each with different values of L/L_c are indicated in the figure.

13.1.3 Reversed $\Delta\beta$ Switch

The simple $\Delta\beta$ switch discussed above suffers from two important limitations. First, in order to have low cross talk for the voltage off or cross state of the switch, the device must be exactly one coupling length (or odd multiples of the coupling length). If the length deviates from the distance of one coupling length, light will "leak" through, reducing the on/off ratio, which will lead to serious degradation in the performance. The discrete operating points on the plot of L/L_c versus $\Delta\beta$ (see Fig. 13.4), mean that an error in the length, L, cannot be corrected with a corresponding change in $\Delta\beta$. Second, due to the typical electrode spacings used in most switches, the coupling length is generally small compared to the electrode length needed to get a sufficient phase shift at an acceptable range in voltages. Thus, since $L/L_c \gg 1$, the switching voltage and, hence, the voltage-speed product will then be unacceptably high.

In order to solve the above two limitations, an approach is used which divides an electrode of length $L/L_c > 1$, into N segments of equal length. Each of the lengths is periodically biased so as to reverse the charge polarity on the electrodes. Hence, this switch is called a "reverse $\Delta\beta$-switch." The response of such a series of multiple reversed devices can be obtained via multiplication of the transfer matrix of each individual segment. This analysis shows that for a switch with a series of N sections having periodic voltage reversal, the overall cross efficiency is given by

$$\eta_N = \sin^2 \kappa_N L \tag{13.14}$$

where

$$\kappa_N = \frac{N}{L} \sin^{-1} \sqrt{\eta_x} \tag{13.15}$$

where η_x is the efficiency of transfer into the cross state for a single element on length L/N, given earlier by (13.5). Finally, note that the required $\Delta\beta L$ for this switch gives $\sqrt{3}\pi$ (see (13.20), even for $L/L_c > 1$,) so long as the number of electrode segments equals approximately the number of coupling lengths.

The simplest multiple electrode switch is the reversed $\Delta\beta$ switch with an overall length of L, shown in Fig. 13.5 Schmidt and Kogelnik (1976). This switch uses two $\Delta\beta$ switches of equal length $\sim L/2$. Because of its simplicity, it can be used to delineate clearly the advantages of reversed $\Delta\beta$ switches. The switches are arranged in a series with $\Delta\beta$ of the first section put in series with—$\Delta\beta$ of the second; the shortest device, which can be voltage tuned in both cross and bar states, has $L > L_c$ or $L_c/2$ for each section.

The most important feature of such a reversed $\Delta\beta$ switch is that it is insensitive to the exact length of the electrode segments since each "state" of the switch can be voltage-tuned for optimum performance. This length insensitivity can be shown clearly by an analysis based on the transfer matrix approach introduced in the previous section. This approach can also be readily applied to a device having multiple sections.

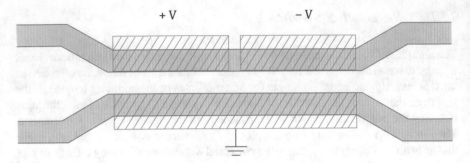

Fig. 13.5 A sketch of the "reversed $\Delta\beta$-switch." Reversed voltages are applied on each half of the coupling region

The overall transfer matrix for a two section device is the product of two matrices for electrodes of length $L/2$ and opposite-biasing polarity. The product matrix is then

$$u^t = u_1^- \cdot u_1^+ = \begin{bmatrix} u_{11}^t & -ju_{12}^t \\ -ju_{12}^{*t} & u_{11}^{*t} \end{bmatrix} \qquad (13.16)$$

where $u_{11}^t = 1 - 2u_{12}^2 = 2u_{11}^* u_{12}$ and where the fact that $u_{11}u_{11}^* = 1$ has been used. The values of these matrix elements can be obtained from (6.16) to (6.17) and the fact that the length of the segment is $L/2$. It is interesting that the nature of the coupling interactions does not cause the two reverse bias sections to "cancel out" their effect. For a single input only in waveguide 1, i.e., $a_2^i = 0$, the total device will be in the cross state when $u_{11}^t = 0$, or when

$$2u_{12}^2 = 1 \qquad (13.17)$$

Substitution of (6.17) in this equation leads to a family of curves, which obey the relation,

$$\sin^2 \frac{\pi}{4} = \left(\frac{\kappa}{\beta_c}\right)^2 \sin^2 \frac{\beta_c L}{2} \qquad (13.18)$$

Note that this same result can be obtained by direct application of the coupled mode equations Haus (1984). Unlike the case of a single section device which has only isolated operating points in the cross state, irrespective of $\Delta\beta$, the two section device allows tuning of $\Delta\beta$ to compensate for an inexact value of L as long as $L > L_c$.

In the bar state, $u_{12}^t = 0$, $a_i^0 = 0$. This condition may be achieved either via $u_{11} = 0$, which implies $\Delta\beta = 0$ and

$$\frac{L}{L_c} = 2(2n + 1) \qquad (13.19)$$

or by $u_{12} = 0$, which implies

$$\left(\frac{L}{L_c}\right)^2 + \left(\frac{\Delta\beta L}{\pi}\right)^2 = (4n)^2 \qquad (13.20)$$

where n is an integer 0, 1, *etc.* The first condition leads to isolated points and is not of interest, while the second leads to a set of curves (circular quadrants) on a plot of L/L_c versus $\Delta\beta/\pi$. The curves also show that in the range of $1 < L/L_c < 3$, both the bar and cross states can be accessed by a simple adjustment of the voltage to change $\Delta\beta$. For example, if $2 < L/L_c < 3$, the device is switched between cross and bar states by, roughly, doubling $\Delta\beta$. Since operating curves, rather than points, exist for both the bar and cross states, the device can, as mentioned above, be tuned to compensate for inexact fabrication of the electrode lengths. Note also in each case, the shift in $\Delta\beta$ and, hence, the value of adjustment voltage is at a value comparable to the operating voltages in a single section $\Delta\beta$ switch and thus does not impose any additional voltage penalty for using multiple electrodes.

A more intuitive understanding of the operation of a reverse $\Delta\beta$ switch can be obtained by considering the operation of this switch at the particular length of $\sim L/L_c = 2$ (Nishihara 1980) for this length. These conditions lead up to a $\sim 50\%$ power transfer at the end of the first section or subcoupler, i.e., at $L/2$. When this split light beam enters the second subcoupler, it experiences the opposite phase mismatch, and the original local mode is reconstructed at the output of the cross waveguide. If the voltage for the bar state is now applied, the signal is switched to the bar state at the end of each subcoupler. Finally, notice that if $L/L_c = \sqrt{2}$ and a voltage of $\Delta\beta = 2\pi$ is added, the polarity of the second section can be merely reversed and the device will be switched to the on the state of uniform $\Delta\beta$ switch. Finally, note that reverse $\Delta\beta$ switches have been made with extremely low cross talk. For example, Bogert et al. (1986) reported a LiNbO$_3$ device with 43 dB extinction ratio.

13.2 Interference-Based Switches

Switching can also be accomplished by using electro-optical phase shifting in one or both arms of a "coupler-equipped" MZ interferometer. For this switch, the necessary two input/output waveguides are provided by 3 dB directional couplers at the input/output (see Fig. 13.6) Martin (1975), Ramaswamy et al. (1978). In this device, the output is either bar or cross, depending on the $\Delta\beta$ induced in the device. The device is thus identical to similar structures used for wavelength filtering. The device switching efficiency from input waveguide one to output waveguide two can be shown to be

$$\eta = \cos^2 \Delta\beta L/2 \qquad (13.21)$$

Notice that the required switching voltage in this case is $\Delta\beta L = \pi$ (i.e., $p = 1$ in (13.2)), and is thus less, by $\sqrt{3}$, than that required for the $\Delta\beta$ directional coupler switches described in the previous section. Also, both switch states are obtained periodically as the voltage is varied.

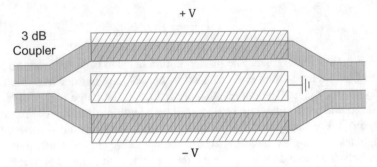

Fig. 13.6 A sketch of the interference-based switch. This is a MZinterferometer consists of two 3 dB coupler and two balanced arms. The voltage is applied to introduce a phase shift. Note that in this case, the phase shift needed to induce is $\Delta\beta = \pi/L_c$, which is smaller than that in a directional coupler case. Also note that by introducing asymmetrical couplers, this switch can also operate as a mode sorter

The MZ interferometer is a related interferometric device using 3 dB Y-branch inputs and outputs. This device does not act like a typical optical switch, since the output is not directed into one of two output ports. However, it can function as a switch in certain integrated circuit layouts by generating an on–off waveform. Because this device is described in detail in Chap. 12, only a few brief comments will be added here. First, for this device, as for a 3 dB-coupler interferometer, the output is fully modulated when $\Delta\beta L = \pi$, i.e., again $p = 1$. Also, it is possible to make a true switch from a MZ device by using an asymmetric branch at the output so as to form a mode sorter. In a conventional device, without the mode sorter, the antisymmetric mode, which is formed in the off, on-transmission, state of the modulator is radiated into the substrate. With the mode sorter, this mode can be directed to one of the output arms of the asymmetric Y-branch Izutsu et al. (1982).

Multimode interference devices have also been used as the splitting or "feeding" elements in interferometric-based switches. Specifically, 1×2 dB MMI couplers can be used in conventional MZ devices because of their good output uniformity and stable relative phase. For example, considering one purely passive device, a pair of 3 dB MMI couplers was used with a MZ structure to make a polarization splitter with a 60 nm wavelength range Soldano et al. (1994). Using the same basic elements plus electro-optic control, MZ interferometer switches have been demonstrated in double heterostructure (DH) Sekiguchi et al. (2012) as well as in multiquantum well (MQW) III–V materials Morl et al. (1998); these devices had extinction ratios of from -10 to -19 dB. The MMI couplers were important in attaining wide bandwidth and polarization-independent operation Bachmann et al. (1993), Zucker et al. (1992).

More complex MZ-like interferometer switches can also be made with MMIs. In such an interferometer, the input signal is split with a $1 \times N$ (or $N \times N$) MMI splitter, fed into individually addressable waveguide phase shifters, and then recombined in an $N \times N$ MMI coupler (see Fig. 13.7). By controlling the phase at the output of each of the N shifters, the input signal(s) can be switched to any one (or certain

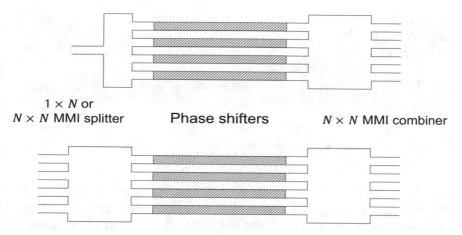

Fig. 13.7 A sketch of the MMI-based switch. It can be a $1 \times N$ or a $N \times N$ switch, with the active tuning capability by adding phase shifters in the middle of two MMIs

sets) of the N output waveguides, although independent control is not possible. The use of MMI couplers for switching, including both the $1 \times N$ and $N \times N$ devices, for the splitter/combiner elements allow the realization of very compact integrated multiway optical switches. For example, 10.6 mm long 1×10 and 13.1 mm long 10×10 switches have been made in a $GaAs/AlGaAs$ epitaxial layers Jenkins et al. (1994). These switches showed ±9% switching uniformity, −10 dB maximum cross talk, and ∼ 6 dB excess loss.

13.3 Modal Interference Switches

Modal interferometric switches rely on interference between two or more modes, which are excited by an input waveguide. Because the input conversion process and the interference do not cause any modal extinction, and hence power loss, these switches can be, in principle, of high efficiency. In general, these devices have not been investigated as extensively as the $\Delta\beta$ switches described earlier.

13.3.1 Intersecting Waveguide Structures

Modal interference in a multimode waveguide has also been used as the basis for switching. An intersecting waveguide switch is one of the earliest versions of these devices Neyer et al. (1986). In this device, sketched in Fig. 13.8, light entering each

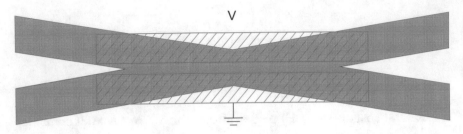

Fig. 13.8 A sketch of a 2 × 2 intersecting waveguide that uses the modal interference to switch

one of the input waveguides excites the two lowest, symmetric and antisymmetric system modes. In the normally off state of the two output waveguides, the symmetric and antisymmetric lowest order modes interfere, with the relative phase depending on the total path length in the mixing region. The modes then propagate within the structure so as to form a single local mode, at one of the waveguide ports. When voltage is applied, the modal propagation constants are changed by the electro-optical effect to interfere in such a way that light exits on the other waveguide. If the path lengths are properly fabricated, the device will switch from one output waveguide to another, according to its voltage state, with low cross talk. This device has a voltage–length requirement similar to that for directional coupler switches and has displayed low cross talk, i.e. 30 dB Neyer et al. (1986), but the requirements on the device geometry are demanding.

13.3.2   MMI Coupler Switches

Switching can also be realized by changing the modal phases in the "free-space" region of an MMI coupler by varying its refractive index. These MMI switches are simpler in structure than those based on phase shifters, which were discussed above, and thus they require simpler controlling circuitry. To illustrate this device, a 1 × 3 switch is shown in Fig. 13.9. The device dimensions, which are defined in the figure, are chosen based on self-imaging principles for each switching function. The required index changes are made in the regions that are shaded. In a semiconductor device, for example, these are realized by current injection, since injection of current can alter the index of a semiconductor as was discussed in Chap. 2. The device in the figure operates as follows: when $n_1 = n_2 = n_g$, the device is a 1 × 3 symmetric MMI splitter; when $n_2 < n_1 = n_g$, it is a 2 × 2 MMI coupler; and when $n_1 = n_2 < n_g$, it is a 1 × 1 MMI coupler. A general 1 × N switch can be constructed using a similar structure Zhao et al. (1998).

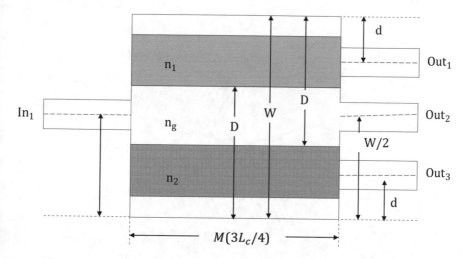

Fig. 13.9 A sketch of a 1 × 3 MMI-based reconfigurable switch. The refractive index of n_1 and n_2. When $n_1 = n_2 = n_g$ due to the "free-space length" is equal to the self-imaging distance, the signal would come out from output port 2; when $n_2 < n_1 = n_g$ lower part can be ignored, the upper part becomes a 2 × 2 MMI, the signal will come out from output port 1

13.3.3   Digital Optical Switch by Mode Sorting

The reverse $\Delta\beta$ switch described earlier has several attractive features, including relative insensitivity to voltage and electrode length. However, it does not clearly operate as a threshold switch, such as is needed, for example, in digital electronics; it is also sensitive to input polarization. Similarly, the intersecting waveguide switch described above also does not operate as a threshold switch and, in fact, has a sinusoidal response.

A truly digital 2 × 2 switch has been described Silberberg et al. (1987), which is based on electro-optical control of mode sorting in two waveguides which intersect at an angle θ. A schematic of this device is shown in Fig. 13.10 for use on z-cut LiNbO$_3$. The device operates by switching the modes entering on either the wider or narrower waveguide via control of the refractive index on the output arms of the switch.

Specifically, the device geometry is such that if light enters on the wider guide, it excites only the lower order mode of the junction. If it enters on the narrower guide, it will excite the second-order mode of the junction region.

This mode sorting occurs if the transition is adiabatic, that is, if

$$\theta \ll \frac{\Delta\beta}{\gamma} \tag{13.22}$$

where γ is the modal decay constant in the cladding region. Recall that this condition was discussed previously in Chap. 7; see (7.24). Note that since the requirement for

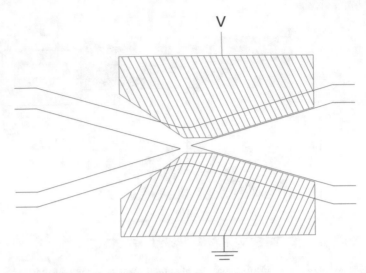

Fig. 13.10 A sketch of a digital 2×2 switch, based on electro-optical control of mode sorting. Note that the upper input waveguide is wider compared to the lower one

mode sorting is that of a simple inequality, it requires only that the angle be sufficiently small.

Once these modes are excited, the light will be directed to a particular output waveguide based on the effective index of that waveguide, which in this section of the device is controlled by the biasing of the output waveguide. For example, the fundamental mode is always directed to the waveguide with the highest index. Reversing the bias reverses the illuminated output guide. With no bias, both output guides are illuminated.

Note that this switch does not rely on modal interference but instead on bias-induced symmetry breaking. The calculated switch output intensity in the two output guides versus induced index is shown in Fig. 13.11. The figure shows a sharp step response in the switching signal with small changes in the effective index, indicating true threshold behavior, for their over biasing does not alter this state.

This digital switch has been fabricated in Ti-diffused, z-cut $LiNbO_3$. The device had a switch voltage for -15 dB cross talk of ± 45 V for the TE mode and $\pm 15 V$ for the TM mode, and a wide wavelength response $\sim 1.3 - -1.5 \mu$m. While the crosstalk values for this switch are practical, the voltage requirement is very high; this high voltage has inhibited greater use of the device. Because of the requirement on adiabaticity in the output arms, the device can be very long. More recently, digital optical switches have also been made in polymeric materials Moosburger et al. (1996); however, these use the thermo-optic effect and are thus discussed in Sect. 13.4.

Fig. 13.11 A plot of output intensity in the two output waveguides verses induced index change. The response is shaped, which means over biasing does not change the device performance

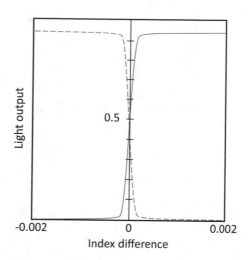

13.4 Thermo-Optical Switches

Thermo-optical-induced changes in the refractive index of materials can also be used to drive switches. Although, in general, devices based on thermo-optical effects are slow compared to electro-optical switches, their response times, which may be as fast as $\sim 1\,\mu s$, are short enough that the switches are useful for many applications. These include reconfigurable PICs, such as delay lines or dispersion compensators, or WDM routers. Thermo-optical switching is also extremely important in a practical sense because it can be used to give some "active-optical" character to a PIC, which is made of normally purely passive materials such as silica, Si, or polymer.

In principle, the operating principle and device layouts for thermo-optical switches can be identical to many of those used in electro-optical devices, except that in this case, a temperature-induced index change is used to provide the needed phase change in the switch. In practice, however, a few simple designs are used most frequently. They include MZ interferometers and digital optical switches. Thermo-optical switches are useful in WDM applications such as add/drop and cross-connect functions since they can switch signals without resorting to optical/electrical signal conversion prior to switching. Notice that the system is transparent to the signal format, either analog or digital.

A typical design for a MZ switch made of SiO_2-on-Si is shown in Fig. 13.12. It consists of a switching unit having two 3 dB MZ devices, each with a heating unit on one arm. For this silica system, the switch time was $2ms$.

More recently, faster devices, with $\sim 1 - 5\,\mu s$ response times, have been achieved in the Si-on-SiO_2 or *SOI* materials system. The silicon waveguide in this system is thermally isolated during switching and is amenable to simple analytic thermal analysis. In this text, we will use this materials system to demonstrate the design considerate for thermo-optical switches. The phase shift in one arm of an interfer-

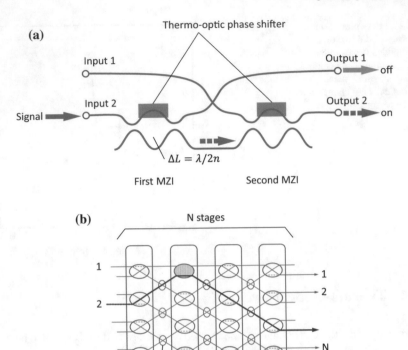

Fig. 13.12 a A sketch of typical MZ-based switch on the platform of SOI. Due to the relatively low speed of thermal–optical modulation, the device speed is about 2 ms. **b** A sketch for the MZ-based switch implementation of a $N \times N$ switch

ometer can be obtained using a modified analytical treatment of the heat flow. This phase shift is

$$\Delta\phi = \frac{2\pi}{\lambda}\left(\frac{dn}{dT}\right)\Delta T L_H \qquad (13.23)$$

where λ is the wavelength, dn/dT is the thermo-optic coefficient of silicon, ΔT is the change in the temperature, and L_H is the heater length. For example, (13.23) shows that a temperature change of $8°K$ is needed to achieve π phase shift in a silicon interferometer, which has a thermo-optical coefficient of $\sim 1.8 \times 10^{-4} K^{-1}$, a length of 700 μm, and $\lambda = 1.5$ μm. Additional useful expressions can also be easily obtained. For example, if the heat flow is corrected for lateral spread, the switching power can be written as

$$P_\pi = \lambda \kappa_{SiO_2} \left(\frac{W_H}{t_{SiO_2}} + 0.88 \right) \left(\frac{dn}{dT} \right)^{-1} \qquad (13.24)$$

where $\kappa_{SiO_2} = 1.4 \, \text{W/cm} \cdot \text{K}$ is the thermal conductivity of SiO_2, t_{SiO_2} is the thickness of the SiO_2, and W_H is the width of the heater. Using the $\Delta T_\pi = 8°\text{K}$ calculated in (13.23), we find that the switching power if $P_\pi = 46 \, \text{mW}$ if the heater is $14 \, \mu\text{m}$ in length and t_{SiO_2} is $1 \, \mu\text{m}$.

The temporal response of the thermo-optic switch can be obtained from an analytical expression, derived by Fischer et al. (1994), relating the cutoff frequency to the switching power and heated area of such a device,

$$f_{cutoff} = \frac{1}{\pi \lambda \rho \epsilon_{th}} \left(\frac{P_\pi}{A} \right) \left(\frac{dn}{dT} \right)_{Si} \qquad (13.25)$$

where P_π is the switching power, A is the heated cross-sectional area, λ is the wavelength, and ρ and ϵ_{th} are the density and specific heat of thermally grown SiO_2, respectively. This equation gives an estimated cutoff frequency, $f_{cutoff} = 49 kHz$, corresponding to a rise time, $\tau_{rise} = 7 \, \mu\text{m}$.

Thermo-optical switches have been made in other materials platforms. For example, a polymer digital–optical switch, which was based on an asymmetric waveguide cross, i.e., the digital optical switch discussed in Sect. 13.3, has been successfully fabricated. The device had low insertion losses, $< 3 \, \text{dB}$, due to the use of an oversize rib waveguide and an extinction ratio of $> 20 \, \text{dB}$ with a $200 \, \text{mW}$ heater Moosburger et al. (1996). In addition, devices based on total internal reflection have also been described Diemeer et al. (1989).

13.5 Optical Switch Arrays

13.5.1 Electro-Optical Switching Arrays

Optical switching arrays in $LiNbO_3$, and more recently Si, comprise some of the most complex working PICs made, and their fabrication has provided a good testbed for examining various PIC technologies. These devices have arrays of directional couplers, and generally have a straightforward cross–bar switch geometry, which requires N^2 switches for an $N \times N$ switch. This switch geometry requires that these devices be arranged in $2N - 1$ stages.

13.5.2   LiNbO₃ Arrays

Historically, LiNbO$_3$ arrays were the first successful array type to be considered for commercial switching fabrics. These large arrays were the product of a long series of advances by, for example, Bell Labs. To achieve low crosstalk for this application, reverse $\Delta\beta$ switches were generally used. When low index-contrast Ti-diffused waveguides are used, bend radii in this material are relatively large. Waveguide loss and crosstalk and operating voltages are the major operating parameters of concern.

The most complex array reported was an 8 × 8 switch, containing 64 directional couplers Granestrand et al. (1986). Another 8 × 8 switch used a rearrangeable switch-path layout of 28 elements to achieve a non-blocking layout. In this case, each switch had a 27 V switching voltage, V_π, with a 1.3 V standard deviation, and a total loss of 5.5 dB at 1.3 μm Duthie and Wale (1988). The switches in this case were $\Delta\beta$ switches, with ~4 mm long interaction lengths. The bends had 40 mm radii.

Smaller, i.e., 4 × 4 LiNbO$_3$ arrays have been examined by many groups and thus their design has been carefully examined. In one example, a 4 × 4 array at 1.3 μm was made using reverse-$\Delta\beta$ cross switches and uniform-$\Delta\beta$ bar switches. In this case, the operating voltages, V_π, were 8 and 13 volts for the bar and cross state, respectively, with a cross talk of −35 dB Bogert et al. (1986). This switch was mounted on a silicon optical bench and pig-tailed to an optical fiber input/output array. In another example, an 4 × 4 crossbar array had a V_π of 15 V, a crosstalk value of 35 dB and a path-dependent loss of ~5 dB McCaughan and Bogert (1985).

Finally, there has been an increasing interest in semiconductor arrays; the initial projects were made using III–V ($GaAs/AlGaAs$) materials. For example, a 4 × 4 array was reported which used reversed $\Delta\beta$ switches. Each switch required 22 volts for switching and had a 0.5 dB path-dependent loss Komatsu et al. (1991). Since this initial work the commercial and advanced development groups have focused on Si materials; see the following section!

13.5.3   Si Arrays: Principally Mach–Zehnder-Based

The very large advance in large-scale Si photonics systems principally to handle large-scale data streams has naturally given rise to research on Si switching arrays. An excellent review of this work has been written by (Fang 2012; Chen 2015) The technological challenge, which drives the remarkable growth of Si photonics, is the need for expanding overall bandwidths of systems for integration of computing and/or signal processing applications; these needs cannot be realized with the traditional metallic interconnects used in the past or even the high-quality arrays using LiNbO$_3$ discussed above. On the other hand, silicon photonics provides a low-cost approach to high-data rate transmission by using well-established planar processing methods to fabricate photonic integrated circuits. In addition, in certain designs, the use of Si ring switches allows ultrasmall on-chip footprints.

A discussion of the technology in switching arrays provides excellent examples of what is possible with silicon photonics. Note that arrays, whether using LiNbO$_3$ or III–Vs have in the past been the photonic-system types, which have paced the development of large-scale photonic circuits; *Si* arrays now have become the same pacing leader for large-scale photonic circuits. As a result, we devote the remainder of this section to describing three examples of *Si* switching arrays; our focus in this discussion will be MZ-based devices due to many recent applications. However, some comments on the emerging use of ring-based arrays will also be included as well.

For example, (Suzuki 2014) has fabricated a silicon photonics 8 × 8 non-blocking optical switch based on double-MZ element switches. These double-MZ switches consisted of a waveguide intersection and two asymmetric MZ switches. The design allowed crosstalk suppression for a large range of spectral bandwidth. In particular, the 88 switches exhibited an average fiber-to-fiber insertion loss of 11.2 dB as well as −20 dB cross talk over a 30 nm spectral bandwidth. As an important system demonstration, the devices were used to show 32-Gbaud dual polarization, quadrature-phase-shift keying, for four-channel wavelength-division-multiplexed signal transmission.

As a second example, Zhang (2016) demonstrated an 8 × 8 × 40-Gbps fully integrated silicon photonic network on a Si chip. This was an integrated photonic network-on-chip circuit and operated using wavelength division multiplexing transceivers.

In a third example, another group has shown the operation of a large-scale low-loss 88 silicon photonic module for a wide variety of add–drop multiplexers. The switch was based on using the thermoelectric effect to carry on the needed device phase shifting. The system incorporated an advanced silicon optical switch chip with spot size converters. The devices resulted in polarization-insensitive and wavelength-insensitive properties over several optical bands. The optical loss in the fabricated devices was particularly low and in fact the modules had only 6 dB excess optical loss for each of the 64 optical paths. These paths also had low polarization-dependent loss and low cross talk. With these modules, a transponder aggregators (TPA) prototype was fabricated having a 100-port optical switch subsystem.

Please note while the majority of this section has focused on MZ-based arrays, there have been major and very attractive array designs that have used ring-based switching arrays (Chen 2015). These arrays are of lower power and with small footprints due to their small diameters (an advantage ultimately based on their high index contrast when the *Si* is grown on SiO$_2$). Discussions of the early design of these arrays are provided in the references given here (Chen 2015).

13.5.4 *Thermo-Optical Switching Arrays*

Because thermo-optical switching uses standard passive-waveguide device material, it can provides a low-cost, high-performance solution to making large-scale arrays. Since thermo-optical switches can be made in several optimized passive-waveguide materials platforms, the arrays can have very low input and propagation loss even on large substrates of these materials. Of course, the switching speed is limited to the speeds in our discussion above on the device level aspects of on thermo-optical switches.

To give one example of a thermo-switching array, consider the matrix–array design shown in Fig. 13.12 for use with silica planar circuits (Miya 2000). The individual switch unit in this case consists of two 3 dB-coupler MZ interferometers arranged in series, with a waveguide intersection. Pairing the interferometers allows any leakage in the first interferometer into an on-chip absorber. The design of the array matrix results in path-independent loss and a shorter total loss than a conventional cross–bar switch. This design was been fabricated into a 16 × 16 array. This large array was measured to have a 6.6 dB average insertion loss, a 55 dB extinction ratio per path, and a total electric operating power of 1.1 W. The switch time for each node is ∼2 ms. Equally impressive arrays have been made in polymeric materials (Eldada and Shacklette 2000).

13.6   Summary

This chapter has discussed one particularly important application area for large-scale PICs, namely switching, typically with the goal of selecting optical paths. Electro-optical switches were one early "contender" for this application because of high speed and electronic control. However, the advantages of thermo-optic switches in low-cost passive materials such as polymers or planar SiO_2 have made this approach very attractive. Large-scale arrays have been made of both of these types of switches. The largest scale optical switch applications use very high I/O count arrays for switching fiber links. In this case, high extinction ratios are very important. Both requirements have driven interest in micromechanical-based (MEMS) switches. These are not discussed in this book, since its focus is in-plane or waveguide devices. Finally, this chapter has also presented devices fabricated from different materials including polymers, SiO_2, $LiNbO_3$, and semiconductors—particularly Si. In fact, in the chapter, several different Si device sections are presented, since the growth of interest in Si switching systems has been significant. Si arrays and devices are easy to fabricate and can be ultrasmall.

Problems

1. Assume single-mode $LiNbO_3$ waveguide with width $w \approx 6 \, \mu m$, y-cut, directional coupler.
   Design a 1 cm-long $\Delta\beta$ switch, assume free space wavelength $\lambda = 1.55 \, \mu m$, electrode spacing $d \approx 2w$ and field overlap factor $\Gamma \approx 0.3$, determine the voltage needed to introduce a π-phase change V_π.

2. Consider a $\Delta\beta$ modulator with a coupling coefficient $\kappa = \frac{1}{200}$ μm^{-1}, plot out the power transfer ratio η_x vs. δ/κ, for $\kappa L = 5$, where L is the length of coupling region.

3. If $L = 2$ cm and $R \approx 60\pi$, determine $\Delta\gamma$ for a lumped circuit modulator and $\Delta\gamma_{\text{transit}}$.

4. Refer to the $\Delta\beta$ switch in Fig. 13.1, answer the questions below:

   a. If the coupling coefficient is κ, what is the minimum length needed for a $\Delta\beta$ switch?

   b. Plot modulation intensity for the cross port of a $\Delta\beta$ switch as a function of $\Delta\beta L/\pi$ for a device of length L, where L is the physical length of the electrodes.

References

Alferness, R. C. (1982). Waveguide electrooptic modulators. *IEEE Transactions on Microwave Theory Techniques*, *30*, 1121–1137.

Bachmann, M., Smit, M., Besse, P., Melchior, L., et al. (1993). Polarization-insensitive low-voltage optical waveguide switch using ingaasp/inp four-port mach-zehnder interferometer. In *Optical Fiber Communication Conference* (pp. TuH3). Optical Society of America.

Bogert, G., Murphy, E., & Ku, R. (1986). Low crosstalk 4× 4 tilinbo 3 optical switch with permanently attached polarization maintaining fiber array. *Journal of Lightwave Technology*, *4*(10), 1542–1545.

Campbell, J., Blum, F., Shaw, D., & Lawley, K. (1975). Gaas electro-optic directional-coupler switch. *Applied Physics Letters*, *27*(4), 202–205.

Chen, C. P., et al. (2015) Performing intelligent power distribution in a 4×4 silicon photonic switch fabric. In *2015 IEEE Optical Interconnects Conference (OI)*. IEEE.

Diemeer, M., Brons, J., & Trommel, E. (1989). Polymeric optical waveguide switch using the thermooptic effect. *Journal of Lightwave Technology*, *7*(3), 449–453.

Duthie, P., & Wale, M. (1988). Rearrangeably nonblocking 8*8 guided wave optical switch. *Electronics Letters*, *24*(10), 594–596.

Eldada, L., & Shacklette, L. W. (2000). Advances in polymer integrated optics. *IEEE Journal of Selected Topics in Quantum Electronics*, *6*(1), 54–68.

Fang, Z., & Zhao, C. Z. (2012) Recent progress in silicon photonics: a review. *International Scholarly Research Notices*.

Fischer, U., Zinke, T., Schuppert, B., & Petermann, K. (1994). Singlemode optical switches based on soi waveguides with large cross-section. *Electronics Letters*, *30*(5), 406–408.

Granestrand, P., Stoltz, B., Thylen, L., Bergvall, K., Döldissen, W., Heinrich, H., et al. (1986). Strictly nonblocking 8× 8 integrated optical switch matrix. *Electronics Letters*, *22*(15), 816–818.

Haus, H. A. (1984). *Waves and fields in optoelectronics*. Prentice-Hall.

Izutsu, M., Enokihara, A., & Sueta, T. (1982). Optical-waveguide hybrid coupler. *Optics Letters*, *7*(11), 549–551.

Jenkins, R., Heaton, J., Wight, D., Parker, J., Birbeck, J., Smith, G., et al. (1994). Novel 1× n and n × n integrated optical switches using self-imaging multimode gaas/algaas waveguides. *Applied Physics Letters*, *64*(6), 684–686.

Kogelnik, H., & Schmidt, R. V. (1976). Switched directional couplers with alternating $\delta\beta$. *IEEE Journal of Quantum Electronics*, *12*(7), 396–401.

Komatsu, K. et al. (1991). 4* 4 gaas/algaas optical matrix switches with uniform device characteristics using alternating delta beta electrooptic guided-wave directional couplers. *Journal of Lightwave Technology*, 9(7), 871–878.

Martin, W. E. (1975). A new waveguide switch/modulator for integrated optics. *Applied Physics Letters*, 26(10), 562–564.

McCaughan, L., & Bogert, G. (1985). 4× 4 ti: Linbo3 integrated-optical crossbar switch array. *Applied Physics Letters*, 47(4), 348–350.

Miya, T. (2000). Silica-based planar lightwave circuits: Passive and thermally active devices. *IEEE Journal of Selected Topics in Quantum Electronics*, 6(1), 38–45.

Moosburger, R., Fischbeck, G., Kostrzewa, C., & Petermann, K. (1996). Digital optical switch based on'oversized'polymer rib waveguides. *Electronics Letters*, 32(6), 544–545.

Morl, L. et al. (1998). A travelling wave electrode mach-zehnder 40 gb/s demultiplexer based on strain compensated gainas/alinas tunnelling barrier mqw structure. In *Conference Proceedings. 1998 International Conference on Indium Phosphide and Related Materials (Cat. No. 98CH36129)* (pp. 403–406). IEEE.

Neyer, A., Mevenkamp, W., and Ctyroky, J. (1986). Single-mode ti: Linbo [sub] 3 [/sub] waveguide crossings and switches: Design rules and applications. In *Integrated Optical Circuit Engineering III* (vol. 651, pp. 169–176). International Society for Optics and Photonics.

Papuchon, M., Combemale, Y., Mathieu, X., Ostrowsky, D., Reiber, L., Roy, A., et al. (1975). Electrically switched optical directional coupler: Cobra. *Applied Physics Letters*, 27(5), 289–291.

Ramaswamy, V., Divino, M., & Standley, R. (1978). Balanced bridge modulator switch using ti-diffused linbo3 strip waveguides. *Applied Physics Letters*, 32(10), 644–646.

Schmidt, R., & Kogelnik, H. (1976). Electro-optically switched coupler with stepped $\delta\beta$ reversal using ti-diffused linbo3 waveguides. *Applied Physics Letters*, 28(9), 503–506.

Sekiguchi, S., Kurahashi, T., Zhu, L., Kawaguchi, K., & Morito, K. (2012). Compact and low power operation optical switch using silicon-germanium/silicon hetero-structure waveguide. *Optics Express*, 20(8), 8949–8958.

Silberberg, Y., Perlmutter, P., & Baran, J. (1987). Digital optical switch. *Applied Physics Letters*, 51(16), 1230–1232.

Soldano, L., De Vreede, A., Smit, M., Verbeek, B., Metaal, E., & Green, F. (1994). Mach-zehnder interferometer polarization splitter in ingaasp/inp. *IEEE Photonics Technology Letters*, 6(3), 402–405.

Suzuki, K., Tanizawa, K., Matsukawa, T., Cong, G., Kim, S.H., Suda, S., Ohno, M., Chiba, T., Tadokoro, H., Yanagihara, M. and Igarashi, Y., Masahara, M., Namiki, S., & Kawashima, H. (2014) Ultra-compact 8 × 8 strictly-non-blocking Si-wire PILOSS switch, Opt. Express 22, 3887–3894 (2014)

Zhang, C., Zhang, S., Peters, J.D., & Bowers, J.E. (2016) 8 × 8 × 40Gbps fully integrated silicon photonic network on chip. Optica **3**, 785–786 (2016)

Zhao, P., Chrostowski, J., & Bock, W. J. (1998). Novel multimode coupler switch. *Microwave and Optical Technology Letters*, 17(1), 1–7.

Zucker, J., Jones, K., Chiu, T., Tell, B., & Brown-Goebeler, K. (1992). Strained quantum wells for polarization-independent electrooptic waveguide switches. *Journal of Lightwave Technology*, 10(12), 1926–1930.

Chapter 14
Numerical Methods

Abstract Due to the complex nature of the light–matter interactions and the ultra-small scale of many photonic components, analytical solutions of the Maxwell's equations in most cases may not exist. Thus, experimental studies rely heavily on numerical analysis to provide guidance both for the design of the photonic components as well as for the interpretation of their performance prior to fabrications. In most cases, one must first develop a quantitative theoretical description of the photonic systems using advanced computational techniques, which requires solving the corresponding partial differential equations numerically. In a broad sense, there are two categories of modeling methods: *finite-difference* method (FDM) and *finite-element* method (FEM), as well as two categories of equation-solving techniques: *frequency-domain* solver, and *time-domain* simulations. In this section, we briefly present an overview of the modeling methods and solving techniques.

14.1 Numerical Methods in Nanophotonics

14.1.1 Finite-Difference Versus Finite-Element

In mathematics modeling, FDM are the popular methods for solving differential equations by approximating them with difference equations, and then use a finite-difference grid to approximate the derivatives. Due to the simple discretization process, the development time for FDM is very short and it is easily understandable and directly follows from the differential equations. The stability and dispersion and inhomogeneous characteristics also follow from a simple, intuitive understanding of the updating procedure. However, the orthogonal grid structure of the FDM implies that the edges of structures modeling have "stair-step" edges, which can become problematic for curved surfaces to achieve sufficient accuracy. Special treatments have been developed to overcome this limitation. These include the nonuniform grids; however, other methods, such as FEM, are generally better suited for complex irregular geometries.

On the other hand, the FEM, in general, subdivides a large problem into smaller, simpler parts that are called finite elements, which are based on triangular or tetrahe-

© Springer Nature Switzerland AG 2021
R. Osgood jr. and X. Meng, *Principles of Photonic Integrated Circuits*,
Graduate Texts in Physics,
https://doi.org/10.1007/978-3-030-65193-0_14

dral sub-regions. The simple equations that model the sub-regions are then assembled into a larger system of equations that models the entire problem. FEM then minimize an associated error function to approximate a solution of the system. Note that developing FEM is not as straightforward as FDM. For example, creating the numerical grid along for FEM could require an entire software package, and understanding of the discretization procedure can be quite convoluted.

14.1.2   Time-Domain Versus Frequency-Domain

The solution based on the time-domain can be computed by time stepping, whereas the same problem in the frequency-domain can be solved only through a linear system of equations. Using the time-domain solver, the time step, at which we advance the solution, is limited by the spatial dimensions. Thus, for simulations with large spaces, the simulation is very computationally expensive. On the other hand, frequency-domain solvers generally require linear algebra or matrix inversions, and thus, there is an inherent limit to the size of the simulation, especially for large three dimensional problems.

The most popular numerical methods for solving Maxwell's equations or the wave equation are the combinations of the discretization methods and solving techniques, as illustrated in Fig. 14.1, namely, Finite-Difference Time-Domain (FDTD), Finite-Difference Frequency-Domain (FDFD), Finite-Element Frequency-Domain (FEFD), and Finite-Element Time-Domain (FETD). In this chapter, we will provide a brief overview of each major technique used in photonics design along with an illustration of each of their use.

Fig. 14.1 Different numerical methods for solving the electromagnetic problems

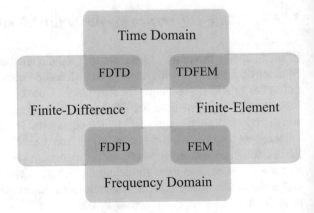

14.2 The Finite-Difference Time-Domain Method

FDTD is a well-known numerical technique in electrodynamics to compute Maxwell's equations. It translates the differential form of Maxwell's equations into difference equations that can be solved numerically. However, before the 1990s, the FDTD method was limited by the need to discretize the simulation space on sub-wavelength scales, with relatively small time steps. Thus, at that time, a typical photonics problem would require a computer memory that would exceed the technology limits at that time. However, since the 1990s, the FDTD has become more computationally affordable with large increases in computer memory and speed.

There are several advantages of FDTD. First, the method is accurate and robust, such that approximations are minimized and detailed solutions are provided with an accuracy determined by the grid resolution. Second, the method naturally includes effects such as polarization, dispersion, and nonlinearities. Furthermore, FDTD is able to readily calculate the full-wave response, which includes the transient behavior of an electromagnetic system.

In this section, we first introduce Yee's unique, yet powerful, algorithm for solving Maxwell's equations. Using a 1D example, we demonstrate the basic principle and formulation of the FDTD method for the analysis of electrodynamic problems. We then discuss stability analysis, boundary conditions, and the extensions to the analysis of 2D/3D problems.

14.2.1 Yee's Algorithm

Yee's algorithm, introduced in 1966, established a set of finite-difference equations for the time-dependent Maxwell's curl equations system (Yee 1966). In this algorithm, the continuous derivatives in space and time are approximated to second-order accuracy with two-point centered difference forms. The resulting finite-difference equations are solved via leapfrog stepping, that is, the electric field vector components in the modeled space are solved at a given instant in time; then the magnetic field vector components in the same spatial volume are solved at the next instant in time using the previously stored electric field data. This process is repeating until the desired transient or steady-state electrodynamic behavior is fully evolved.

The fundamental unit of the 3D grid, known as the Yee lattice, is shown in Fig. 14.2, which discretizes and solves the six components of E and H fields that satisfy the six-coupled scalar Maxwell curl equations in free space:

Fig. 14.2 Position of electric and magnetic vector components in a 3D staggered unit cell known as Yee lattice. The vectors are placed at the point in the mesh at which they are defined and stored

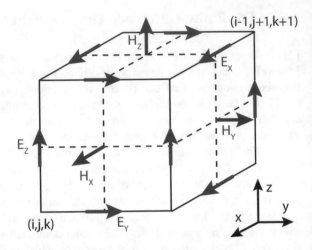

$$\frac{\partial E_x}{\partial t} = \frac{1}{\epsilon_0}\left(\frac{\partial H_z}{\partial y} - \frac{\partial H_y}{\partial z}\right) \qquad \frac{\partial H_x}{\partial t} = \frac{1}{\mu_0}\left(\frac{\partial E_y}{\partial z} - \frac{\partial E_z}{\partial y}\right)$$

$$\frac{\partial E_y}{\partial t} = \frac{1}{\epsilon_0}\left(\frac{\partial H_x}{\partial z} - \frac{\partial H_z}{\partial x}\right) \qquad \frac{\partial H_y}{\partial t} = \frac{1}{\mu_0}\left(\frac{\partial E_z}{\partial x} - \frac{\partial E_x}{\partial z}\right) \qquad (14.1)$$

$$\frac{\partial E_z}{\partial t} = \frac{1}{\epsilon_0}\left(\frac{\partial H_y}{\partial x} - \frac{\partial H_x}{\partial y}\right) \qquad \frac{\partial H_z}{\partial t} = \frac{1}{\mu_0}\left(\frac{\partial E_x}{\partial y} - \frac{\partial E_y}{\partial x}\right)$$

Rather than solving for the electric field alone with a wave equation, the Yee algorithm solves the coupled Maxwell's curl equations directly. In which way, both electric and magnetic material properties can be easily modeled. This is especially convenient when modeling the inhomogeneous materials and a full-wave response of a dispersive medium.

Here, we demonstrate an example of using the FDM to calculate the fundamental TE mode profile of a channel waveguide. In detail, we illustrate how the computing grid size can affect the calculated mode profile. Figure 14.3a shows the structure we calculated with a Si channel waveguide sitting on top of the SiO$_2$ substrate. The channel waveguide is 400 nm in width and 200 nm in height. Figure 14.3b, c, d shows the calculated TE mode profile with different grid sizes of 50 nm, 20 nm, and 5 nm, respectively.

14.2.2   FDTD: 1D Example

In this section, we illustrate the basic implementation of FDTD method for a 1D case, including details on the discretization of Maxwell's equations. To solve a specific problem of pulse propagation, we demonstrate a 1D model with details on the discretization of Maxwell's equation. Taking the advantages of the simplicity of 1D

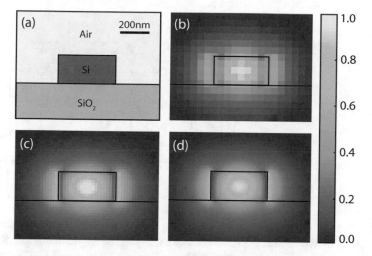

Fig. 14.3 Demonstration of the calculated mode profile with different computing grid size: **a** The structure of a Si channel waveguide (400 nm in width and 200 nm in height) on top of the SiO₂ substrate. The fundamental TE mode profile was calculated using different grid size of **b** 50 nm, **c** 20 nm, and **d** 5 nm. In all cases, $\lambda = 1550$ nm

example, we also illustrate the stability criterion for the FDTD simulations. This discretized equation and stability criterion can be easily expanded to both 2D and 3D models.

In one dimension, the medium extends to infinity in the y-direction and z-direction. This translational symmetry then leads to $\frac{\partial}{\partial x} = \frac{\partial}{\partial y} = 0$. For a free-space situation, Maxwell's curl equations take the form:

$$\frac{\partial H_z}{\partial t} = -\frac{1}{\mu_0} \frac{\partial E_y}{\partial x} \quad \text{and} \quad \frac{\partial E_y}{\partial t} = -\frac{1}{\epsilon_0} \frac{\partial H_z}{\partial x} \tag{14.2}$$

In order to introduce 1D Yee discretization, we use the following notation:

$$E_y(m \cdot \Delta x, n \cdot \Delta t) \equiv (E_y)_m^n$$
$$H_z(m \cdot \Delta x, n \cdot \Delta t) \equiv (H_z)_m^n \tag{14.3}$$

where m and n are the index of the spatial and temporal grids, respectively. Using the central difference approximation with second-order accuracy for space and time derivatives, the equations can be discretized as

$$\frac{(E_y)_m^{n+\frac{1}{2}} - (E_y)_m^{n-\frac{1}{2}}}{\Delta t} = -\frac{1}{\epsilon_0} \frac{(H_z)_{m+\frac{1}{2}}^n - (H_z)_{m-\frac{1}{2}}^n}{\Delta x}$$

$$\frac{(H_z)_{m+\frac{1}{2}}^{n+1} - (H_z)_{m+\frac{1}{2}}^n}{\Delta t} = -\frac{1}{\mu_0} \frac{(E_y)_{m+1}^{n+\frac{1}{2}} - (E_y)_m^{n+\frac{1}{2}}}{\Delta x} \tag{14.4}$$

Fig. 14.4 Visual illustration
of the numerical
dependencies in the 1D
FDTD method

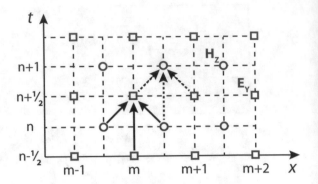

In a practical sense, for simplicity of the calculations, physical constants are omitted from mathematical expressions. For example, the implementation of FDTD uses the speed of light c, which is exactly dimensionless 1, i.e., $c = 1/\sqrt{\epsilon_0 \mu_0} \equiv 1$. By redefining the electric field as a typical approach in many theoretical electromagnetic computations, $E_y \equiv \sqrt{\epsilon_0/\mu_0} E_y$ (14.4) can be expressed as

$$
\begin{aligned}
(E_y)_m^{n+\frac{1}{2}} &= (E_y)_m^{n-\frac{1}{2}} - \frac{\Delta t}{\Delta x}\left[(H_z)_{m+\frac{1}{2}}^{n} - (H_z)_{m-\frac{1}{2}}^{n}\right] \\
(H_z)_{m+\frac{1}{2}}^{n+1} &= (H_z)_{m+\frac{1}{2}}^{n} - \frac{\Delta t}{\Delta x}\left[(E_y)_{m+1}^{n+\frac{1}{2}} - (E_y)_m^{n+\frac{1}{2}}\right]
\end{aligned}
\tag{14.5}
$$

Figure 14.4 is the visual illustration of (14.5), which indicates the numerical dependencies in the 1D FDTD formulation. The value of a field at any point is determined by three previous values: two from the neighbors of opposite field at the previous half time step; one from the same position at the previous one time step.

Note that this discretization procedure can be easily extended into 2D and 3D space (Wartak 2013). However, for example, if 3D problem would require N grid cells in each dimension, the total grid cells are N^3. With a minimum of six fields to compute in double precisions, it can easily take gigabytes of memory with billions of operations. Thus, FDTD is a computationally intensive method. However, advances in CPU speed and memory and the emergence of inexpensive parallel systems with parallel computing technology enable a full 3D FDTD simulation without any constraints.

Despite the requirements for spatial grid size in order to maintain the numerical accuracy, the time step must be small enough so that it satisfies the Courant condition (Press 1996) in order to achieve a convergence while solving partial differential equations numerically. A detailed mathematical discussion can be found in (Taflove and Hagness 2005). The physical stability criterion is that the speed of numerical propagation should not exceed the physical speed. Thus, the lattice speed $\Delta x/\Delta t$ must be less than the physical velocity v_p in the medium with refractive index of n, where $v_p = c/n$. A summary of stability criteria for various dimensions is presented in Table 14.1.

Table 14.1 Stability criterion for FDTD

Dimensionality	Criterion
1D	$v_p \Delta t \leq \Delta x$
2D	$v_p \Delta t \leq \left(\frac{1}{\Delta x^2} + \frac{1}{\Delta y^2} \right)^{-\frac{1}{2}}$
3D	$v_p \Delta t \leq \left(\frac{1}{\Delta x^2} + \frac{1}{\Delta y^2} + \frac{1}{\Delta z^2} \right)^{-\frac{1}{2}}$

14.2.3 Boundary Conditions

One of the major challenges of using FDTD method for solving unbounded electro-magnetic problems is to employ a finite computational domain. Thus, our simulation domain needs to be terminated with proper boundary conditions. This boundary termination can be accomplished by introducing an artificial layer to enclose the domain of interest. However, in order to duplicate the original open-space environment, the artificial boundary layer has to treat the field incident on top of the layer, so as to eliminate the artificially reflected fields. There are several approaches to achieve this implementing a mathematical boundary condition (i.e., absorbing boundary condition) or a fictitious absorbing material layer (perfect matching layer).

14.2.3.1 Absorbing Boundary Condition

An absorbing boundary condition (ABC) is a mathematical technique such that it acts to estimate the missing field outside the FDTD domain, and thus emulates an infinite space. This is normally done by assuming an incident plane wave. Unfortunately, in many cases, the incident wave at the boundary is usually not a plane wave with a well-defined angle of incidence. Thus, ABC is a general approximation, which reflects some of the waves back into the FDTD space. An advanced treatment is available, such as arbitrary wave can be decomposed into many place waves incident (Jin 2011).

14.2.3.2 Perfectly Matched Layer

To correctly model the boundary condition, one can define artificial thin layers as absorbers solely for simulation purposes. A popular absorber model has been proposed by Berenger (1994) for the FDTD simulation, which is named as the perfectly matched layer (PML). Regardless of the frequency, polarization, and angle of incidence of a plane wave incident upon its interface, PML creates no reflected fields.

As an example to illustrate the idea of PML, the modified source-free Maxwell's curl equation for electric field is shown below:

$$\nabla_s \times E = -j\omega\mu H \qquad (14.6)$$

where ∇_s is defined by

$$\nabla_s = \hat{x}\frac{1}{s_x}\frac{\partial}{\partial x} + \hat{y}\frac{1}{s_y}\frac{\partial}{\partial y} + \hat{z}\frac{1}{s_z}\frac{\partial}{\partial z} \qquad (14.7)$$

and ∇_s can be considered as the standard ∇ operator in Cartesian coordinates whose x-, y-, and z-axes are stretched by a complex numbers of s_x, s_y, and s_z, respectively. Throughout the simulation domain, the complex diagonal tensor is the identity tensor, but inside the PML, it has the following form (Gedney 1996):

$$\overset{\leftrightarrow}{s} = \begin{bmatrix} \dfrac{s_y s_z}{s_x} & 0 & 0 \\ 0 & \dfrac{s_z s_x}{s_y} & 0 \\ 0 & 0 & \dfrac{s_x s_y}{s_z} \end{bmatrix} \qquad (14.8)$$

Note that in a computation, when a material property changes abruptly and the spatial discretization is not sufficiently dense, undesirable numerical reflections may occur. One approach to avoiding this problem is to vary the material parameters smoothly within the PML (Chew and Jin 1996), thus, we have

$$s_{x,y,z} = 1 - j\left(\frac{\alpha - L}{L}\right)\delta_{x,y,z} \qquad (14.9)$$

where $\delta_{x,y,z}$ is the loss tangent in dimension x, y and z, α is the distance from the edge, and L is the thickness of the PML, which is terminated at the simulation domain edge with a perfect electrical conductor (PEC) boundary condition. Figure 14.5 illustrates the effect of the PML for a point source in free space. The calculated spatial profile is shown at time $t = 5t_0$, $t = 7t_0$, and $t = 10t_0$, respectively. As we can see, for the simulation domain with PML, the spherical wave generated by the point source is perfectly absorbed at the simulation domain boundary, while for the case without the PML, the reflection from the boundary significantly interfaces with the spherical wave generated from the point source.

14.3 Finite-Difference Frequency-Domain (FDFD)

As we discussed earlier, time-domain methods such as FDTD are also extremely useful for transient behavior analysis. However, when looking into a steady-state solution at a single frequency, the time-domain method is relatively time-consuming. Instead, the frequency-domain method, such as the FDFD is highly applicable, since it maintains the finite-difference spatial features, but removes time stepping (Kunz

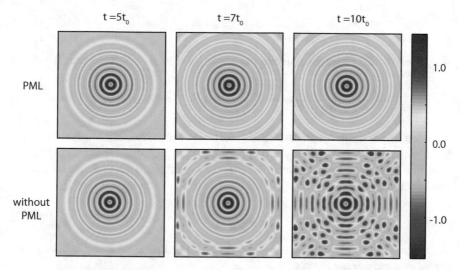

Fig. 14.5 Visual illustration of dipole source in free space at time $t = 5t_0$, $t = 7t_0$, and $t = 10t_0$, with and without PML, respectively

and Luebbers 1993). The steady-state solution is found at a single frequency through a matrix inversion process. In addition, the FDFD method has the advantage of treating dispersive materials. In FDTD, implementing dispersive materials requires either convolution terms or auxiliary equations, but in FDFD, only one simple set of values of material properties are needed at the frequency of interest.

14.3.1 FDFD Using the Wave Equations

In this section, by taking the advantages of the simplicity of 1D example, we briefly illustrate the frequency-domain representation of the wave equation and the possible implementations through Maxwell's equations. Thus, in practice, FDFD normally utilizes the frequency domain wave equation for its setup (Joannopoulos et al. 2011). The wave equation has a more compact form and does not require interleaving or a Yee lattice. Meanwhile, only solving one wave equation is necessary; after solving for E, for example, H can be calculated directly follows from the frequency-domain Maxwell's equations.

In the source-free space, the wave equations can be simplified to

$$\nabla^2 E + k^2 E = 0 \quad \text{and} \quad \nabla^2 H + k^2 H = 0 \qquad (14.10)$$

where $k = \omega\sqrt{\epsilon\mu}$ is the wavevector. To illustrate how these equations are solved, we will consider the 1D example with a current source term J_z for generality

$$\frac{\partial^2 E_z}{\partial x^2} + k^2 E_z = j\omega\mu J_z \tag{14.11}$$

We proceed to discretize the above equation using a second-order centered difference:

$$\frac{(E_z)_{m+1} - 2(E_z)_m + (E_z)_{m-1}}{(\Delta x)^2} + k^2(E_z)_m = j\omega\mu(J_z)_m \tag{14.12}$$

Note that, using the notation defined in (14.3), we need to discretize in space as there is no time dependence. Thus, there is no superscript n, compared with the notion used in FDTD. The wavevector k is a complex constant. This difference equation can be rearranged as

$$(E_z)_{m-1} + A(E_z)_m + (E_z)_{m+1} = b(J_z)_m \tag{14.13}$$

where $a = [k^2(\Delta x)^2 - 2]$ and $b = j\omega\mu(\Delta x)^2$. The entire system can then be represented in a tridiagonal matrix:

$$\underbrace{\begin{pmatrix} a & 1 & 0 & 0 & \cdots \\ 1 & a & 1 & 0 & \cdots \\ 0 & 1 & a & 1 & \cdots \\ \vdots & & & \cdots & \\ \cdots & 0 & 1 & a & 1 \\ \cdots & 0 & 0 & 1 & a \end{pmatrix}}_{M} \underbrace{\begin{pmatrix} (E_z)_0 \\ (E_z)_1 \\ (E_z)_2 \\ \vdots \\ (E_z)_{l-1} \\ (E_z)_l \end{pmatrix}}_{[E_z]} = b \underbrace{\begin{pmatrix} (J_z)_0 \\ (J_z)_1 \\ (J_z)_2 \\ \vdots \\ (J_z)_{l-1} \\ (J_z)_l \end{pmatrix}}_{[J_z]} \tag{14.14}$$

where the total number of spatial cells in our 1D system is $l + 1$. The solution of this equation can be found by simple matrix inversion as

$$[E_z] = [M]^{-1}b[J_z] \tag{14.15}$$

Note that this 1D example of a homogeneous medium can be easily replaced with inhomogeneous and frequency-dependent materials, by replacing a, b with a_m and b_m, where it takes consideration of the material constant such as ϵ, μ and σ for each grid point. From this above example, we see that the solution from FDFD involves taking an inverse of the matrix M, which can be easily done in 1D simulation. The discretized FDFD equations in 2D and 3D follow straightforwardly from the 1D example. However, in a higher dimensional setup, this matrix M can be very large, thus advanced linear algebra techniques are required for efficient calculation. In general, the Laplace matrices and the **Kronecker** products are introduced to assist the FDFD setup in 2D and 3D models. Detailed discussion can be found in Inan and Marshall (2011).

Finally, in the 2D and 3D formulation of this material, directly solving the wave equation of E field may suffer from a problem with the current source. This is due to the assumption that the divergence of E and H are both equal to zero. Clearly, this is not valid at the position of the source. For this reason, it is common to solve the wave equation of the H field, which can easily include the curl J component due to the source, while $\nabla \cdot H = 0$ still holds.

14.3.2   FDFD from Maxwell's Equations

The FDFD method can also be formulated using Maxwell's equations. Here, the frequency-domain Maxwell's equations are as follows:

$$
\begin{aligned}
\nabla \times E &= -j\omega\mu H - M \\
\nabla \times H &= j\omega\epsilon E + J
\end{aligned}
\tag{14.16}
$$

In the 1D case, we consider an x-directed propagation with E field polarized along y direction, we have

$$
\begin{aligned}
\frac{\partial E_y}{\partial x} &= -j\omega\mu H_z - M \\
\frac{\partial H_z}{\partial x} &= j\omega\epsilon E_y + J
\end{aligned}
\tag{14.17}
$$

Discretizing using the leapfrog method, we have the finite difference equations:

$$
\begin{aligned}
\frac{(E_y)_{m+1} - (E_y)_m}{\Delta x} &= -j\omega\mu(H_z)_{m+\frac{1}{2}} - M \\
\frac{(H_z)_{m+\frac{1}{2}} - (H_z)_{m-\frac{1}{2}}}{\Delta x} &= j\omega\epsilon(E_y)_m + J
\end{aligned}
\tag{14.18}
$$

The matrix system can then be formulated into $[M][F] = [S]$, where the column vector $[F]$ includes each of the E components followed by each of the H components and $[S]$ is the source of the system. This linear system can be rearranged as follows:

$$
\begin{aligned}
a_h(H_z)_{m+\frac{1}{2}} + (E_y)_{m+1} - (E_y)_m &= -\Delta x M \\
a_e(E_y)_m + (H_z)_{m+\frac{1}{2}} - (H_z)_{m-\frac{1}{2}} &= \Delta x J
\end{aligned}
\tag{14.19}
$$

Note that in this setup based on Maxwell's equations, the matrix M, which must be inverted, is no longer tridiagonal; this means the inversion process will be more computationally intensive. In addition, the vector of field values $[F]$ will double in length compared with the wave equation setup (14.14), since we are simultaneously solving for E and H. Thus, the doubling of the field vector results in a quadrupling of the matrix M, which increase the computational cost considerably. Therefore, the FDFD method from the wave equation, in general, is more attractive than the FDFD method from Maxwell's equations.

14.4 Beam Propagation Method

Among the many numerical methods available for modeling optical propagation in integrated and fiber-optic photonic devices, the Beam Propagation Method (BPM) is the most commonly used technique for larger photonic systems. BPM is an approximation technique for simulating the propagation of light in slowly varying optical waveguides (Obayya 2011). It solves the well-known parabolic or paraxial approximation of the *Helmholtz* equation. There are several reasons for using the BPM over other numerical methods. First, it is a conceptually straightforward technique and is easily implemented even in three dimensions. Second, it is a very efficient method with an optimal computational complexity, i.e., the computational effort is proportional to the number of grid points used in the simulation. Overall, the BPM is a very flexible method and requires less intensive computing power compared with other methods such as FDTD.

14.4.1 Paraxial Formulation

In this section, we demonstrate the simplest version of BPM, where one assumes scalar electric field E with paraxial approximations. These treatments restrict its applicability to the fields propagating at small angles with respect to the axis of the waveguide (Saleh et al. 1991). We define this axis as z axis. To illustrate the method, we start from the monochromatic wave equation. Assuming a scalar field, ϕ, and paraxiality, the wave equation is written in the form of the Helmholtz equation,

$$\frac{\partial^2 \phi}{\partial x^2} + \frac{\partial^2 \phi}{\partial y^2} + \frac{\partial^2 \phi}{\partial z^2} + k^2(x, y, z)\phi = 0 \qquad (14.20)$$

where the spatially varying wavenumber is $k(x, y, z) = k_o n(x, y, z)$, and the free-space wavenumber is $k_o = 2\pi\lambda$. The refractive index $n(x, y, z)$ solely defines the geometry of the problem. Considering that the most rapid variation in the field ϕ is the phase variation due to propagation predominantly along the z direction, it is beneficial to factor out this rapid variation by introducing a slowly varying field u,

$$\phi(x, y, z) = u(x, y, z)e^{i\beta z} \qquad (14.21)$$

where β is a free parameter called the reference wavenumber and is frequently expressed in terms of a reference refractive index, n_0, via $\beta = k_0 n_0$. Here, n_0 can be the refractive index of the substrate or cladding. Substituting (14.21) into (14.20) gives the equation for the envelope of the field:

$$\frac{\partial^2 u}{\partial z^2} + 2i\beta\frac{\partial u}{\partial z} + \frac{\partial^2 u}{\partial y^2} + \frac{\partial^2 u}{\partial x^2} + (k^2 - \beta^2)u = 0 \tag{14.22}$$

By assuming that the variation of u with z is sufficiently slow such that

$$\left|\frac{\partial^2 u}{\partial z^2}\right| \ll \left|2\beta\frac{\partial u}{\partial z}\right| \tag{14.23}$$

the above equation reduces to

$$\frac{\partial u}{\partial z} = \frac{i}{2\beta}\left[\frac{\partial^2 u}{\partial x^2} + \frac{\partial^2 u}{\partial y^2} + (k^2 - \beta^2)u\right] \tag{14.24}$$

which is known as a Fresnel or paraxial equation. This approximation eliminates of the second-order derivative term in z reduces the second-order boundary value problem to a first-order initial value problem, which can be solved by simple integration of the above equation along the propagation direction z. In addition, the efficiency is enhanced by the fact that the longitudinal grid can be much coarser than the wavelength for many problems.

14.4.2 Finite-Difference BPM

Here, we present the detailed implementation of BPM based on FDM (Scarmozzino et al. 2000; Scarmozzino and Osgood 1991). The above differential equation can be numerically integrated in the forward z direction using the Crank–Nicolson scheme, which is a finite-difference approach and is the most widely used. In this numerical scheme, the field in the transverse x–y plane is denoted as discrete points on a grid, and at discrete points along the longitudinal propagation direction z. Given the field at one z plane, the field at the next z plane can be determined. The stepping process is repeated to account for the propagation throughout the structure. Assuming a 2D BPM case, if we let u_i^m denote the field at the transverse grid point i and longitudinal plane m and assume the the grid points and planes are equally spaced by Δx and Δz apart. Thus, in the Crank–Nicolson scheme, (14.24) is represented at the midplane between the known plane m and the unknown plane $m + 1$ as follows:

$$\frac{u_i^{m+1} - u_i^m}{\Delta z} = \frac{i}{2\beta}\left[\frac{\delta^2}{\Delta x^2} + (n^2 k_0^2 - \beta^2)\right]\frac{u_i^{m+1} + u_i^m}{2} \tag{14.25}$$

where δ^2 is the second-order difference operator, $\delta^2 u_i = [u_{(i+1)} + u_{(i-1)} - 2u_i]$, and n is the averaged refractive index between the two planes. The above equation can be rearranged into the form of a standard tridiagonal matrix equation for the unknown field u in the plane $(n + 1)$ in terms of known quantities, resulting in

$$au_{i-1}^{n+1} + bu_i^{n+1} + cu_{i+1}^{n+1} = d \qquad (14.26)$$

where the expressions for the coefficients a, b, c, and d above are readily derived and can be found in Chung and Dagli (1990).

Boundary conditions. Since the field can only be represented on a finite computational domain, the above equation requires an appropriated boundary condition, which completes the system of equations. Transparent boundary condition (TBC) is commonly used, which assumes the field behaves as an outgoing plane wave near the boundary. The TBC is very effective in terminating undesired reflections and details on implementations are given in Hadley (1992b).

14.4.3 BPM Expansions of the Method

There are a few limitations on the traditional BPM that are based on the paraxial approximations. For example, the fields must propagate primarily along the z axis, i.e., fields are paraxial and limited to a small angular spread in wavenumber. This places a restriction on geometries with large and abrupt perturbations along the z-axis. Also, the gradient of the refractive index must be small. In addition, the elimination of the second-order derivative term eliminates the possibility of a backward propagating wave solution; thus, devices relying on large-angle reflections cannot be modeled. In this section, we briefly introduce a few techniques, that is, to eliminate or significantly relax these limitations.

14.4.3.1 Wide-Angle BPM

The physical limitation of the above BPM approach results from the parabolic approximation to Helmholtz equation, which implies a paraxiality condition on the primary direction of propagation. This restriction and the related restrictions on index contrast can be addressed using the wide-angle BPM, which is an approach to incorporate the effect of the second-order derivative term that was neglected in the basic BPM, thus a more accurate approximation to Helmholtz equation. The most popular formulation is based on Padé approximants (Hadley 1992c). In general, larger angles, higher index contrast, and more complex mode interference can be analyzed in both guided-wave and free-space problems as the Padé order increases. Detailed discussion for using this technique can be found in Hadley (1992a).

In addition. various bidirectional BPM techniques have been considered to include the backward traveling wave, with most focusing on the reflection of a wave along z-direction. For example, the guided-wave propagation can be divided into regions that are uniformed along z. Both forward and backward waves can exist at any point along the structure. The essential idea is to employ a transfer matrix M' which describes the entire structure that is composed of propagation and interface matrices (Kaczmarski and Lagasse 1988; Rao et al. 1999).

14.4.3.2 Full-Vector BPM

The basic BPM approach discussed above results from the assumption of scalar waves, which prevents the polarization effects from being considered. This limitation can be overcome through a full-vector BPM technique, which is to recognize the electric field as a vector and solving from the vector wave equation rather than the scalar Helmholtz equation. This approach can be found with more details in Huang and Xu (1993), Xu et al. (1994).

14.5 Finite-Element Methods

The FEM was originally developed for mechanical and structural analyses in the 1950s. It became popular in solving the vector electromagnetic problems after an important breakthrough occurred in the 1980s with the development of an edge-based vector element (Nédélec 1980; Barton and Cendes 1987). There is a major difference between the FDM and the FEM. From the discussions earlier in this chapter, we note that, in principles, FDM finds an *approximation to the differential operators*, and then use these difference equations to solve for the fields at each grid.

In the sections below, we illustrate the basic principle of the FEM by briefly introducing the methods for solving the boundary-value problems in mathematical modeling. Then we present the formulation procedure of the FEM to solve the electromagnetic problems in the frequency-domain.

14.5.1 Boundary-Value Problems

Boundary-value problems have long been a major topic in mathematical modeling. A typical boundary-value problem can be defined by a governing differential equation in a domain Ω, with boundary conditions specified on the boundary that encloses the domain: $\mathcal{L}\psi = f$, where $\mathcal{L}$ is a differential operator, f is the source function, and ψ is the unknown quantity. In electromagnetics, the form of the governing differential equation ranges from a simple Poisson equation to complicated vector wave equations.

To solving the boundary-value problems, various approximate methods have been developed, and among them, the Ritz and Galerkin's methods have been used most widely (Axelsson and Barker 2001). The **Ritz method** is a direct method to find an approximate solution for boundary-value problems. It is a variational method which starts from the variational representation, which is referred to as functional, of the boundary-value problem. The minimum of the functional corresponds to the governing differential equation under the given boundary conditions. The approximate solution is then obtained by minimizing the functional with respect to its variables. On the other hand, **Galerkin's method** belongs to the family of weighted residual methods, which start directly from the partial differential equation of the boundary-value problem and seek the solution by weighting the residual of the differential equation. In this method, a continuous operator problem is converted into a discrete problem, then characterizes the discrete space with a finite set of basis functions, i.e., the weighting function, used for the expansion of the approximate solution. A brief review of the Ritz and Galerkin's methods and a detailed illustration of their solution procedures to a simple boundary-value problem can be found in Jin (2015).

14.5.2   Implementation of FEM

In general, it is a very challenging step in the Ritz and Galerkin's methods to find a trial function defined over the entire solution domain, which is capable of representing the true solution to the problem. This is particularly true for 2D and 3D problems. To simplify the problem, we can divide the entire domain into small subdomains and employ the trial functions defined over each subdomain. These trial functions are usually in a much simpler form compared with original equations in the entire domain since the subdomains are small. Therefore, the principle of the FEM is to replace an entire continuous domain by a number of subdomains, in which the unknown function is represented by simple interpolation functions with unknown coefficients. A system of equations is then obtained by applying the Ritz variational or Galerkin's procedure and the solution of the boundary-value problem is achieved by solving the system of equations. The basic steps for a finite element method can be summarized as follows:

- *Domain discretization.* The first and perhaps the most important step in the FEM is to discretize the domain over which the solution is desired. An effective discretization with proper numbering for each element can significantly affect the computing time, memory usage, as well as accuracy of the numerical results (Jin 2011). Note that the linear line segments, triangles, and tetrahedral are the most frequently used subdomain elements for 1D, 2D, and 3D modeling, due to their simplicity and suitability for domains with arbitrary shape and volume. We demonstrate two examples in Fig. 14.6 showing the finite element discretization of a 2D and a 3D domains.
- *Select interpolation functions.* The approximation of the unknown solution is assumed to take a specific functional form over each small element. In general, the interpolation is usually selected to be the linear (first-order polynomial) or parabolic (second-order) functions. These functions are then matched to the adjacent cells to ensure continuity across the cell boundaries

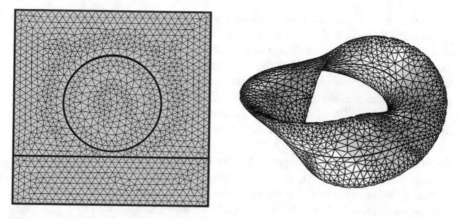

Fig. 14.6 Visual illustration of the finite element mesh grid for 2D (left) and 3D (right) object

- *Formulate a system of equations.* In this step, each elemental equation can be formulated using either the Ritz variational or Galerkin's method. The system of equations can then be set up by summing the elemental equations over the entire domain. The boundary conditions are then imposed to obtain the final form of the system of equations. This system includes information about boundaries and sources, also have the added constraint of continuity across element boundaries.
- *Solve the system of equations.* The resultant system, in general, has one of the following two forms:

$$[M][\phi] - [S] \tag{14.27}$$

or

$$[A][\phi] = \lambda[B][\phi] \tag{14.28}$$

In electromagnetics, (14.27) is corresponding to the wave equations with the known vector $[S]$ as the source. Equation 14.28 represents the eigenvalue systems that associated with source-free problems. In this case, the source vector $[S]$ vanishes and the matrix $[K]$ can be written as $[A] - \lambda[B]$, where λ denotes the unknown eigenvalues. Similar to FDFD, solving these systems becomes purely a linear algebra problem and the truncation the infinitely large solution domain into a finite computation domain is taking cared by setting up an artificial mesh layer with either ABC or PML.

14.6 Summary

In this chapter, we have briefly described the basic principle and formulation of a few major computational methods for the numerical analysis of electromagnetic fields for photonics applications. These include the FDTD method, FDFD method, beam propagation method, and the FEM. We started the discussion with the construction of the finite differencing formulae and demonstrated their applications in solving 1D Maxwell's equations. Taking the advantage of the simplicity of 1D problems, we then discussed the working principles, stability criterion, and boundary conditions in the time-domain simulation. Further, the FDM was extended to frequency-domain, where the linear system of equations can be formulated. In addition, we have illustrated the basic principle and steps of FEM with possible applications to electromagnetic problems. These four methods are chosen because they represent the fundamental and popular approaches for numerical analysis of photonics engineering design. The reader is also encouraged to consult more advanced books and references listed in the end of this chapter, for a more comprehensive understanding of these methods with a variety of advanced treatments and applications.

References

Axelsson, O., & Barker, V. A. (2001). *Finite element solution of boundary value problems: Theory and computation*. Philadelphia: SIAM.

Barton, M., & Cendes, Z. (1987). New vector finite elements for three-dimensional magnetic field computation. *Journal of Applied Physics, 61*(8), 3919–3921.

Berenger, J.-P. (1994). A perfectly matched layer for the absorption of electromagnetic waves. *Journal of Computational Physics, 114*(2), 185–200.

Chew, W., & Jin, J. (1996). Perfectly matched layers in the discretized space: An analysis and optimization. *Electromagnetics, 16*(4), 325–340.

Chung, Y., & Dagli, N. (1990). An assessment of finite difference beam propagation method. *IEEE Journal of Quantum Electronics, 26*(8), 1335–1339.

Gedney, S. D. (1996). An anisotropic perfectly matched layer-absorbing medium for the truncation of FDTD lattices. *IEEE Transactions on Antennas and Propagation, 44*(12), 1630–1639.

Hadley, G. R. (1992a). Multistep method for wide-angle beam propagation. *Optics Letters, 17*(24), 1743–1745.

Hadley, G. R. (1992b). Transparent boundary condition for the beam propagation method. *IEEE Journal of Quantum Electronics, 28*(1), 363–370.

Hadley, G. R. (1992c). Wide-angle beam propagation using padé approximant operators. *Optics Letters, 17*(20), 1426–1428.

Huang, W., & Xu, C. (1993). Simulation of three-dimensional optical waveguides by a full-vector beam propagation method. *IEEE Journal of Quantum Electronics, 29*(10), 2639–2649.

Inan, U. S., & Marshall, R. A. (2011). *Numerical electromagnetics: The FDTD method*. Cambridge: Cambridge University Press.

Jin, J.-M. (2011). *Theory and computation of electromagnetic fields*. Hoboken: Wiley.

Jin, J.-M. (2015). *The finite element method in electromagnetics*. Hoboken: Wiley.

Joannopoulos, J. D., Johnson, S. G., Winn, J. N., & Meade, R. D. (2011). *Photonic crystals: Molding the flow of light*. Princeton: Princeton University Press.

Kaczmarski, P., & Lagasse, P. (1988). Bidirectional beam propagation method. *Electronics Letters, 24*(11), 675–676.

Kunz, K. S., & Luebbers, R. J. (1993). *The finite difference time domain method for electromagnetics*. Boca Raton: CRC Press.

Nédélec, J.-C. (1980). Mixed finite elements in r3. *Numerische Mathematik, 35*(3), 315–341.

Obayya, S. (2011). *Computational photonics*. Hoboken: Wiley.

Press, W. H. (1996). *FORTRAN numerical recipes: Numerical recipes in FORTRAN 90*. Cambridge: Cambridge University Press.

Rao, H., Scarmozzino, R., & Osgood, R. M. (1999). A bidirectional beam propagation method for multiple dielectric interfaces. *IEEE Photonics Technology Letters, 11*(7), 830–832.

Saleh, B. E., Teich, M. C., & Saleh, B. E. (1991). *Fundamentals of photonics* (Vol. 22). New York: Wiley.

Scarmozzino, R., Gopinath, A., Pregla, R., & Helfert, S. (2000). Numerical techniques for modeling guided-wave photonic devices. *IEEE Journal of Selected Topics in Quantum Electronics, 6*(1), 150–162.

Scarmozzino, R., & Osgood, R. (1991). Comparison of finite-difference and Fourier-transform solutions of the parabolic wave equation with emphasis on integrated-optics applications. *JOSA A, 8*(5), 724–731.

Taflove, A., & Hagness, S. C. (2005). *Computational electrodynamics*. Boston: Artech House.

Wartak, M. S. (2013). *Computational photonics: An introduction with MATLAB*. Cambridge: Cambridge University Press.

Xu, C., Huang, W., Chrostowski, J., & Chaudhuri, S. (1994). A full-vectorial beam propagation method for anisotropic waveguides. *Journal of Lightwave Technology, 12*(11), 1926–1931.

Yee, K. (1966). Numerical solution of initial boundary value problems involving Maxwell's equations in isotropic media. *IEEE Transactions on Antennas and Propagation, 14*(3), 302–307.

Appendix
Problem Set Solutions

Chapter 1

Chapter 2

1. According to ITU-T G.652, the FWHM of the mode is $8.6 \sim 9.5\,\mu m$. Its mode distribution is approximately Gaussian (in fact, it is in the shape of J_0 Bessel function).

2. According to batop.de

$\lambda/\mu m$	x	Δn
1.3	0.085	0.04
0.9	0.045	0.04

3. Consider at point $(x, y) = (0, 0)$, index change can be obtained by

$$\Delta n_0 \approx b \cdot C_0$$

where

$$C_0 = \frac{2}{\sqrt{\pi}} \frac{\tau}{d} \mathrm{erf}\left(\frac{w}{2d}\right)$$

and

$$
\begin{aligned}
(\lambda) &= 0.552 + 0.065/\lambda^2 \\
&= 0.552 + 0.065/1.3^2 \\
&= 0.59
\end{aligned}
$$

Plug in b and C_0, we have

$$\Delta n_0 = \frac{1.18}{\sqrt{\pi}} \frac{\tau}{d} \mathrm{erf}\left(\frac{w}{2d}\right)$$

© Springer Nature Switzerland AG 2021
R. Osgood jr. and X. Meng, *Principles of Photonic Integrated Circuits*,
Graduate Texts in Physics,
https://doi.org/10.1007/978-3-030-65193-0

assume $\dfrac{w}{2d} \gg 1$, we have

$$\Delta n_0 = \frac{1.18}{\sqrt{\pi}} \frac{\tau}{d}$$

Plug in $\Delta n_0 = 10^{-3}$ and $\tau = 900$ Å, we have

$$d = 59.9\,\mu\text{m}$$

Remind that $d = 2\sqrt{Dt}$ and $D = D_0 \exp(-T_0/T)$, thus,

$$t = \frac{d^2}{4D_0 \exp(-T_0/T)} = \frac{(59.9 \times 10^{-4})^2}{4 \times 2.5 \times 10^{-4} \times \exp(-2.5 \times 10^4/1323.15)}$$
$$= 5761640\,\text{s} = 1600\,\text{h}$$

We will need 67 days to obtain an index change of 0.001!

4. Helmholtz equation:
$$\nabla^2 \vec{E}_x + (k_o n_x)^2 \vec{E}_x = 0$$

In homogeneous material, we have $n_x = n = \sqrt{\varepsilon/\varepsilon_0}$ and $k_o = \omega/c = \sqrt{\varepsilon_0 \mu_0}\omega$, where ε_0 is the vacuum permittivity. Thus,

$$\frac{\partial^2}{\partial z^2} \vec{E}_x + (nk_o)^2 \vec{E}_x = 0$$

The solution to the equation above is simply

$$\vec{E}_x = Ae^{-j(nk_o)z} + Be^{j(nk_o)z}$$

where the propagation in the direction of $+z$ requires that $B = 0$. Considering the wave having a frequency of ω_o, the full solution for $\vec{E}_x$ is

$$\vec{E}_x = Ae^{j[\omega_o(t - \sqrt{\varepsilon \mu_0}z)]}$$

5. (a) Lithorgraphy.
   (b) Diffusion.

Chapter 3

1. From batop.de, we have $n_f(\lambda_o = 0.84\,\mu\text{m}) = 3.65$, $n_c(\lambda_o = 0.84\,\mu\text{m}) = 3.59$. Thus,

$$V = k_o d \sqrt{n_f^2 - n_s^2}$$

$$= \frac{2\pi}{\lambda_o} d \sqrt{n_f^2 - n_s^2}$$

$$= \frac{2 \times 3.14}{0.84} \times 0.2 \times \sqrt{3.65^2 - 3.59^2}$$

$$\approx 0.985$$

Remind that phase equation for waveguide mode

$$V\sqrt{1-b} = m\pi + \arctan\sqrt{\frac{b}{1-b}} + \arctan\sqrt{\frac{b+a}{1-b}}$$

where $m = 0$ for TE_0 mode and $a = 0$ for symmetric waveguide.
Plug V in the equation above and solve for b yields $b \approx 0.185$. Therefore,

$$n_{\text{eff}} = \sqrt{b(n_f^2 - n_c^2) + n_c^2} \approx 3.60$$

Thus,

$$\gamma = k_o \sqrt{n_{\text{eff}}^2 - n_c^2} \approx 2.12 \times 10^6 \, \text{m}^{-1}$$

2. (a) The diffusion coefficient can be calculated by

$$D = D_0 \exp(-T_0/T) \approx 7.4 \times 10^{-13} \, \text{cm}^2/\text{s}$$

and therefore the diffusion distance is

$$d = 2\sqrt{Dt_0} - 2\sqrt{7.4 \times 10^{-17} \times 6 \times 3600} \approx 2.53 \, \mu\text{m}$$

The concentration at $y = 0.6 \, \mu\text{m}$ is then

$$C(y) = \frac{2}{\sqrt{\pi}} \frac{\tau}{d} \exp\left\{-\frac{y^2}{d^2}\right\} = \frac{2}{\sqrt{3.14}} \times \frac{0.08}{2.53} \times \exp\left(-\frac{0.6^2}{2.53^2}\right) \approx 0.034$$

Therefore, the index change would be

$$\Delta n(\lambda) = b(\lambda)C(y) = (0.552 + 0.065/\lambda^2) \times 0.034 = 0.0188 + 0.00221/\lambda^2$$

(b) At $\lambda = 1.3 \, \mu\text{m}$, $b = 0.552 + 0.065/1.3^2 - 0.59$
Thus, the index change of the material along y is

$$\Delta n(y) = b(\lambda)C(y) = 0.0211 \exp(-y^2/d^2)$$

and

$$n_s = 2.2204$$

$$n_f = n_s + \Delta n(0) = 2.2204 + 0.0211 = 2.2415$$

For graded index guide, the phase equation becomes

$$V_c^{g,m} = \sqrt{2\pi}\left(m + \frac{3}{4}\right)$$

where

$$V_d = \frac{2\pi}{\lambda} d\sqrt{n_f^2 - n_s^2} = 3.752$$

Thus,

$$m \le \frac{V_d}{\sqrt{2\pi}} - \frac{3}{4} = 0.747$$

which means that only 1 mode ($m = 0$) is supported.

3. Given that $x = 0.9$, $\lambda = 0.84\,\mu$m:

$$n_f = n(\text{GaAs}) = 3.645$$
$$n_s = n(\text{Ga}_{0.9}\text{Al}_{0.1}\text{As}) = 3.590$$
$$n_c = n(air) = 1$$

**Sketch of mode profile

$$V = k_o d\sqrt{n_f^2 - n_s^2} \approx 0.94$$

$V = 0.94 < \pi$ tells us that only the fundamental mode is supported.
by solving $V\sqrt{1-b} = \arctan\sqrt{\dfrac{b}{1-b}} + \arctan\sqrt{\dfrac{b+a}{1-b}}$ we can obtain b, thus,
we have n_{eff}, and furthermore, β, also including δ, γ.

$$T_{\text{eff}} = \frac{1}{\gamma} + \frac{1}{\delta} + d$$

Therefore, the percentage is $\dfrac{\gamma^{-1}}{\gamma^{-1} + d + \delta^{-1}}$

4. xpressing the relations of δ, γ, and κ in terms of β and κ, i.e., (3.12), and inserting them into (3.16), we obtain an eigenvalue equation for the allowed βs.

5. The simplest way to do this would be to use the normalized frequency V. According to the definition of V, it implies how many modes a specific waveguide geometry can support. In this case, according to (3.19), we obtain

$$V = \frac{2\pi d}{\lambda}\sqrt{n_f^2 - n_s^2} = 45.5 \tag{S.3.1}$$

Use (3.27), then gives us the highest order mode index m

$$m = \left[\frac{V}{\pi}\right] = 14 \tag{S.3.2}$$

where the symbol "$[a]$" means to get the largest integer not larger than a. Recall the total mode number is $m + 1$, so the total allowed mode number is 15.

Another approach is using the physical sense of ray picture of wave propagation in waveguides, as shown below.

At the core and cladding interface, the light must have a total internal reflection, which means

$$n_f \sin\theta > n_s \tag{S.3.3}$$

where θ is the angle between the light and the interface, as shown in the picture below. We then obtain the following relation using (S.3.3):

$$\kappa = \frac{2\pi n_f}{\lambda}\cos\theta$$
$$< \frac{2\pi}{\lambda}\sqrt{n_f^2 - n_s^2} = 3.03\,\mu m^{-1} \tag{S.3.4}$$

We know that there should be a wave in the transverse direction with integer number of periods, as shown in the following picture (Fig. A.1).

So the largest order mode this waveguide can support is

$$m = \left[\frac{3.03\,\mu m^{-1} \times 15\,\mu m}{\pi}\right] = 14 \tag{S.3.5}$$

Again, we get the same result. Note that the form of κd is the same as normalized frequency V, which should remind us of what the physical meaning of V is.

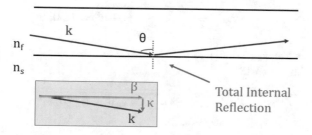

Fig. A.1 Ray picture of wave propagation in a waveguide

6. According to (3.20),

$$\kappa w = (m+1)\pi - \tan^{-1}\left(\frac{\kappa}{\gamma}\right) - \tan^{-1}\left(\frac{\kappa}{\delta}\right) \tag{S.3.6}$$

where the parameters are defined as in textbook. In the case where $n_f \gg n_c$, which means a good mode confinement, we obtain

$$\tan^{-1}\left(\frac{\kappa}{\gamma}\right) = \tan^{-1}\left(\frac{\kappa}{\delta}\right) = \frac{\pi}{2} \tag{S.3.7}$$

Plug this into (S.3.6), it turns out that

$$\kappa = \frac{m\pi}{w} \tag{S.3.8}$$

Then plugging this into (3.13) yields

$$\beta_m = \sqrt{\left(\frac{2\pi n_f}{\lambda_0}\right)^2 - \left(\frac{m\pi}{w}\right)^2} \tag{S.3.9}$$

7. (a) This is a symmetric waveguide with $a = 0$, (3.27) yields

$$V_c^m = m\pi \tag{S.3.10}$$

So, for $m = 1$, the cutoff normalized frequency is $V_c^{(1)} = \pi$. Plugging this into (3.19) yields

$$d_{design} = 0.5 d_{cutoff} = 0.5 \frac{V_c^{(1)}\lambda}{2\pi\sqrt{n_f^2 - n_s^2}} = 4.84\,\mu\text{m} \tag{S.3.11}$$

(b) Calculate b first. From (a), we have $V = \pi/2 = 1.57$, from the normalized diagram, we obtain $b \approx 0.3$. Plugging this into (3.24) gives

$$N_{eff} = \sqrt{b(n_f^2 - n_s^2) + n_s^2} = 1.5006 \tag{S.3.12}$$

So, $\beta = N_{eff}k_0 = 6.29\,\mu\text{m}^{-1}$.

(c) According to (3.13),

$$\delta = \gamma = \sqrt{\beta^2 - (n_s \cdot k_0)^2} = 0.29\,\mu\text{m}^{-1} \tag{S.3.13}$$

The decay length is $L_{decay} = \dfrac{1}{\delta} = 3.45\,\mu m$. Note this is a quite large decay length, as we can see in this case, the mode is not well confined in the waveguide.

8. Consider low index contrast, the TM modes can be treated as TE modes in effective index method. First calculate the asymmetry parameter a:

$$a = \frac{n_s^2 - n_c^2}{n_f^2 - n_s^2} = 76.88 \approx \infty \qquad (S.3.14)$$

Plug this into (3.27), which yields

$$V_c^m = m\pi + \tan^{-1}\sqrt{a} = \frac{3}{2}\pi \qquad (S.3.15)$$

Then we could easily get the desired thickness d

$$d = \frac{\lambda}{2\pi\sqrt{n_f^2 - n_s^2}} \cdot 0.9 V_c^m = 1.83\,\mu m \qquad (S.3.16)$$

where the normalized frequency $V = 1.5\pi \times 0.9 = 4.24$. According to the normalized parameter diagram, we have $b = 0.64$. Plug into (3.24), we obtain

$$n_{eff1} = \sqrt{b(n_f^2 - n_s^2) + n_s^2} = 3.393 \qquad (S.3.17)$$

Then the effective waveguide structure is shown below. To get the 10% reduction to the cutoff frequency, $V = 0.9\pi$, similar to the previous procedure, we get

$$w = \frac{0.9\pi \cdot \lambda}{2\pi\sqrt{n_{eff1}^2 - n_s^2}} = 1.517\,\mu m \qquad (S.3.18)$$

Also, we obtain from the normalized parameter diagram that

$$b = 0.6 \qquad (S.3.19)$$

So, eventually

$$n_{eff} = \sqrt{b(n_{eff1}^2 - n_s^2) + n_s^2} = 3.388 \qquad (S.3.20)$$

Chapter 4

1. Step one: we divide the waveguide section along y into three regions (Figs. A.2 and A.3):

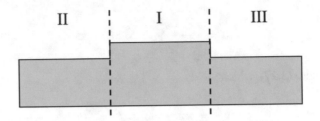

II   I   III

Fig. A.2 Define the waveguide regions

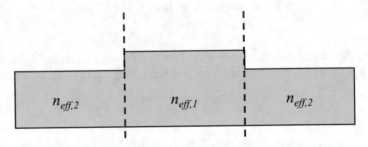

$n_{eff,2}$ $n_{eff,1}$ $n_{eff,2}$

Fig. A.3 Calculated the effective index for each region

In Region I:

$$V_I = k_o d \sqrt{n_f^2 - n_s^2} = \frac{2\pi}{\lambda_o} d \sqrt{n_f^2 - n_s^2} = 1.26d$$

$$a = \frac{n_f^4}{n_c^4} \frac{n_s^2 - n_c^2}{n_f^2 - n_s^2} = 3.41^4 \times \frac{3.41^2 - 1^2}{3.41^2 - 3.4^2} \approx 2.1 \times 10^4$$

By solving the phase equation,

$$V\sqrt{1-b} = m\pi + \arctan\sqrt{\frac{b}{1-b}} + \arctan\sqrt{\frac{b+a}{1-b}}$$

and plug in $b = 0$, $m = 1$ we obtain that

$$V = \pi + \arctan\sqrt{a} \approx \frac{3}{2}\pi$$

Therefore, $d_c = \dfrac{V}{1.26} \approx 3.74\,\mu m.$

$$d_{design} = 0.8 d_c \approx 3\,\mu m$$

$$V_I = 1.26 d_{design} = 3.78$$

Plugging V_I in the phase equation yields

$$b_I \approx 0.58$$

Therefore,

$$n_{eff,1} = \sqrt{n_s^2 + b_I(n_f^2 - n_s^2)} \approx 3.4$$

In Regions II and III,

$$V_{II} = V_{III} = k_o h\sqrt{n_f^2 - n_s^2} = \frac{2\pi}{\lambda_o} d\sqrt{n_f^2 - n_s^2} = 1.26\,h$$

Similarly, we have $b_{II} = b_{III} \approx 0.28$ and $n_{eff,2} = n_{eff,3} = 3.40$.
Step two: where $a = 0$, $V_{c,1} = \pi$. By solving for w_c from

$$V_{c,1} = \frac{2\pi}{\lambda_o} w_c \sqrt{n_{eff,1}^2 - n_{eff,2}^2} = 0.69 w_c,$$

we obtain that $w_{design} = 0.8 \times \dfrac{\pi}{0.69} = 3.64\,\mu m.$

2. Similarly, we can divide the waveguide into three parts (Fig. A.4):
Step one, in Region I,

$$V_I = k_o t\sqrt{n_f^2 - n_s^2} = \frac{2\pi}{\lambda_o} t\sqrt{n_f^2 - n_s^2} = 6.27 t$$

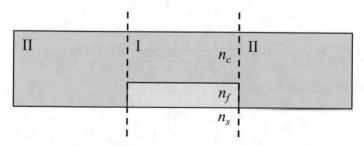

Fig. A.4 Define the waveguide regions

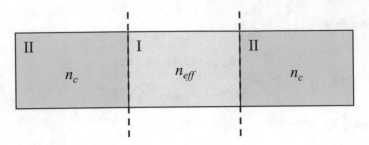

Fig. A.5 Calculated the effective index for each region

$$a = \frac{n_f^4}{n_c^4} \frac{n_s^2 - n_c^2}{n_f^2 - n_s^2} = \frac{1.96^4}{1.39^4} \times \frac{1.47^2 - 1.39^2}{1.96^2 - 1.47^2} \approx 0.54$$

$$V_{c,I} = \pi + \arctan \sqrt{a} \approx 3.78$$

$$t = \frac{3.78}{6.27} \times 0.8 = 0.48 \,\mu\text{m}$$

Solve for b yields $b = 0.597$, thus $n_{\text{eff}} = \sqrt{n_s^2 + b(n_f^2 - n_s^2)} = 1.78$
Step two (Fig. A.5):
Note that we have $a = 0$, $V_c = \pi$. Therefore,

$$V_c = \frac{2\pi}{\lambda_o} w \sqrt{n_f^2 - n_s^2} = 4.84w$$

$$w = \frac{0.8V_c}{4.84} = \frac{0.8\pi}{4.84} = 0.52 \,\mu\text{m}$$

3.

$$d_x = \frac{V_I^g}{\frac{2\pi}{\lambda} \sqrt{n_{fx}^2 - n_s^2}}$$

where $V_{I,c}^g = \frac{7}{4}\sqrt{2\pi}$. Thus $d_x = \frac{0.8V_I^g}{\frac{2\pi}{\lambda} \sqrt{n_{fx}^2 - n_s^2}} \approx 6.28 \,\mu\text{m}$

The phase equation for graded index guides is

$$2V^g \int_0^{\frac{x_t}{d}} \left[f\left(\frac{x}{d}\right) - b \right]^{\frac{1}{2}} d\left(\frac{x}{d}\right) - \frac{3}{2}\pi = 2m\pi$$

By solving the equation, we obtain that $b \approx 0.29$. You can use numerical method or graphical method to solve for b. Then,

$$n_{\text{eff}} = \sqrt{n_s^2 + b(n_f^2 - n_s^2)} \approx 2.2206$$

Therefore, $V_{II}^{2g} = 0.8 \times 3\sqrt{\dfrac{11}{8}} = 1.504$.

$$d_y = \frac{1.504}{\dfrac{2\pi}{\lambda_o}\sqrt{2.2206^2 - 2.22^2}} = 6.96\,\mu\text{m}.$$

4.

$$n_{\text{eff}} = \sqrt{n_s^2 + b(n_f^2 - n_s^2)}$$
$$= n_s + b\left(\frac{n_f}{n_s}\right)(n_f - n_s) + \dots \qquad (4.21)$$
$$\approx n_s + b(n_f - n_s)$$
$$\text{Q.E.D.}$$

Chapter 5

1.

$$\kappa_{12} = \frac{j2\gamma k_x^2}{\beta(k_x^2 + \gamma^2)(2/\gamma)}e^{-\gamma(h-d)}$$

$$\begin{cases} k_x = k_o\sqrt{n_f^2 - n_{\text{eff}}^2} \\[2mm] \gamma = k_o\sqrt{n_{\text{eff}}^2 - n_s^2} \\[2mm] \beta = k_o n_{\text{eff}} \end{cases}$$

For V being 20% below cutoff for the $m = 1$ mode, $V = 0.8\pi$.

$$k_o d\sqrt{n_f^2 - n_s^2} = 0.8\pi$$

Solving the above equation for d yields $d \approx 1.16\,\mu\text{m}$. Also by solving the normalized phase equation, we can obtain $b \approx 0.554$. Therefore, $n_{\text{eff}} = \sqrt{n_s^2 + b(n_f^2 - n_s^2)} \approx 3.32$.

Plug n_{eff} into the equations above for k_x, γ, we can obtain h by

$$h = -\frac{1}{\gamma} \log \frac{\kappa_{12}\beta(k_x^2 + \gamma^2)(2/\gamma)}{2\gamma k_x^2} + d \approx 2.73\,\mu m$$

where $\kappa_{12} = \dfrac{\pi}{2L_c}$.

2. (a)

Similar to Problem 1: $V = 0.8(\pi + \arctan\sqrt{a}) \approx 0.8 \times \dfrac{3}{2}\pi = \dfrac{6}{5}\pi$. Also note

that $V = k_o t \sqrt{n_f^2 - n_s^2}$ yields that $t \approx 9.4\,\mu m$

Solve for b yields that $b_I \approx 0.293$, and $n_{\text{eff},I} = \sqrt{n_s^2 + b(n_f^2 - n_s^2)} \approx 1.52788$.

(b)

$$V = 0.8\pi = k_o w \sqrt{n_{\text{eff},I}^2 - n_s^2}$$

Solve for w yields that $w \approx 11.57\,\mu m$.

(c)

Plugging in $V = 0.8\pi$, $a = 0$ into normalized phase equation, we can obtain that
$b_{II} = 0.554$, thus $n_{\text{eff},II} \approx 1.52749$, and

$$\begin{cases} k_x = k_o\sqrt{n_f^2 - n_{\text{eff}}^2} \approx 1.40 \times 10^5\,\text{m}^{-1} \\ \gamma = k_o\sqrt{n_{\text{eff}}^2 - n_s^2} \approx 1.56 \times 10^5\,\text{m}^{-1} \\ \beta = k_o n_{\text{eff}} \approx 6.19 \times 10^6\,\text{m}^{-1} \end{cases}$$

Thus,

$$h = -\frac{1}{\gamma} \log \frac{\kappa_{12}\beta(k_x^2 + \gamma^2)(2/\gamma)}{2\gamma k_x^2} + w \approx 14.94\,\mu m$$

3. For a codirectional coupler, the power transfer is $\dfrac{|a_1(z)|^2}{|a_1(0)|^2} = 1 - \left(\dfrac{\kappa}{\beta_c}\right)^2 \sin^2(\beta_c z)$, where

$$\beta_c = \sqrt{\kappa^2 + \left(\frac{\beta_2 - \beta_1}{2}\right)^2} = \frac{\sqrt{13}}{2}\kappa$$

Maximum power transfer is

$$1 - \left(\frac{\kappa}{\beta_c}\right)^2 = 1 - \left(\frac{2}{\sqrt{13}}\right)^2 = \frac{9}{13}$$

4. Equation (5.42):

$$\kappa = j\frac{\pi}{\lambda}\frac{\Delta d}{T_{\text{eff}}}\frac{n_f^2 - n_{\text{eff}}^2}{n_{\text{eff}}^2}$$

From Problem 1, we have $\gamma = 1.55 \times 10^6 \text{ m}^{-1}$, thus, $T_{\text{eff}} = d + \dfrac{2}{\gamma} \approx 2.45\,\mu\text{m}$.
Therefore,

$$\Delta d = \frac{\kappa \lambda T_{\text{eff}} n_{\text{eff}}^2}{\pi(n_f^2 - n_{\text{eff}}^2)} \approx 217.7\,\text{nm}$$

5. (a)   i. If $\beta_1 = \beta_2$, the power transfer between the two waveguides becomes perfectly matched with detuning $\delta = 0$, according to (5.18), we obtain

$$\begin{cases} P_1 = P\cos^2 \kappa z \\ P_2 = P\sin^2 \kappa z \end{cases} \tag{5.22}$$

where P is the incident power from waveguide 1, κ is the coupling coefficient.

   ii. If $\beta_1 \gg \beta_2$, recall when the two waveguides have a detuning δ, the power transfers more frequently with less power evolved in this process. According to (5.18), the power from waveguides 1 and 2 can be expressed as

$$\begin{cases} P_1 = P\dfrac{\Delta^2}{\kappa^2 + \Delta^2}\cos^2 \sqrt{\kappa^2 + \Delta^2}z \\ P_2 = P\dfrac{\kappa^2}{\kappa^2 + \Delta^2}\sin^2 \sqrt{\kappa^2 + \Delta^2}z \end{cases} \tag{5.23}$$

where $\Delta = \frac{\beta_1 - \beta_2}{2}$ is the propagation constant difference in the two waveguides.

In the limit where $\beta_1 \gg \beta_2$, we get $\Delta \approx \beta_1/2$. In a waveguide, $\kappa \ll \beta$. Plug these into (5.23), we obtain $P_1 = P$ and $P_2 = 0$.

(b) According to (5.15),

$$\beta'_{1,2} = \frac{\beta_1 + \beta_2}{2} \pm \sqrt{\frac{\beta_1 - \beta_2}{2}^2 - \kappa^2} \tag{5.24}$$

where $\beta_{1,2}$ and $\beta'_{1,2}$ are the modes in a single waveguide and the coupled mode of the system, respectively. Here, "+" for the even and "−" for the odd system mode (Figs. A.6 and A.7).

6. (a)

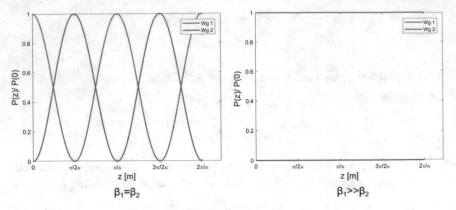

Fig. A.6 Power transfer *versus* propagation length

Fig. A.7 Even and odd
modes in directional coupler

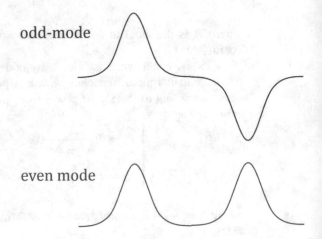

$$d = \frac{\lambda}{2\pi} \frac{0.8 V_c^1}{\sqrt{n_f^2 - n_s^2}}$$

$$= \frac{1.55}{2\pi} \frac{0.8\pi}{\sqrt{3.4^2 - 3.387^2}} \qquad (5.25)$$

$$= 2.087 \, \mu m$$

(b) According to (5.18), $\beta = \kappa$ in this case,

$$L_{\min} = \frac{\pi}{2\kappa} = \frac{\pi}{2 \times 50 \, cm^{-1}} = 0.0314 \, cm = 31.4 \, \mu m \qquad (5.26)$$

For all possible length, we could express it as $L = (2m + 1)L_{\min}$, where m
is an integer number.

(c) When $d = 2.087\,\mu\text{m}$, $V = 0.8\pi$, from the normalized parameter diagram, we obtain that $b \approx 0.55$. Plugging this into (3.24) yields

$$n_{eff} = \sqrt{0.55(3.4^2 - 3.387^2) + 3.387^2} = 3.394 \qquad (5.27)$$

So the propagation constant for waveguide 1 is

$$\beta_1 = \frac{2\pi}{\lambda} n_{eff} = 13.758\,\mu\text{m}^{-1} \qquad (5.28)$$

Similarly, we can obtain propagation constant β_2 of waveguide 2 as

$$\beta_2 = \frac{2\pi}{\lambda} \cdot \sqrt{b_2(n_f^2 - n_s^2)} = 13.74\,\mu\text{m}^{-1} \qquad (5.29)$$

In order to use (5.18), we determine β_c first

$$\beta_c = \sqrt{\kappa^2 + \left(\frac{\beta_2 - \beta_1}{2}\right)^2} = 0.064\,\text{cm}^{-1} \qquad (5.30)$$

According to (5.18), we then obtain

$$L'_{\min} = \frac{\pi}{2\beta_c} = \frac{\pi}{2 \times 64\,\text{cm}^{-1}} = 0.025\,\text{cm} = 25\,\mu\text{m} \qquad (5.31)$$

So all the lengths that have a maximum power transfer can be expressed as $L' = (2m + 1)L'_{\min}$, where m is an integer. Also note that in such case, the power that can be transferred reduces by a factor of $\left(\frac{\kappa}{\beta_c}\right)^2 \approx 0.61$ due to the asymmetry structure of such directional coupler.

Chapter 6

Chapter 7

1. (a) First, we determine n_{eff} using normalized parameter

$$V = \frac{2\pi}{\lambda} w \sqrt{n_f^2 - n_s^2} = \frac{2\pi}{1.5\,\mu\text{m}} \times 4\,\mu\text{m} \times \sqrt{3.4^2 - 3.38^2} = 6.17 \quad (7.32)$$

In this case, $a = 0$ as $n_s = n_c$. From the normalized parameter diagram, we can get $b \approx 0.85$. We then obtain

$$n_{\text{eff}} = \sqrt{n_s^2 + b(n_f^2 - n_s^2)} = 3.397 \qquad (7.33)$$

According to (3.30), the field has the form of

$$
\begin{cases}
Ae^{-\delta x} & , x > 0 \\[2mm]
A\left(\cos \kappa x - \dfrac{\delta}{\kappa}\sin \kappa x\right) & , -d \le x < 0 \\[2mm]
A\left(\cos \kappa d + \dfrac{\delta}{\kappa}\sin \kappa d\right) e^{\gamma(x+d)} & , x < -d
\end{cases}
\tag{7.34}
$$

where

$$
\begin{cases}
\kappa = \dfrac{2\pi}{\lambda}\sqrt{n_f^2 - n_{\text{eff}}^2} = 0.598\,\mu\text{m}^{-1} \\[3mm]
\gamma = \delta = \dfrac{2\pi}{\lambda}\sqrt{n_{\text{eff}}^2 - n_{s,c}^2} = 1.422\,\mu\text{m}^{-1}
\end{cases}
\tag{7.35}
$$

According to (3.33),

$$
A^2 = \frac{4\kappa^2 \omega \mu_0 P}{|\beta|(\kappa^2 + \delta^2)\left(d + \dfrac{1}{\gamma} + \dfrac{1}{\delta}\right)}
\tag{7.36}
$$

Thus, we obtain that $A = 3.51V/\,\mu\text{m}$.

(b) According to (7.41),

$$
u = R_t \ln(r/R) \approx -R + r
\tag{7.37}
$$

(c) According to (7.44), we obtain

$$
\alpha \approx \frac{2\gamma_t^{22}\exp(\gamma w)}{(N_{\text{eff}}^2 - n_s^2)k^2\beta(2 + \gamma w)}\exp\left\{\left(-\frac{2\gamma^3}{3\beta^2}R\right)\right\} = 1.22 \times 10^{-42}\,\mu\text{m}^{-1}
\tag{7.38}
$$

Note this is a quite small value for the loss constant α.

2. In this case, the mode is well confined to the waveguide. So the width of waveguide is $4 \times 1.5\,\mu\text{m} = 6\,\mu\text{m}$. According to (7.13),

$$
\frac{(3\alpha/4\pi)^2}{(3\alpha/4\pi)^2 + 1} = 0.2 \quad \Rightarrow \alpha = 1.5\pi
\tag{7.39}
$$

Also we know

$$
\lambda_g = \frac{2\pi}{\beta_0} = 0.44\,\mu\text{m}
\tag{7.40}
$$

Thus, according to (7.12), we obtain

$$
w = \left(16 + \frac{4}{3}\pi \times 0.44 \times L\right)^{1/2} = 6 \quad \Rightarrow L = 12.24\,\mu\text{m}
\tag{7.41}
$$

3. Here are several approaches to this; here, we show one using the ray optics picture. Refer to the picture below, the whispering gallery mode (*WGM*) appears when the light can only hit on one side of the ring waveguide.

According to the geometry relation, the following condition needs to be satisfied when there is a *WGM*:

$$w > R - R \sin \theta \tag{7.42}$$

Plug in the following expression

$$\sin \theta = \frac{\beta}{k} = \frac{n_{\text{eff}}}{n_f} \tag{7.43}$$

One obtains

$$w > \frac{n_f - n_{\text{eff}}}{n_f} R \tag{7.44}$$

Since $\delta n = n_f - n_s > n_f - n_{\text{eff}}$, we obtain

$$\frac{\delta n}{n_s + \delta n} > \frac{n_f - n_{\text{eff}}}{n_f} \tag{7.45}$$

For all the mode in the ring waveguide to be *WGM*, we need to make the highest order mode to satisfy (7.44). The higher order mode will have a refractive index more close to n_s, which means a strict condition for all the modes are *WGM's* should be

$$w > \frac{\delta n}{n_s + \delta n} R \tag{7.46}$$

4. According to (7.6),

$$
\begin{aligned}
t &= \frac{4}{\left(\frac{w_x}{a} + \frac{a}{w_x}\right)\left(\frac{w_y}{a} + \frac{a}{w_y}\right)} T \\
&= \frac{N_f N_g}{N_f + N_g} \frac{4}{\left(\frac{w_x}{a} + \frac{a}{w_x}\right)\left(\frac{w_y}{a} + \frac{a}{w_y}\right)} \\
&= 0.382 \\
\Rightarrow \quad t^2 &= 0.146
\end{aligned}
\tag{7.47}
$$

It yields the loss l in *dB* is

$$l = 10 \log_{10}(0.146) = -8.36 \text{ dB} \tag{7.48}$$

5. (a) The modes are shown in Fig. A.9 with label.

Note that at the transition region, the mode of straight waveguide needs to couple to the mode of curved waveguide. But these two have a mode shape mismatch, which leads to a loss.

(b) Usually, we will have a displacement to allow better mode matching, such as in Fig. 7.27.

6. Instead of an abrupt transition design, we would prefer to use an adiabatic design, which means to change gradually from one waveguide to the other. Intuitively, we can understand that this will give more "time" and "space" for the modes to evolve from mode in waveguide 1 to that in waveguide 2 (Fig. A.8).

Let's first consider the difference between the above two structures. According to (7.6), we have the transmission rate $t_{1,2}$ of cases 1 and 2 are

Fig. A.8 Whispering gallery modes

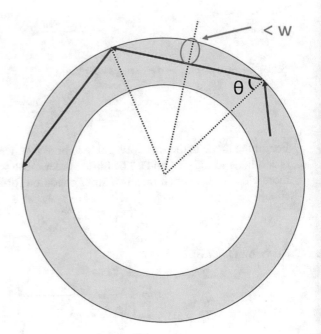

Fig. A.9 Mode shapes in straight and curved waveguides

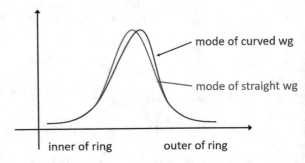

$$\begin{cases} t_1 = \dfrac{2}{w/5w + 5w/w} = 0.38 \\[4mm] t_2 = \dfrac{2}{w/3w + 3w/w} \dfrac{2}{3w/5w + 5w/3w} = 0.53 > t_1 \end{cases} \qquad (7.49)$$

This provides us a special case, where we add a small intermediate section in between that can increase the transmission rate.

Now, let's consider in a more general way, where we have two waveguides to connect, with width w_1 and w_2. We compare the case where we abruptly connect them together and put a waveguide with width $w_1 + \Delta w$ in between.

In the abrupt case, we have

$$t_1 = \frac{2}{w_1/w_2 + w_2/w_1} \qquad (7.50)$$

In the case with one transition waveguide in between, we have

$$t_2 = \frac{2}{\dfrac{w_1 + \Delta w}{w_1} + \dfrac{w_1}{w_1 + \Delta w}} \cdot \frac{2}{\dfrac{w_1 + \Delta w}{w_2} + \dfrac{w_2}{w_1 + \Delta w}} \qquad (7.51)$$

We want to prove: $\forall w_1, w_2, \Delta w$ satisfying $0 < w_1 < w_2$ and $0 < \Delta w < w_2 - w_1$, it holds that $t_1 < t_2$. As t_1 and t_2 are positive. Consider

$$\begin{aligned} \frac{t_1}{t_2} &= \frac{2(w_1/w_2 + w_2/w_1)}{\left(\dfrac{w_1 + \Delta w}{w_2} + \dfrac{w_2}{w_1 + \Delta w} \right)\left(\dfrac{w_1 + \Delta w}{w_1} + \dfrac{w_1}{w_1 + \Delta w} \right)} \\[4mm] &= \frac{2}{\left[\left(\dfrac{w_1 + \Delta w}{w_1} \right)^2 + 1 \right]\left[\left(\dfrac{w_2}{w_1 + \Delta w} \right)^2 + 1 \right]} + \frac{2}{\left[\left(\dfrac{w_1}{w_1 + \Delta w} \right)^2 + 1 \right]\left[\left(\dfrac{w_1 + \Delta w}{w_2} \right)^2 + 1 \right]} \end{aligned} \qquad (7.52)$$

We change the variable and define $f = t_1/t_2$ to have the following form:

$$f(A, B) = \frac{t_1}{t_2} = \frac{2}{(A+1)(B^{-1}+1)} + \frac{2}{(A^{-1}+1)(B+1)} \qquad (7.53)$$

where

$$\begin{cases} A = \left(\dfrac{w_1 + \Delta w}{w_1} \right)^2 > 1 \\[4mm] B = \left(\dfrac{w_2}{w_1 + \Delta w} \right)^2 > 1 \end{cases} \qquad (7.54)$$

We only need to prove $f < 1$. Take the partial derivative of f with respect to A

$$\frac{\partial f}{\partial A} = \frac{1}{(A+1)^2} \frac{1 - B}{B + 1} < 0 \qquad (7.55)$$

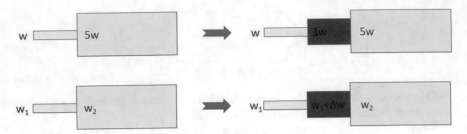

Fig. A.10 Abrupt (left) and adiabatic (right) transition

Due to the rotational symmetry of (7.53), we know $\partial f/\partial B < 0$ for $A > 1$. So the maximum value of f

$$f_{max} < f(1, 1) = 1 \tag{7.56}$$

Q.E.D.

Note that we have proved the adiabatic case is always better than the abrupt one. And to make the argument even stronger, the more intermediate sections, the better the performance, because we can always use the conclusion at each transition and insert a new waveguide in between. In the infinite transition limit, we will get a continuously changing transition such as described in the textbook. Here, we won't discuss which shape is the best, but readers should be able to explore in a similar fashion (Fig. A.10).

Chapter 8

Chapter 9

1. (a) According to (9.31), the spatial resolution ρ is approximately

$$\rho \approx \frac{W_e}{m} = \frac{20\,\mu m}{30} = 0.67\,\mu m \tag{9.57}$$

(b) According to 9.34, the tolerance of the length can be obtained by

$$\frac{\delta L}{L} = 2\frac{\delta W_e}{W_e}$$

$$\Rightarrow \delta L = 2L\frac{\delta W_e}{W_e} = 2 \times 200 \times \frac{0.1}{20} = 2\,\mu m \tag{9.58}$$

2. According to Table 9.1, for a symmetric MMI, the first N-fold image distance is $3L_\pi/4N$. For a 1×2 MMI (where $N = 2$), one obtains

$$L = \frac{3L_\pi}{8} = \frac{4n_f W_e^2}{8\lambda_0} = 246.77\,\mu m \tag{9.59}$$

3. This depends on how many ports there are. Consider we have N output ports in total, also assume the outputs are uniform, then ideally we should have

$$P = \frac{P_{total}}{N} \tag{9.60}$$

at each output port. However, due to the nonuniform performance of a star coupler in general, the outer ports will have a large loss, causing a decrease in the output power.

4. (a) In this case, $a = 0$, we obtain

$$V_c^1 = \pi \tag{9.61}$$

Plug this into (3.19), and we get

$$d = \frac{0.8V_c^1\lambda}{2\pi\sqrt{n_f^2 - n_s^2}} = 1.07\,\mu m \tag{9.62}$$

According to the normalized parameter diagram, $b = 0.55$. Plug this into (3.24)

$$n_{eff1} = \sqrt{b(n_f^2 - n_s^2) + n_s^2} = 3.378 \tag{9.63}$$

Then the waveguide is reduced to the following effective structure.
Again, we use the same technique to calculate the width of this waveguide, where $V_2 = 0.8\pi$. We obtain

$$w = \frac{0.8V_2\lambda}{2\pi\sqrt{n_{eff1}^2 - n_s^2}} = 1.43\,\mu m \tag{9.64}$$

To conclude

$$\begin{cases} d = 1.07\,\mu m \\ w = 1.43\,\mu m \end{cases} \tag{9.65}$$

(b) i. If the wavevector in the transverse direction is κ, according to the maximum mode number $m = 25$, we obtain:

$$\kappa_{max} W_{fs} \approx (25 + 1)\pi \tag{9.66}$$

Also, we have $\kappa_{max} = \frac{2\pi}{\lambda}\sqrt{n_f^2 - n_s^2} = 1.76$. These two equations yield

$$w_{fs} \approx \frac{26\pi}{\kappa_{max}} = 46.43\,\mu m \qquad (9.67)$$

And $w_e \approx w_{fs}$, according to (9.6), we get

$$L_\pi = \frac{4n_f W_e^2}{3\lambda_0} = 6264.18\,\mu m \approx 0.63\,cm \qquad (9.68)$$

ii. According to (9.13), for a double image to occur, the minimum length is given by

$$L_{min} = \frac{1}{2} \times 3L_\pi \approx 0.95\,cm \qquad (9.69)$$

Chapter 10

1. The reflectivity at $z = 0$ is given by

$$R = \frac{(\kappa/\beta_d)^2 \sinh^2 \beta_d L}{1 + (\kappa/\beta_d)^2 \sinh^2 \beta_d L}$$

where

$$\beta_d = \sqrt{\kappa^2 - \delta^2}$$

Use the relationship above and plot for R versus δ.

2.

$$d = \frac{\lambda}{n_{eff} - \sin\phi_c}$$

By solving the phase equation, we can obtain that $n_{eff} \approx 3.433$. Therefore,

$$\phi_c = \arcsin\left(n_{eff} - \frac{\lambda}{d}\right) \approx 10.54°$$

Chapter 11

1. (a) According to (10.7),

$$R = \frac{\sinh^2(\kappa L)}{1 + \sinh^2(\kappa L)} = 0.99 \qquad (11.70)$$

Solve for κ, it yields

$$L = \frac{5.29}{200 \, \text{cm}^{-1}} = 0.0215 \, \text{cm} = 21.5 \, \mu\text{m} \tag{11.71}$$

(b) According to (11.20),

$$\Delta\omega_s = 2v_g\kappa \approx 1.2 \times 10^{13} \, \text{Hz} \tag{11.72}$$

2. According to (11.8), one can use transfer matrix to evaluate the system parameters by cascade the transfer matrix of each components

$$\begin{bmatrix} a^o \\ b^o \end{bmatrix} = [t_c(L_4)] \cdots [t_D][t_c(L_2)][t_D][t_c(L_1)] \begin{bmatrix} a^i \\ b^i \end{bmatrix} \tag{11.73}$$

where in this case, the input vector, the transfer matrices $[t_D]$ and $[t_c(L_i)]$ are given by

$$\begin{bmatrix} a^i \\ b^i \end{bmatrix} = \begin{bmatrix} a^i \\ 0 \end{bmatrix}$$

$$[t_c(L_i)] = \frac{\sqrt{2}}{2} \begin{bmatrix} 1 & -i \\ -i & 1 \end{bmatrix} \tag{11.74}$$

$$[t_D] = \begin{bmatrix} \exp(i\pi/4) & 0 \\ 0 & \exp(-i\pi/4) \end{bmatrix}$$

Plug these expressions into (11.92), one obtains

$$\begin{bmatrix} a^o \\ b^o \end{bmatrix} = \frac{\sqrt{2}}{2} a^i \begin{bmatrix} (1+i) \\ -(1+i) \end{bmatrix} \tag{11.75}$$

3. (a) According to (11.22), assuming we are calculating the central wavelength, it yields

$$\Delta L = \frac{m\lambda_s}{N_{\text{eff}}^f} = \frac{150 \times 1.5 \, \mu\text{m}}{1.4513} = 155.03 \, \mu\text{m} \tag{11.76}$$

(b) According to (11.28), we obtain

$$\Delta\nu = \frac{d_r}{R_a} \left(\frac{m\lambda^2 n_g}{N_{\text{eff}}^s d_a c N_{\text{eff}}^f} \right)^{-1}$$

$$= \frac{30 \, \mu\text{m}}{2 \, \text{cm}} \left(\frac{150 \times (1.5 \, \mu\text{m})^2 \times 1.4752}{1.4513 \times 30 \, \mu\text{m} \times 3 \times 10^8 \times 1.4529} \right)^{-1} \tag{11.77}$$

$$= 5.717 \times 10^{12} \, \text{Hz}$$

$$= 517 \, \text{GHz}$$

(c) According to (11.36),

$$Lu = 8.7\theta_{max}^2/\theta_0^2 \tag{11.78}$$

where

$$\begin{cases} \theta_{max} = x_{max}/R_a = \dfrac{8 \times 30\,\mu m}{2\,cm} = 1.2 \\[2mm] \theta_0 = \dfrac{\pi\lambda}{N_{eff}^s w_0} = \dfrac{\pi \times 1.5\,\mu m}{1.4529 \times 4.5\,\mu m} = 0.72 \end{cases} \tag{11.79}$$

Thus,

$$Lu = 8.7 \times \left(\frac{1.2}{0.72}\right)^2 = 24.2\,dB \tag{11.80}$$

(d) According to (11.39), we obtain the loss l is

$$l = 17\exp(-2\pi^2 w_0^2/d_a^2) = 17\exp\left(-\frac{2\pi^2 4.5^2}{30^2}\right) = 10.9\,dB \tag{11.81}$$

4. The diffraction equation to get start is

$$N_{eff}^s d_a(\sin\theta_i + \theta_o) + N_{eff}^f \Delta L = m\lambda \tag{11.82}$$

According to the text description above (11.29), we then obtain another expression for $\nu' = \nu + FSR$ as below.

$$\left(N_{eff}^s - \frac{FSR}{\nu}\frac{dN_{eff}^s}{d\lambda}\lambda\right)(\sin\theta_i + \sin\theta_o)d_a + \left(N_{eff}^f - \frac{FSR}{\nu}\frac{dN_{eff}^f}{d\lambda}\lambda\right)\Delta L = \left(1 - \frac{FSR}{\nu}\right)(m+1)\lambda \tag{11.83}$$

Note that the diffraction order becomes $m' = m + 1$ and the wavelength $\lambda' = \lambda(1 - \frac{FSR}{\nu})$. Substrate (11.82) from (11.83), it yields

$$\frac{FSR}{\nu}\frac{dN_{eff}^s}{d\lambda}\lambda(\sin\theta_i + \sin\theta_o)d_a + \frac{FSR}{\nu}\frac{dN_{eff}^f}{d\lambda}\lambda\Delta L = \frac{FSR}{\nu}m\lambda - \left(1 - \frac{FSR}{\nu}\right)\lambda \tag{11.84}$$

We plug (11.82) into the right side of this equation and arrange the terms, it yields

$$\frac{FSR}{\nu}\left(N_{eff}^s - \lambda\frac{dN_{eff}^s}{d\lambda}\right)(\sin\theta_i + \sin\theta_o)d_a + \frac{FSR}{\nu}\left(N_{eff}^f - \lambda\frac{dN_{eff}^f}{d\lambda}\right)\Delta L = \left(1 - \frac{FSR}{\nu}\right)\lambda \tag{11.85}$$

Note that

$$\begin{cases} \lambda = \dfrac{c}{\nu} \\[2mm] n_g = N_{eff}^{f,s} - \lambda\dfrac{dN_{eff}^{f,s}}{d\lambda} \end{cases} \tag{11.86}$$

We then simplify (11.85) as

$$FSR \cdot n_g(\sin\theta_i + \sin\theta_o)d_a + FSR \cdot n_g\Delta L = (\nu - FSR)\frac{c}{\nu} \approx c \tag{11.87}$$

Note that usually the operation frequency ν is much larger than *FSR*, the right side is reduced to c. Thus,

$$(FSR)^{-1} = \frac{1}{c}[n_g(\Delta L + d_a \sin\theta_i + d_r \sin\theta_o)] \tag{11.88}$$

5. 1. According to (10.7),

$$R = \frac{\sinh^2(\kappa L)}{1 + \sinh^2(\kappa L)} = 0.99 \tag{11.89}$$

Solve for κ, it yields

$$L = \frac{5.29}{200\,\mathrm{cm}^{-1}} = 0.0215\,\mathrm{cm} = 21.5\,\mu\mathrm{m} \tag{11.90}$$

2. According to (11.20),

$$\Delta\omega_s = 2v_g\kappa \approx 1.2 \times 10^{13}\,\mathrm{Hz} \tag{11.91}$$

6. According to (11.8),

$$\begin{bmatrix} a^o \\ b^o \end{bmatrix} = [t_c(L_4)] \cdots [t_D][t_c(L_2)][t_D][t_c(L1)] \begin{bmatrix} a^i \\ b^i \end{bmatrix} \tag{11.92}$$

where in this case, the matrices $[t_D]$ and $[t_c(L_i)]$ are given by

$$[t_c(L_i)] = \frac{\sqrt{2}}{2}\begin{bmatrix} 1 & -i \\ -i & 1 \end{bmatrix}$$

$$[t_D] = \begin{bmatrix} \exp(i\pi/4) & 0 \\ 0 & \exp(-i\pi/4) \end{bmatrix} \tag{11.93}$$

$$\begin{bmatrix} a^i \\ b^i \end{bmatrix} = \begin{bmatrix} a^i \\ 0 \end{bmatrix}$$

Plug these expressions into (11.92), one obtains

$$\begin{bmatrix} a^o \\ b^o \end{bmatrix} = \frac{\sqrt{2}}{2}a^i\begin{bmatrix} (1+i) \\ -(1+i) \end{bmatrix} \tag{11.94}$$

Chapter 12

1. (a) According to (12.44),

$$L = \frac{\lambda d}{2\Gamma n_e^2 r_{33} V_\pi} = \frac{1.55\,\mu m \cdot 2\,\mu m}{2 \times 0.75 \times 2.135^2 \times 30.8\,pm/V \times 10\,V} = 689\,\mu m$$

$$(12.95)$$

   (b) Notice that

$$V_\pi/\Delta\nu = \frac{V_\pi L}{\Delta\nu L} = \frac{10\,V \cdot 6890\,\mu m}{10\,GHz \cdot cm} = 0.0689\,V/GHz \qquad (12.96)$$

   (c) You could for example use Thermal Optics or Acoustic Optics Effect.

Chapter 13

1. According to (13.3), one obtains

$$V_\pi = \frac{p\lambda d}{n^3 r\Gamma L} = \frac{\sqrt{3} \times 1.55\,\mu m \times 2 \times 6\,\mu m}{2.286^3 \times 30.8\,pm/V \times 0.3 \times 1\,cm} = 2.9 \times 10^3\,V \quad (13.97)$$

2. One need to plot the cross-power-transfer efficiency η_x versus δ/κ as shown in Fig. A.11 according to (13.5)

$$\eta_x = \left(\frac{\kappa}{\beta_c}\right)^2 \sin^2 \beta_c L$$

$$= \frac{1}{(\delta/\kappa)^2 + 1} \sin^2(\sqrt{(\delta/\kappa)^2 + 1} \cdot \kappa L) \qquad (13.98)$$

   Note we can see an envelope which is decreasing as the δ/κ increases. This emphasizes the effect when the two waveguides are detuned with each other, meaning they have different propagation constants β's, less power transfer is allowed.

3. 1. According to (5.19), the minimum length is

$$L_{min} = \frac{\pi}{2\kappa} \qquad (13.99)$$

   Notice this is different from the reversed-$\Delta\beta$ switch, which at $V = 0$ is at bar state.

   2. Refer to Fig. 13.2, which plot for (13.5)

Fig. A.11 Power transfer ratio *versus* detuning

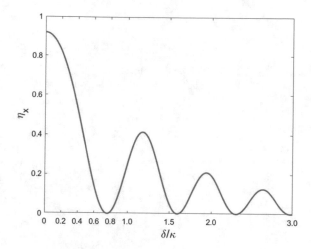

$$\eta_x = \left(\frac{\kappa}{\beta_c}\right)^2 \sin^2 \beta_c L \qquad (13.100)$$

Note that at $V = 0$, $\beta_c = \kappa$. With the increase of voltage, β_c increases, which results in the power allowed to transfer decreases by a factor of $(\kappa/\beta_c)^2$ and the transfer period (the voltage needed to transfer from one maximum to another) decreases (Fig. A.11).

Index

© Springer Nature Switzerland AG 2021
R. Osgood jr. and X. Meng, *Principles of Photonic Integrated Circuits*,
Graduate Texts in Physics,
https://doi.org/10.1007/978-3-030-65193-0

Printed in the United States
by Baker & Taylor Publisher Services